FUNDAMENTAL RESEARCH ON CREEP AND SHRINKAGE OF CONCRETE

FUNDAMENTAL RESEARCH ON CREEP AND SHRINKAGE OF CONCRETE

Edited by

F. H. WITTMANN

1982

MARTINUS NIJHOFF PUBLISHERS
THE HAGUE / BOSTON / LONDON

Distributors

for the United States and Canada
Kluwer Boston, Inc.
190 Old Derby Street
Hingham, MA 02043
USA

for all other countries
Kluwer Academic Publishers Group
Distribution Center
P.O. Box 322
3300 AH Dordrecht
The Netherlands

Library of Congress Cataloging in Publication Data
Main entry under title:

Fundamental research on creep and shrinkage of concrete.

Papers presented at a conference held at the Swiss Federal Institute of Technology in Lausanne in Sept. 1980.
Includes index.
1. Concrete--Creep--Congresses. 2. Concrete--Expansion and contraction--Congresses. I. Wittmann, F. H. (Folker H.)
TA440.F86 620.1'3633 81-18990
AACR2

ISBN-13: 978-94-010-3718-1 *e-ISBN-13: 978-94-010-3716-7*
DOI: 10.1007/978-94-010-3716-7

PREFACE

Today research on creep and shrinkage of concrete is diversified to such a degree that specialists working in different areas sometimes find it difficult to understand one-another. Materials scientists are mainly interested in processes on a microstructural level but they do not necessarily understand the relevance of time dependent deformation in structural design. On the other hand engineers who apply simplified model laws in non-elastic structural analysis are not always in the position to judge the limitations implied in their approach. It is generally realized that further development can be stimulated by a more effective exchange of results and ideas among the different groups involved.

In an attempt to bridge this obvious gap in September 1980 there was a Conference organized at Swiss Federal Institute of Technology in Lausanne. The papers presented at this meeting covered the wide range starting with microstructural aspects and mechanisms and including constitutive modelling and structural creep analysis. These contributions together with summaries of two panel discussions are being published in this volume.

All serious of the meeting have been introduced by invited lectures. These papers will be published in a special volume "Creep and Moisture Effects in Concrete". This special volume is rather to be a general survey of the different areas covered while the present conference proceedings provide a unique selection of research papers.

Nowadays time-dependent deformation of concrete can be taken into consideration realistically by computerized structural analysis. So far, however, we do not have appropriate model laws to describe the complex materials behaviour. But several interesting and promising approaches exist in this new branch of materials science which is generally called concrete mechanics. It is hoped that this compilation of papers will help to clarify the common aim and stimulate further research work.

It is a pleasure for me to thank all my collaborators at the Laboratory for Building Materials Science at Swiss Federal Institute of Technology. Without their continuous help during the organization of the meeting and the preparation of the conference proceedings it would have been impossible to achieve the goal. At the same time I would like to thank Mr. Schrijver from Sijthoff and Noordhoff for the efficient collaboration during the editing process.

Lausanne, May 1981 F.H. Wittmann

TABLE OF CONTENTS

Part I
Mechanisms of Creep and Shrinkage

A MODEL OF HARDENED CEMENT PASTE FOR LINKING SHRINKAGE AND CREEP PHENOMENA

Max J. Setzer

Technische Universität München,
Abteilung für Werkstoffphysik
Baumbachstr. 7,D-8000 München 60

ABSTRACT:
It is shown that swelling of hardened cement paste is influenced by two different mechanisms:
1) In the lower vapour pressure region (r.h.< 38%) a reduction of the surface energy of the gel particles leads to an expansion of the single particles.
2) In the upper vapour pressure region the whole structure is widened by a change of interparticle attraction in the small gel pores. (disjoining pressure)
Both effects can clearly be distinguished. The creep velocity of specimens stored at constant but different humidities increases at values above 40% r.h. This increase can be correlated to that part of the swelling which results from the expansion of the structure of hcp. A theoretical explanation is made through adapting the "Extended Munich Model" and using the "thermodynamic approach".

Zusammenfassung:
Es wird gezeigt, daß das Quellen von Zementstein von zwei verschiedenen Mechanismen beeinflußt wird:
1) Im unteren Feuchtebereich (r.h.< 38%) sinkt die freie Oberflächenenergie der Gelpartikel ab. Dies führt zu einer Expansion der einzelnen Partikel.
2) Im oberen Feuchtebereich wird die gesamte Struktur aufgeweitet durch eine Änderung der Anziehung der Teilchen untereinander. (disjoining pressure oder Spaltdruck)
Beide Effekte können gut getrennt werden. Die Kriech-

geschwindigkeit von Zementsteinproben, die bei konstanter und unterschiedlicher Feuchte vorgelagert waren, erhöhte sich bei über 40% r.F. Dieser Zuwachs der Kriechgeschwindigkeit kann zu dem Teil des Quellens korreliert werden, der aus der Expansion der Gelstruktur herrührt. Theoretisch werden die Ergebnisse mit einem verbesserten "Münchener Modell" und der "Thermodynamischen Näherung" erklärt.

Introduction

It is a well known fact that creep velocity of concrete and hardened cement paste depends on the content of physically bound water. Therefore drying out of concrete leads not only to shrinkage but also to a diminished deformation under external load. Several models have been proposed and discussed by different authors to explain and link these two phenomena which lead to the inelastic behaviour of hardened cement paste and concrete.

Powers (1,2) describes swelling as a hindered adsorption of water in fine gel pores. Due to this, a disjoining pressure is generated which expands the solid. In Powers' opinion creep is caused by a diffusion of a load bearing water, because an external load changes the free energy of the adsorbed water.
Therefore, according to Powers, creep and shrinkage are two different names for the same phenomenon. Bazant (3,4) used surface thermodynamics to improve these considerations.

Feldman and Sereda (5,6) measured simultaneously adsorption isotherms and swelling and deduced a different model. In their opinion water is inserted between the layers of gel particles, which leads to a swelling of the body. They regard creep as a displacement of the interlayer water and as a sliding of the gel layers. Feldman and Sereda have improved their model (7,8). They distinguish between mono-layer water, intracrystalline water and intercrystalline water. Hope and Brown (9) used this model to explain their findings of combined creep and shrinkage.

Both models have been discussed by different authors e.g. (10-12) Wittmann (12) pointed out that creep should be zero if hcp is dried out completely. Although his own measurement shows a remarkable creep even of fully dried specimen. Creep can not be explained by a

pure diffusion of water. Even if shrinkage and creep are influenced by the physically bound water, they are caused by different mechanisms. Parrott (13,14) found some correlation between creep, shrinkage and polysilacate formation of hardened cement paste. Young et. al. (15-19) distinguish between a capillary drying shrinkage and a gel drying shrinkage. Gel drying as well as microsharing under load leads to a structural change in the silicate framework. By drying as well as under load silicate polymerisation increases. The Munich group (12,20-23) has developed a model of xerogel to describe the interaction of solid gel particles and physically bound water due to surface forces. The model has been described in (24). In (25) the thermodynamic background for an extended model is outlined in detail, as well as its application to the mechanical behaviour and its expansion to low temperature studies.
In this paper the adoption of these thermodynamics to creep and shrinkage of hcp shall be outlined.

Theory

1) Swelling
A concept to explain the swelling of ultrafine particles by a change of the free-surface-energy F has been proposed by Bangham (26-28). Any small particle is compressed by surface energy. If surface energy is lowered, the compression is diminished and the particles are enlarged by $\Delta l/l$

$$\frac{\Delta l}{l} = - \kappa \quad \Delta F \cdot \qquad (1)$$

is a constant of proportionality depending on the internal surface, Young's modulus, density, and the geometry of the gel particles. The change of surface forces by adsorbed water films can be calculated by thermodynamics. In a first approximation using a nonporous system, it is:

$$\Delta F = RT \quad rd \quad \ln p/p_o \qquad (2)$$

To evaluate this, absolute temperature T, relative humidity p/p_o and the thickness of the adsorbed film must be measured. (R being the gas constant)

In a material which contains very fine pores, as for example in a hcp, water vapour condensation is present

in addition to adsorbed water. Therefore the 2. equation can not be applied to hcp in an unmodified form. The "Thermodynamic Approach" has been developed (20, 25,29) to seperate adsorption isotherms. Using this method the change in free-surface-energy can be appropriately calculated by an interaction method for hcp in a fair manner.
In porous systems, such as in hcp, swelling is not only caused by an expansion of particles but also by an expansion of the small pores. This means for the whole structure

$$\frac{\Delta l}{l} = \frac{\Delta l_p}{l} + \frac{\Delta l_s}{l} \qquad (3)$$

The two phenomena can be separated.

2) Creep
Creep of materials has generally been discussed by Niklas (30). He adopts the theory of reaction kinetics to creep. In this case the creep centers N which have reacted at time t can be calculated by

$$N = N_o \quad (1-e^{-kt}) \qquad (4)$$

N_o is the number of centers at t=0, and k is given by a modified Arrhenius equation:

$$k = K \quad \exp\left(- \frac{E - E_m}{RT} \right) \qquad (5)$$

where K = constant, E = energy of activation, E_m = mechanical energy due to external stress, and T = absolute temperature.

If one connects to every elementary creep process a medium deformation creep is proportional to N. However, since local energy is distributed according to a distribution function f (E_m), an integral equation for creep ε is found

$$\varepsilon \ (t) = \zeta \int_0^\infty f \ (E_m) \quad 1 - e^{-k\{E_m,T\} \cdot \ t} \ dE_m . \qquad (6)$$

In a previous paper (31) it has been shown that experimental results of creep measurements of hcp can be fitted the Shank (32) expression.

$$\varepsilon = \alpha t^n \ . \qquad (7)$$

(α ,n beeing constant). Creep velocity is then

$$v = \dot{\varepsilon} = \alpha n t^{n-1}. \quad (8)$$

Niklas has calculated a distribution function using this equation:

$$v(t) = nt^{n-1} = \zeta \int_0^\infty \frac{1}{c(n)} k^{(1-n)} \exp(-kt)\, dE_m. \quad (9)$$

c (n) is a constant depending on the exponent n. It is defined by the integral equation (5). For instance, if the gel pores of hcp are widened by disjoining pressure, the interaction of energy E between the particles diminishes.

This energy of interaction between the particles is a function of the distance d between the surfaces i.e. the pore size and its distribution

$$E = E\ (d) \quad (10)$$

Depending on the bonding type (van der Waals interaction, polymerisation bonding, hydrogen bridging, double layers and structured water etc.) we find different types of functional descriptions (33). However if the distance changes are as small as in swelling phenomena then equation 10 can be approximated by

$$E \cong E_o \left(1 - m \frac{\Delta l_s}{l}\right). \quad (11)$$

In addition, this approximation can be adopted in an equivalent way to a small weakening of primary bonds. Using this it can be shown (25) that creep velocity is diminished by widening of small pores in a linear way:

$$v(t) = v_o(t) + \frac{\Delta l_s}{l} v_1\ (t) \quad (12)$$

$$v_o(t) = \zeta \int_0^\infty \frac{1}{c(n)} k_o^{1-n} \exp(-k_o t)\ dE_m \quad (13)$$

$$v_1(t) = \zeta \int_0^\infty m \frac{1}{c(n)} k_o^{1-n} \quad 1 + \exp \quad \frac{E_o(1-n)}{RT} \quad e^{-k_o t} dE_m. \quad (14)$$

Eq. 13 is equivalent to eq. 9. Both, eq. 13 and eq.14 do not contain Δ ls/l.

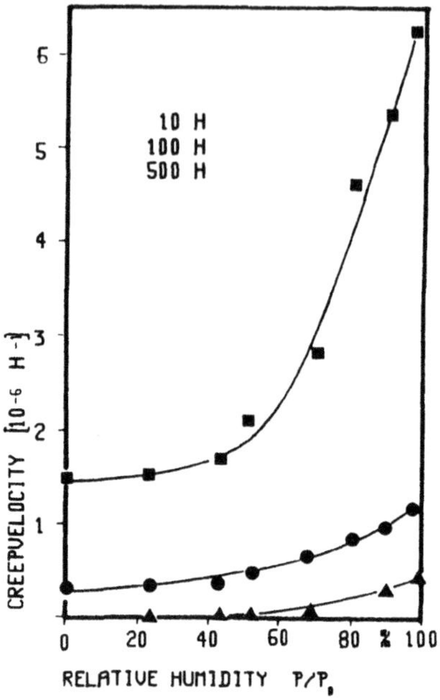

FIG. 1
Creepvelocity vs. relative humidity of hcp (Wittmann/12/)

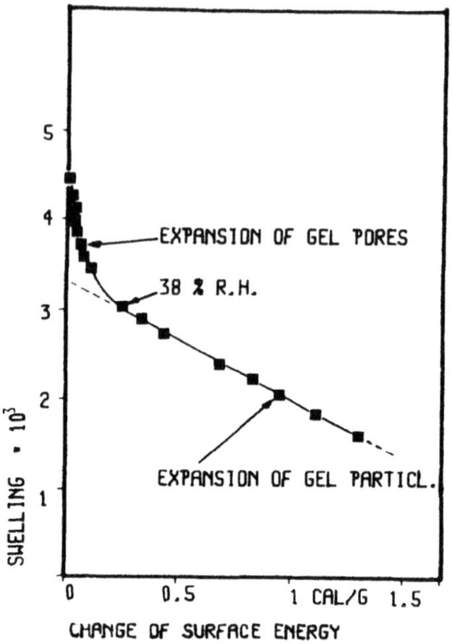

FIG. 2
Total swelling of hcp vs. change of free surface energy

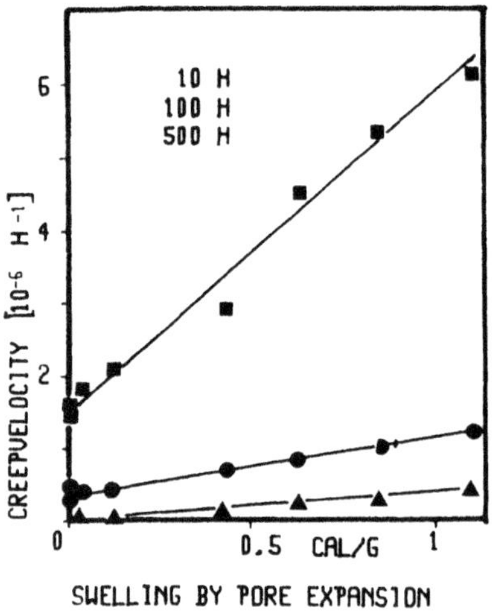

FIG. 3
Creepvelocity vs. swelling of hcp caused by expansion of gel pores l_s/l

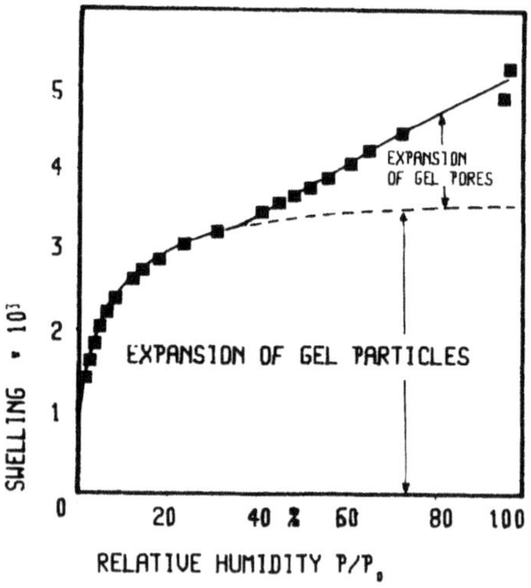

FIG. 4
Total swelling of hcp vs. relative humitidy

Results

Using "Thermodynamic approach" the change of free-surface-energy has been calculated using the adsorption isotherm published by Feldman (6). By this not only strength change due to changes in water content but also swelling can be explained (29). In fig. 2 total swelling, following the data in (6) is plotted versus the calculated change in free-surface-energy. The change of free-surface-energy is related to the value of 100%rh. A fair linear correlation is found in the water vapour pressure region between 2% and 38% r.h. Above 38% r.h. - exquivalent to very small values of free-surface-energy change - the swelling curve deviates from the straight line.
This additional swelling is caused by the disjoining pressure, which increasingly expands the small gel pores at high relative humidity. In fig. 2 the swelling data used in fig. 4 are plotted versus relative humidity.

From this plot the selling due to an expansion of pores can be separated from that due to the expansion of particles following eq. 3. If the creep measurements of Wittmann (12) (fig. 1) are explained by eq. 12, then one should find straight lines if creep velocity is plotted versus the swelling due to pure expansion. Fig. 3 gives appropriate proof to support this statement.

Discussion

The foregoing results lead to the conclusion that shrinkage and creep are based not only on a diffusion of water but on several different mechanisms in the ultramicroscopic scale:
a) Expansion of single gel particles, b) expansion of pores, c) displacement of gel particles.

Since, in the region of nanometers, surface effects can not be neglected because of bulk effects, water, which is physically bound by these surface effects, changes the mechanisms and this leads to shrinkage and creep. The idea of Powers (1,2) and Bazant (3,4) to use thermodynamics, is very helpful. But for a full description of surface interaction the semi macroscopic thermodynamics has to be improved (20,25). Thus not only the behaviour of physically bound water, but also the changes in the gel particles and in the gel pores are better understood.

Feldman and Sereda (5-8) on the other hand discussed the particle interaction and the changes in structure. However, they introduced the somewhat misleading term of interlayer water, which is well defined in mineralogy. Measurements for instance, of x-ray diffraction, sorption isoterms and Mössbauer effect made with hcp lead to quite different results compared with those measurements made with substances containing a remarkable and defined amount of interlayer water (20). In addition, we compared (33) water vapour sorption on hcp and on raw mix, which is inert and contains no interlayer water at all. We measured weight changes of adsorption isotherms, divided by different internal surfaces of the materials. This leads to exactly the same results in the vapour pressure region below 40% rh. Adsorbed water is essentially different from water in bigger pores, as has been proved by low temperature DTA (34). In this paper, we differenciate between the above mentioned highly disturbed water and water in pores of $3<R_H<10$nm and in pores above 10nm. This is in good agreement with the statements of Young et.al. (15-19), who distinguish between micropores below 2,6 nm diameter, mesopores between 2,5 and 50nm, and macropores >50nm following a suggestion of IUPAC (35) and adopted in (20,25), too.

In (33) it will be shown that the postulated disturbance of water near the surfaces and the measured adsorbed films can be described by a simple comprehensive formula, which can explicitly be explained by physical mechanisms. From these findings we conclude that water in pores smaller than 2,5nm is dominant in generation of disjoining pressure. This, combined with the presented model, is in astonishingly good aggreement with some of the hypotheses of Young et.al. The findings of Hope and Brown (9) can also be interpreted in this way, since water and methanol are disturbed in quite a different way by surface forces. Finally, it should be noted that Parrott in (13) finds a linear correlation between creep and shrinkage during initial drying to 60% rh of heat cured specimen. An explanation could be that the gel porosity is diminishe by heat curing (36).

Conclusions

All this leads to the conclusion that swelling is mostly influenced by:

1) swelling of particles due to a diminished surface energy in the lower water vapour pressure region (rh. 40%) and by

2) expansion of small pores due to disjoining pressure above 40% r.h.

Creep is changed by the widening of small pores since particle interaction is diminished.
The findings are in good aggreement with very recent results of the Munich group about the behaviour of water which is highly disturbed by surface forces.

Acknowledgement

The work of the Abteilung für Werkstoffphysik at the Technical University of Munich receives considerable financial support by the Deutsche Forschungsgemeinschaft.

References:

1. Powers, T.C.: in Proc. "International Conference on the Structure of Concrete", A.E. Brooks and K. Newman eds., Cem. Concr. Ass. London, 1965,p. 319
2. Powers, T.C.: Material Structure, 1, 487 (1968)
3. Bazant, Z.P.: Cem.Conr. Res. 2, 1 (1972)
4. Bazant, Z.P.: Nucl. Engeneering Design 20, 477 (1972)
5. Feldman, R.F.; Sereda, P.J.: Materials & Structure 1, 509 (1968)
6. Feldman, R.F.: 5th International Symposium on the Chemistry of Cement, Tokyo, 1968, Vol 2, p.53
7. Feldman, R.F.: Cem. Concr. Res. 2, 521 (1972)
8. Sereda, P.J.; Feldman, R.F.; Ramachandran, V.S.: 7th International Congress on the Chemistry of Cement, Paris, 1980, Vol I., VI-1/3
9. Hope, B.B.; Brown, N.H.: Cem. Concr. Res. 5, 577, (1975)
10. Brunauer, S.; Odler,I.; Judenfreund, M.: Highway Res. Record No. 328, p.89, 1976
11. Kondo, R.; Diamond, M.: 6th International Congress on Chemistry of Cement, Moscow, 1974
12. Wittmann, F.H.: State Institute of Technical Research, Rep. Ser.2, Vol. 167, Finland, 1971
13. Parrott, L.J.: Proc. of Conf. "Hydraulic Cement Paste", Cement Concr. Ass., Sheffield, 1976,p.189
14. Parrott, L.J.: Cem.Concr.Res. 7, 597 (1977)
15. Mindess, S.; Young, S.F.; Lawrence, F:V.: "Creep and Drying, Shrinkage of Cacium Silicate Pastes. I Specimen Preperation and Mechanical Properties". Cem. Concr.Res. 8, 591 (1978)
16. Bentur, A.; Milestone, N.B.; Young, J.F.: "Creep and Drying Shrinkage of Calcium Silicate Pastes, II Induced Microstructural and Chemical Change" Cem. Concr. Res. 8, 721 (1978)
17. Bentur, A.; Berger, R.L.; Lawrence, F.H.; Milestone, N.B.; Mindess, S.; Young, J.F.: "Creep and Drying Shrinkage of Calcium Silicate Pastes. III A Hyporhesis of Irreversible Strains". Cem. Concr.Res. 9, 83 (1979)
18. Bentur, N.B.; Milestone, N.B.; Young, J.F.; Mindess, S.: Cem.Concr.Res. 9, 161 (1979)
19. Bentur, A.; Berger, R.L.; Young, J.F.; Milestone N.B.; Mindess, S.; Lawrence, F.V.: 7th International Congress on the Chemistry of Cement, Paris, 1980, Vol. III, IV-26
20. Setzer,M.J.: "Surface Energy and Mechanical Behaviour of hardened Cement Paste", Thesis, Technical

University of Munich, 1972
21. Splittgerber, H.; Wittmann, F; Surface Sci. 41, 504 (1974)
22. Setzer, M.J.; Wittmann, F.H.: Applied Phys. 3, 403 (1974)
23. Übelhack, H.; Wittmann, F.H.: Proc. Conference on Mössbauer Spectroskopy, Krakow
24. Wittmann, F.H.: Schriftreihe DAfStb, Heft 290, p.43 (1977)
25. Setzer, M.J.: Schriftreihe DAfStb, Heft 280, p. 43 (1977)
26. Bangham, D.H.: Transaction Faraday Soc. 33, 805 (1937)
27. Bangham, D.H.; Fakhoury, N.J.: J. chem.Soc. 1324 (1931)
28. Bangham, D.H.; Fakhoury, N.J.: Proc. Roy. Soc. A130, 81 (1931)
29. Setzer, M.J.: Cem. Concr. Res. 6,37 (1976)
30. Niklas, H.: Z. angew. Phys. 6, 470 (1967)
31. Wittmann, F.H.; Setzer, M.J.: Cem. Concr. Res. 1, 679 (1971)
32. Shank, J.R.: Bulletin No. 91, Ohio State University, Engeneering Experimental Station, 1935
33. Badmann, R.; Stockhausen, N.; Setzer, M.J.: to be published in J.Coll. Interface Sci.
34. Stockhausen, N.; Dorner, H.; Zech, B.; Setzer, M.J.: Cem. Concr. Res. 9, 783, (1980)
35. IUPAC, Manual of Symbols and Terminology, Appendix 2, Part 1, Pure Appl. Chem. 31, 578 (1972)

THE USE OF ACTIVATION SPECTRUM THEORY TO DESCRIBE CREEP IN CONCRETE

R. L. DAY* and B. R. GAMBLE**

*Dept. of Civil Engg., The Hatfield Polytechnic, Hatfield, Herts. England
**Dept. of Civil Engg., University of Calgary, Calgary, Alta. Canada

ABSTRACT

The paper discusses the use of activation spectrum theory to produce a practical model of concrete creep behaviour. The model is shown to be attractive because it naturally allows for the complex and readily changeable concrete microstructure; yet there is no need for a detailed description of the microstructure. The validity and practical potential of the theory are demonstrated by: (1) the realistic description of qualitative creep behaviour over all time ranges, (2) the production of an accurate equation for irrecoverable creep at constant temperature. The theory's potential to produce comprehensive creep prediction equations, and some of the theoretical refinements necessary to achieve it, are discussed.

هذه المقاله تناقش استعمال نظرية " الطيف التنشيطى " لتحصيل نموذج عملى لطبيعة الزحف فى الخرسانه • هذا النموذج مرغوب لانه ياءخذ فى الاعتبار طبيعة الخرسانه المتغيره وفى نفس الوقت لا يحتاج لتوضيح تفصيلى لاجزاء الخرسانه الداخليه •
يمكن تمثيل التطبيق والوضع العملى للنظريه بالتالى —
١ — تفصيل واقعى لنوعيه طبيعه الزحف فى الخرسانه فى اى وقت •
٢ — استنتاج معادله دقيقه للزحف الغير عكوس عند ثبوت درجه الحراره •
المقاله تناقش امكانيه استنتاج معادلات شامله للزحف المتوقع وكذلك التصليحات النظريه الضروريه لاستنتاج هذه المعادلات •

1 INTRODUCTION

Although creep in concrete is believed to be intimately associated with microstructural change, progress in determining the details of this association, so as to develop practical relationships, has been hampered by the inadequacy of experimental equipment and technique; microscopic changes related to creep have not yet been observed or measured directly.

This lack of progress encourages the development of methods of creep analysis which do not depend on microstructural details. One such method uses activation spectrum theory; behaviour is determined from the basic consideration that the material as a whole is altered, along very specific lines, by rate process. Activation spectrum theory and its application to concrete creep is the topic of this paper.

The paper is organised according to its three purposes: (1) to outline the rudiments of the theory, (2) to show that the simple theory can be used to describe accurately creep in concrete, (3) to discuss the potential of the theory for producing comprehensive, yet fundamentally based, creep prediction equations.

2 ACTIVATION SPECTRUM THEORY

2.1 The Basic Creep Process

It has been hypothesized [1] that within the concrete microstructure there are regions, or elements, where the activation process results in the spontaneous movement of matter to new positions. When no directed energy potential exists, activation of 'creep-elements' occurs randomly--in all directions. However, when an energy potential such as an applied stress is present, activation is biased in the direction of the potential; the material rearrangements associated with this activation are revealed on the macroscale as creep.

The basic working equation [1,2] expresses the chance (p) of a creep element activating and contributing to creep strain within time dt:

$$p = 2 \cdot \nu(T) \cdot \exp\{-F/RT\} \sinh\{E_s/RT\} \quad (1)$$

where F is the activation free energy of the element, E_s is the energy contribution of the applied stress to the activation process if the element activates, $\nu(T)$ is the frequency of thermal vibration at absolute temperature T, and R is the universal gas constant.

2.2 The Activation Energy Spectrum

Consider a simple material which at time t contains Y creep-elements, all of which have approximately the same activation free energy F_i. This material can be depicted by a spectrum (Figure 1) where the ordinate is an energy density function (Q); the area under the function represents a number of elements, thus $Y = Q_i \cdot dF$. At time t the rate of activation is given by; $r = p \cdot Q_i \cdot dF$ (rate of activation = chance of activation x no. of centres).

In concrete, it is expected that many elements, having a wide range of activation energies, are present so that Q is virtually a continuous function of F (Figure 2). Thus, for this general case, at time t the rate of activation,

$$r = \int_0^\infty p \cdot Q(F)\, dF \qquad (2)$$

where p is a function of F (Equation 1). If b is defined as the average contribution to creep of a single activation, then the creep rate at any given time is simply $b \cdot r$ (assuming b is not a function of F).

In real materials the activation spectrum changes with time; elements, once activated, are likely to require a different energy for a second activation.

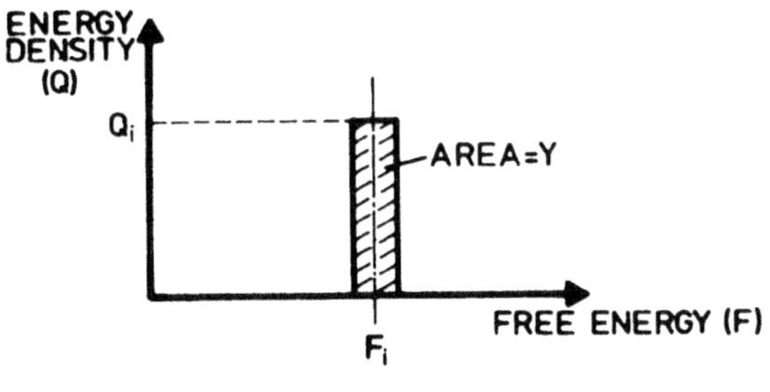

FIG. 1
A Simple Activation Spectrum

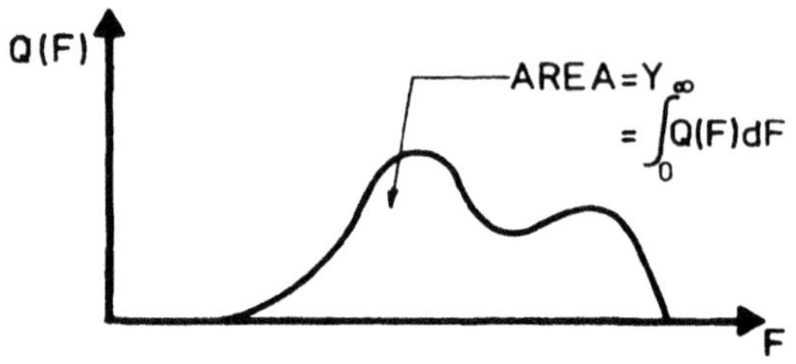

FIG. 2
General Activation Spectrum

To proceed towards a practical creep equation an estimate must be made of post-activation behaviour. The assumption adopted here--similar to that found in papers on the creep of metals [3-6] and soils [2]--is that an element is exhausted after it activates; it cannot activate again. Wyatt [6] showed that a more complicated assumption has little effect on the final outcome of the analysis. In any case, the validity of the assumption can best be judged by the success of the theory in describing creep.

Because of the assumption: the rate of activation of elements in energy range dF = the rate of reduction of elements in this range = the rate of change of Q(F,t)dF. Thus

$$\frac{d[Q(F,t)dF]}{dt} = -p \cdot Q(F,t)\,dF \tag{3}$$

which, upon integration, gives

$$Q(F,t) = Q(F,0) \cdot \exp\{-p \cdot t\} \tag{4}$$

where Q(F,0) represents the state of the material at the time of loading; it is the 'initial' spectrum at t=0.

Since Q(F,t)dF represents a number of elements, the elements which activate between t=0 (time of load) and any later time are depicted by the area between Q(F,t) and Q(F,0), which is proportional to creep. Thus, the creep at time t

$$\epsilon = \int_0^\infty b \cdot Q(F,0)\,[1-\exp\{-p \cdot t\}]\,dF \tag{5}$$

the double exponential function $\exp\{-p \cdot t\}$ is termed a 'decay-operator' (D) which acts on the initial spectrum Q(F,0) to produce the Q(F,t) spectrum.

2.3 General Description of Creep Behaviour

The creep process is illustrated in Figure 3; the Q(F,0) function (Fig. 3.a) is assumed rectangular for simplicity of explanation. During the time under load the decay-operator (D), which represents the activation process, sweeps along the F-axis from left to right (Fig. 3.b). The shape of this function varies only with temperature [6], and its position at any time is a function of t, ν and E_s. The amount of creep which

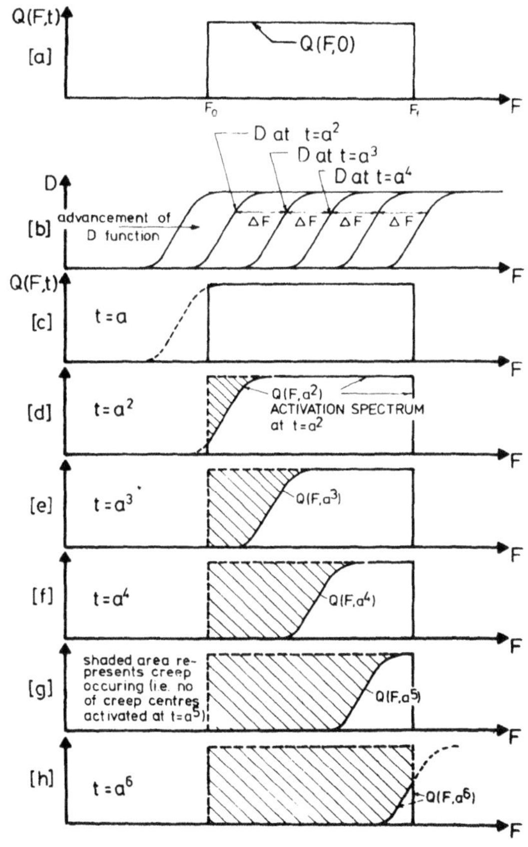

FIG. 3
Illustration of the Creep Process

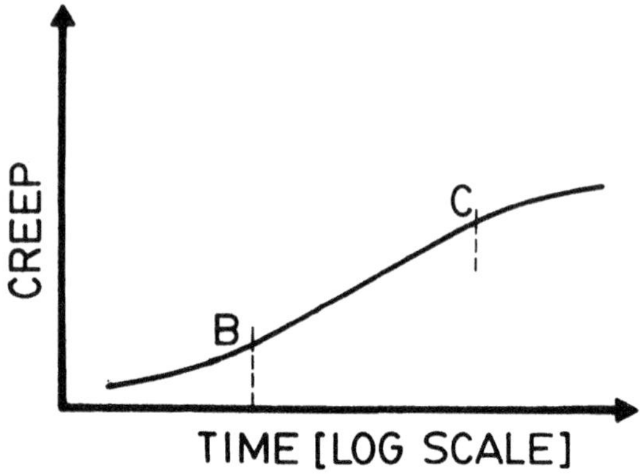

FIG. 4
Qualitative Creep Behaviour

occurs after time t is proportional to the area that the decay-operator sweeps from the original spectrum (shaded).

If Q(F,0) were constant over all F, creep would be linear with logarithmic time (D travels equal increments in equal logarithmic time). Practically, this cannot be the case; there must be an upper and lower level for Q(F,0)--at F_o and F_f say. Therefore, at earlier and later times (Figs. 3.c,d,h) $d\epsilon/d(\log t)$ must be smaller than at intermediate times (Figs. 3.e,f,g).

Figure 4 shows the qualitative creep behaviour at constant temperature which is predicted from the theory.

2.4 Description of Creep in Concrete

Figure 4 is a realistic qualitative description of concrete creep behaviour over all time ranges. This conclusion is made because (a) the transition to linear behaviour at point B is a common aspect of the relevant creep literature and (b) the theory agrees with long term creep tests [7,8] which show a reduction in $d\epsilon/d(\log t)$ after about 2 years loading.

2.5 Quantitative Creep Description

To quantify the spectrum approach, it is necessary to assume a realistic function for Q(F,0). Since the process described in Figure 3 will also occur in an unstressed material in all directions (the material ages with time as thermal fluctuations cause elements to activate in random directions) the initial spectrum is likely to be akin to the $Q(F,a^3)$ curve of Figure 3.e.

Thus, the assumed initial spectrum (Fig. 5.a) is given by:

$$Q(F,0) = s \cdot \exp[-X \cdot \exp\{-F/RT\}] \tag{6}$$

The parameter s, which may be a function of F, represents the density of the creep elements in the material and depends on how the elements originally form. Here, s is assumed to be a constant; elements are presumed not to prefer a particular range of activation free energies. The parameter X is a function of curing conditions and the specimen's environmental and stress history. In effect, X locates the 'front' of the initial spectrum along the free energy axis. The tail of the initial spectrum is not modelled; very long term creep behaviour (i.e. beyond point C, Fig. 4) is not described by the following theory.

Figure 5.a shows the initial spectrum and that at time t after load. Since the area between curves is proportional to creep, and since the decay-operator does not change shape with time (T=constant), then A_1 and A_2 (Fig. 5.b) and creep

$$\epsilon = b \cdot s \cdot [F_t - F_i] \tag{7}$$

where F_t and F_i are defined at any arbitrarily chosen fraction of s, say c. Substituting for Q = c in

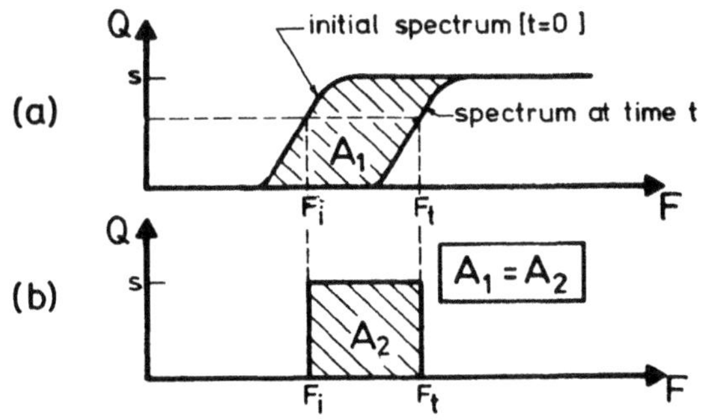

FIG. 5
Creep Behaviour Predicted from Spectrum Movement

equations 4 and 6, rearranging, and substituting into equation 7, gives:

$$\epsilon = C_1 \cdot \ln(1 + C_2 \cdot t) \tag{8}$$

where

$$C_1 = b \cdot s \cdot R \cdot T \tag{9}$$

and

$$C_2 = \frac{2 \cdot \nu(T) \cdot \sinh(E_s/RT)}{X} \tag{10}$$

The new parameters C_1 and C_2 are determined by the state of the material at the time of loading and by the effects of temperature and stress. Since it is not possible to determine directly values for C_1 and C_2 (the proper experiments to measure b, s, X, ν, and E_s have not yet been developed) the suitability of the activation approach is examined by determining how well equation (8) describes relevant creep behaviour which has been reported in the literature.

Figure 6 shows the results of McDonald [9] which were analysed by Wittman & Setzer [10], who fit a power curve to the data. Note that equation (8) fits long term creep well, but total creep at early times is underestimated; this was found in many experimental examinations. If, however, equation (8) is fitted to flow (irrecoverable creep) rather than total creep results, the theory agrees very well with experiment (Fig. 7). This good agreement was also found for the few other experiments in which flow could be determined [12,13,14].

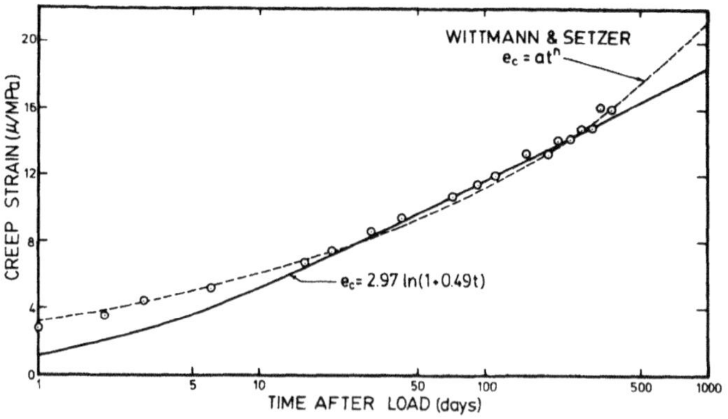

FIG. 6
Creep Results of McDonald [9]

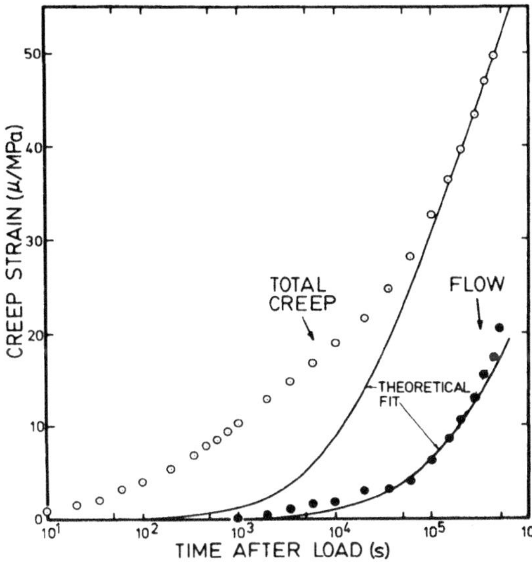

FIG. 7
Agreement of Theory with Flow Results. Data from [11]

3 DISCUSSION: THE VALIDITY AND POTENTIAL OF THE ACTIVATION SPECTRUM APPROACH

The preceding analysis has indicated that spectrum theory can be used to provide realistic and accurate descriptions of some aspects of concrete creep behaviour. Because of this success, because the theoretical concepts are simple, and because the theory is based upon deformation processes which are known to generally occur, it is suggested that the theory may be successfully used to produce comprehensive creep prediction equations.

It is clear that future success depends upon an accurate estimate of the general state of the material at the time of loading (i.e. the initial spectrum). The very simple assumption that creep-elements prefer no particular activation energy range was adequate for the description of irrecoverable creep. The analysis of more complicated behaviour, however, requires a more detailed estimate of the initial spectrum. For example, early time creep (mostly recoverable [15,11]) is generally associated with water movement, which has fairly well defined activation energies [16,17]; some energy ranges are preferred. When the shape of the initial spectrum is estimated, these preferred ranges must be accounted for if total creep is to be modelled.

For spectrum theory to become a truly predictive method, not only the shape, but the position of the initial spectrum along the F-axis must be established. This requires a knowledge of the variables on which C_1 and C_2 depend (Eqtns. 9,10). Phenomenological tests may be useful to establish how the parameters vary with stress and environmental history, mix proportions, curing conditions, etc. Also important is the development of experimental methods which will allow the general state of the microstructure to be monitored; polysilicate, low angle x-ray scattering and nitrogen sorption tests may hold promise in this regard.

Once the initial spectrum is defined, rate process not only determines creep behaviour, but also determines how the spectrum (i.e. state of the material) changes with time. It is anticipated that this extra information can simplify the analysis of behaviour under a complicated series of environmental and stress changes. The analysis can be performed in steps which correspond to the times at which the conditions change because the state of the material is known when each change occurs.

As the theory is developed further, it must be made to account for other subsidiary phenomena which may affect creep. For example, where significant hydration occurs during the loading period or where a temperature (or humidity) change itself initiates new mechanisms of creep, new creep-elements may come into play during the loading period. It is anticipated, however, that proper assumptions can be made without detracting unduly from the simplicity of the method.

REFERENCES

1. Gamble, B.R. and Illston, J.M. "Rate of Deformation of Cement Paste and Concrete During Regimes of Variable Stress, Moisture Content and Temperature", Proc. of Conference on Hydraulic Cement Pastes: Their Structure and Properties, Univ. of Sheffield, Apr. 8-9 (1976)

2. Mitchell, J.K., Campanella, R.G. and Singh, A. "Soil Creep as a Rate Process", Journ. Soil Mech. Div. ASCE, January (1968)

3. Smith, C.L. "A Theory of Transient Creep in Metal", Proc. Physical Soc., 61, Part 3, September (1948)

4. Davis, M. and Thompson, N. "Creep in a Precipitation Hardened Alloy", Proc. Phys. Soc., B63, pp. 847-860 (1950)

5. Cottrell, A.H. "Dislocations and Plastic Flow in Crystals", Clarendon Press, Oxford (1953)

6. Wyatt, O.H. "Transient Creep in Pure Metals", Proc. Phys. Soc., 66B, p.459 (1953)

7. Troxell, G.E., Raphael, J.M. and Davis, R.E. "Long Time Creep and Shrinkage Tests of Plain and Reinforced Concrete", ASTM Proc., 58, pp. 1101-1120 (1958)

8. Hummel, A., Wesche, K. and Brand, W. "Einfluss der Zementart, des Wasser Zement-Berhaltnisses und des Belasturgsalters auf das Kriechen von Beton", Deutscher Ausschuss fur Stahlbeton, No. 146, 133 pp (1962)

9. McDonald, J.E. "An Experimental Study of Multiaxial Creep in Concrete", Amer. Concr. Inst. SP-34: Concrete for Nuclear Reactors, Berlin, October (1970)

10. Wittman, F. and Setzer, M. "Comparison of Some Creep Functions with Experimental Results", Cem. Concr. Res., 1, No. 6, pp. 679-690 (1971)

11. Day, R.L. "An Examination of the Relationship Between Creep and Microstructural Change in Hardened Cement Paste", Ph.D Thesis, Univ. of Calgary, Calgary, Alta. Canada (1979)

12. Parrott, L.J. "Lateral Strains in Hardened Cement Paste Under Short and Long-Term Loading", Magazine of Concr. Res., 26 No. 89, December (1974)

13. Illston, J.M. and Sanders, P.D. "Characteristics and Prediction of Creep of a Saturated Mortar Under Variable Temperature", Magazine of Concr. Res., 26, No. 88, September (1974)

14. Marechal, J.C. "Creep of Concrete as a Function of Temperature", Amer. Concr. Inst. SP-34: Concrete for Nuclear Reactors, Berlin, October (1970)

15. Illston, J.M. "The Components of Strain in Concrete Under Sustained Compressive Stress", Magazine of Concr. Res., 17, No. 50, March (1965)

16. Parrott, L.J. "Load Induced Dimensional Changes of Hardened Cement Paste", Ph.D Thesis, Univ. of London (1973)

17. Sabri, S. "The Response of Hardened Cement Paste to Changes in Ambient Relative Humidity and Temperature", Ph.D Thesis, Univ. of London (1980)

DEFORMATION KINETICS OF CONCRETE.

Jaap Luijerink

D.U.T. Stevinlab.
Department of Civil Engineering
Delft, The Netherlands

ABSTRACT
A method based on the rate theory is developed to describe the total creep deformation, the total strain rate and the apparent free energy of activation curves.
At the time where the creep curve starts its tertiary stage, the strainrate-time curve shows a minimum and the apparent free energy of activation has a maximum. This means that during the primary and secondary creep the material is able to build up higher energy barriers against the work contributed by the external effective stress. During the tertiary stage however the material does not have this capacity and starts to break. No assumption is made concerning the micro structural mechanisms of the rate processe, but it can be concluded that during this third stage the propagation of cracks takes place.
It was also possible to describe the total σ,ε diagram.

ZUSAMMENFASSUNG
Mit einer Methode die basiert ist auf der Verformungskinetik, ist es möglich die totale Kriechverformung, die Kriechgeschwindigkeit und die Freie Enthalpie der Activierung zu beschreiben wenn eine Relation zwischen der Activierungsenergie und der Zeit angenommen wurde: $\exp\{(Q-Q_0)/kT\}=(1+t/d)^m$. Wenn die Kriechverformung mit ihren dritten Stadium anfängt, zeigt die Kriechgeschwindigkeit eines Minimum und die Freie Enthalpie eines Maximum. Das Material hat während der erster und zweiter Stufe die Möglichkeit höhere Energien auf zu bauen gegen die ausseren Arbeit; diese Eigenschaft existiert nicht mehr während die dritte Stufe und das Material fängt an zu brechen. Kein Kriechmechanismus ist vorausgesetzt worden.
Es ist auch möglich einen σ,ε Diagramm zu beschreiben.

INTRODUCTION

In this paper a method based on the rate theory will be discussed. This theory deals with micro structural particles that have to overcome energy barriers, see figure 1.
Instead of keeping the true activation free energy Q and the work per volume W constant, in this contribution they are time dependent. During the deformation of the material inside micro structural displacements take place and therefore the energy barriers Q will change.
By doing so it is possible to describe the total creep curve as well as the total σ,ε diagram of load controlled experiments.

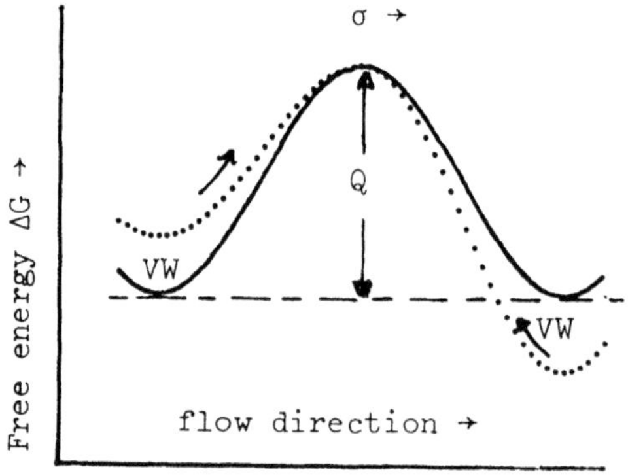

fig 1 Energy barrier

THEORY

In fundamental deformation kinetics {1} the strain rate of creep deformation can be described by the well accepted equation

$$\delta\varepsilon/\delta t = A_f.\exp\{-(Q-VW)/kT\} - A_b.\exp\{-(Q+VW)/kT\} \tag{1}$$

or

$$\dot{\varepsilon} = \{\alpha.\sinh(VW/kT) + \beta.\cosh(VW/kT)\}.\exp(-Q/kT) \tag{2}$$

For a load controlled experiment where $\dot{\sigma}=K$ is the strain rate in the origin (W=0) given by

$$K/E_0 = \beta.\exp(-Q_0/kT) \tag{3}$$

Equation (2) can now be written with (3) as

$$\dot{\varepsilon} = \{\alpha.e^{-\frac{Q_0}{kT}}.\sinh(\frac{VW}{kT}) + \frac{K}{E_0}.\cosh(\frac{VW}{kT})\}.e^{-\frac{Q-Q_0}{kT}} \tag{4}$$

In these equations is

- Q (Nm) the true activation free energy
- V (m^3) the activation volume
- A_f , A_b , α , β and k are constants
- kT (Nm) has the usual meaning
- E_0 (N/m^2) the modulus of elasticity
- $W=\int \sigma \delta\varepsilon$ (N/m^2) the work per volume done on the specimen and contributed by the effective compressive stress σ.

The relation between W and σ is for a sustained load σ_0 given by

- $W = \sigma_0(\varepsilon-\varepsilon_0) + W_0 = \sigma_0.\varepsilon_{creep} + W_0$

The subscript 0 indicates the values at the time t=0.
It is easy to show that in this form of (2) and (4) for creep $\beta=0$.
To solve (2) with $\beta=0$ both sides have to be multiplied by $V\sigma_0/kT$.
Rearranging and integrating with respect to the time t gives the solution

$$\operatorname{arcsinh}\{\frac{1}{\sinh(\frac{VW}{kT})}\} = C - \frac{\alpha V\sigma_0}{kT} \int_0^t \exp(-Q/kT).\delta t \qquad (5)$$

In (5) is
$$\sinh(C) = 1/\sinh(VW_0/kT) \qquad (6)$$

The strain rate at the time t=0 is given by

$$\dot{\varepsilon}_0 = \alpha.\exp(-Q_0/kT).\sinh(VW_0/kT) \qquad (7)$$

We can write (5) in the next form

$$\sinh\{\frac{V(\sigma_0\varepsilon_{cr} + W_0)}{kT}\} = \frac{1}{\sinh\{C - \frac{\alpha V\sigma_0}{kT} \int_0^t \exp(-Q/kT).\delta t\}} \qquad (8)$$

It is easy to see that a material will break when the denominator in the right hand side of (8) is zero. The argument of this denominator can be considerd as a damage function.
The strain rate can now be written with (4) and (8) as

$$\dot{\varepsilon} = \frac{\dot{\varepsilon}_0.\sinh(C).\exp\{-(Q-Q_0)/kT\}}{\sinh\{C - \frac{\alpha V\sigma_0}{kT} \int_0^t \exp(-Q/kT).\delta t\}} \qquad (9)$$

We have now to find a relation between the true activation free energy Q and the time t. If we consider VW/kT to be very small or in (9)

$$C \gg \frac{\alpha V\sigma_0}{kT} \int_0^t \exp(-Q/kT).\delta t \qquad (10)$$

then equation (9) reduces to the simple and well known relation

$$\dot{\varepsilon} = \dot{\varepsilon}_0.\exp\{-(Q-Q_0)/kT\} \tag{11}$$

During the propagation of the creep deformation we can suppose that a material undergoes transitions from a low energy stage to a higher state. This means that we can expect a positive rate of the activation energy Q. If we assume that there is a relation betwee the energy rate and the time given by

$$\frac{\delta Q}{\delta t} = \frac{m.kT}{t + d} > 0 \tag{12}$$

then the solution of (12) is

$$_0\int^t \exp\{-(Q-Q_0)/kT\}\delta t = {_0\int^t} \left(\frac{d}{t + d}\right)^m.\delta t = \frac{d}{n}\{\left(\frac{t+d}{d}\right)^n - 1\} \tag{13}$$

It has to be mentioned here that the constant $m=1-n > 0$ and that the experimental results show that $m<=1$. So the material is able to build up higher energy barriers against the work contributed by the external effective stress σ_0 {2}.
If the sustained load $\sigma_0 = 0$ then no changes in the activation energy Q has to be expected and hence it follows that $d=\infty$. If the load is very high then d has to be small compared to the time t. The results show a relation between d and σ_0: $d=d_0/\sinh(b.\sigma_0)$. This relation is derived theoretical in {3}.
Equations (8) and (9) with (13) can describe the total primary-secondary-tertiary creep deformation and the strain rate respectively.
A special point of interest on the creep curve is the time that the creep deformation starts its tertiary or accelerating stage. At this moment t_1 the strain rate has a minimum. The equation to calculate this extreme can be derived on the usual way and gives $\dot{Q}_1 = V\sigma_0\dot{\varepsilon}_1$. This time t_1 can be calculated for

m=1 by $$\ln(L) = \{C - \text{arcsinh}(M)\}/M \tag{14 a}$$

m≠1 by $$L^n = n.\{C - \text{arcsinh}(M.L^n/(1-n))\}/M + 1 \tag{14 b}$$

with $L = t_1/d + 1$ and $M = V\sigma_0\dot{\varepsilon}_0 d.\sinh(C)/kT$

The apparent free energy of activation $\Delta G = Q - VW$ associated with the forward creep deformation shows on the same moment a maximum and (14 a,b) satisfies also this extreme condition, see figure 8. In {4} is Q replaced by the crack propagation parameter $2\gamma a$ and VW by $\pi\sigma^2 a^2/2E$. So at this maximum the classical Griffith theory define the equilibrium crack size (surface). This means that during the primary and secondary creep deformation depends on microstructural displacements and during the tertiary stage

(micro-)crack propagation occurs.
There is no physical reason to expect a maximum of the strain rate when the deformation starts to creep. During loading the strain rate increases but after the sustained load is reached the creep deformation decelerates. Hence the strain rate should have a maximum at time t=0. Equations (14) now become

$$\frac{V}{kT}\,\frac{\sigma_0 . \dot{\varepsilon}_0 d}{m} = 1 \qquad (15)$$

Figures 2,3,4,5,6 and7 show the influence on the creep curve by the parameters $\dot{\varepsilon}_0$,d,σ_0,V,W_0 and n. In the fig. 2,3,4 and 5 the relevance parameters decrease from left to right with 25%. Fig.6 shows that a change of 4 times of W_0 does not influence the creep curve so much. This is also in fig.7 where the value of n changes 16 times. Figures 2 and 3 show that $\dot{\varepsilon}_0$ and d influence also the primary stage.

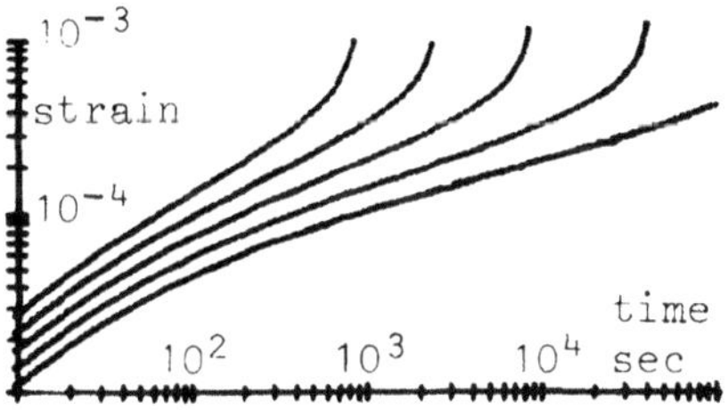

fig 2 influence of $\dot{\varepsilon}_0$

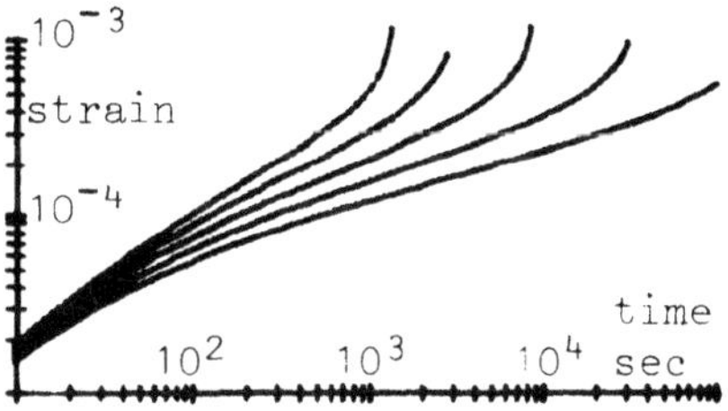

fig 3 influence of d

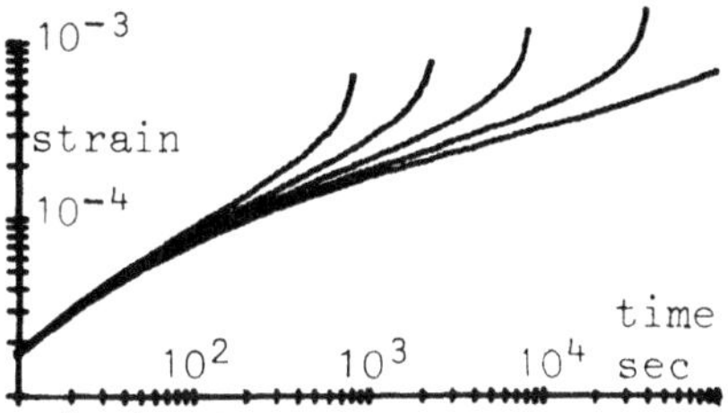

fig 4 influence of σ_0

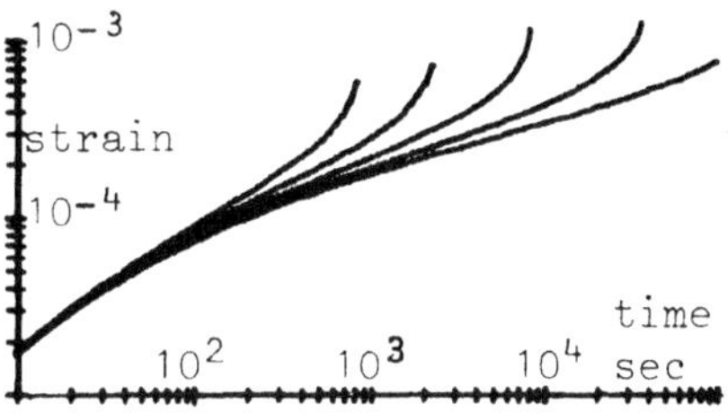

fig 5 influence of V

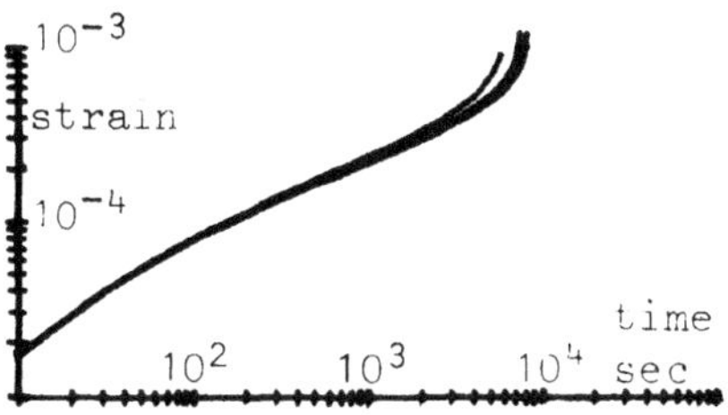

fig 6 influence of W_0

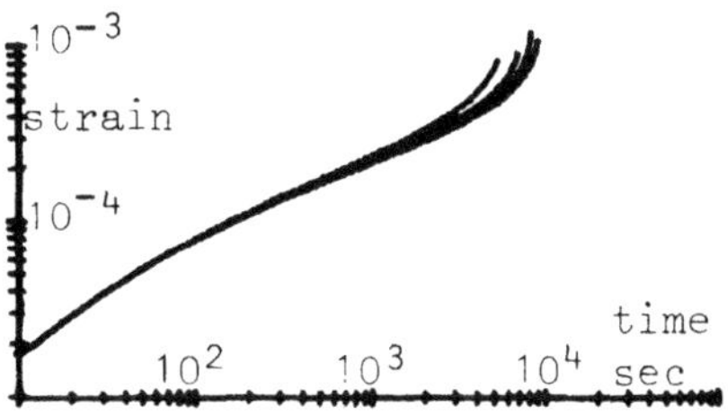

fig 7 influence of n

Finally figure 8 shows versus the time: the creep deformation $\varepsilon(t)$ according (8) with (13), the strain rate $\dot{\varepsilon}(t)$ according (9) with (13) and the relation of the apparent free energy of activation: $\Delta G - \Delta G_0 = Q - Q_0 - V\sigma_0\varepsilon_{cr}$.

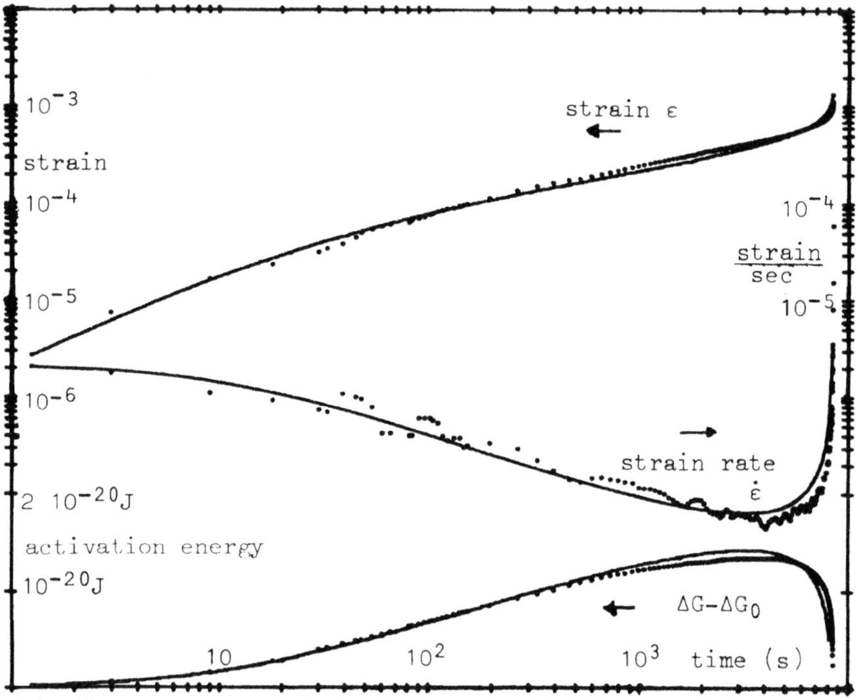

fig 8 Creep deformation

EXPERIMENTS AND RESULTS

The creep experiments were carried out as usual. To obtain the initial condition W_0 precautios were taken to measure the σ,ε values during loading.
The environmental temperature was 20 ^{0}C, so $kT = 4.045.10^{-21}$ J.
Although not allowed the humidity and the aging effect of the specimens were not taken into account.
After the calculation of the parameters $\dot{\varepsilon}_0$, d, V and n several combinations of these parameters does not depend on σ_0.
The results are summarised in table 1.
The value $\alpha.\exp(-Q_0/kT)$ and Q_0 can be obtained from (7). The mean value over all specimens was $2.53.10^{-6}$(1/s) resp. $Q_0 = 52\ 10^{-21}$ J.
In the literature the combination $V\sigma_0$ is known as to be constant and according (15) $\dot{\varepsilon}_0 d/m$ does not depend on σ_0, see column 4. The average of this value is $52\ 10^{-6}$ and gives $V\sigma_0 = mkT/\dot{\varepsilon}_0 d = 7.8.10^{-17}$J. This is in good agreement with {5}. This means that the creep

deformation take place in the hardened cement paste.

Table 1 Experimental results

σ_0	d	m	$\frac{\dot{\varepsilon}_0 d}{m}$	$\frac{V\sigma_0 . \dot{\varepsilon}_0 d}{kT \quad m}$	$\frac{Q_1-Q_0}{kT}$	t_1	t_r
$10^6 \frac{N}{m^2}$	sec		10^{-6}			sec	sec
9.54	10800	1	26.9	18.10^{-6}	64.8	$4.3\ 10^{33}$	$12\ 10^{33}$
15.26	5720	1	46.4	$73\ 10^{-4}$	31.5	$9.4\ 10^{24}$	$26\ 10^{24}$
28.32	38.3	0.991	55.8	0.10	7.29	73700	$20\ 10^4$
28.67	38.9	0.99	81.2	0.26	3.65	9374	$25\ 10^3$
29.51	22.3	0.996	39.6	0.16	5.11	3734	10^4
31.61	63.1	0.988	58.7	0.12	6.24	35200	$96\ 10^3$
31.68	8.1	0.985	60.7	0.27	2.65	226	661

In spite of this column 5 is very discouraging.
If the loading is very fast then W_0 can be replaced by $\sigma_0{}^2/2E_0$.
The young's modulus is $E_0=4\ 10^{10} N/m^2$ and $b=V\sigma_0/2kTE_0=2.4\ 10^{-7} m^2/N$.
The values of d and σ_0 have been fitted and give the relation:

$$d = \frac{9000}{\sinh(6.55\ 10^{-8}.\sigma_0)} \tag{16}$$

The correlation coefficient r=0.997.
The value b is now $6.55\ 10^{-8}\ m^2/N$, that means that in this case $V\sigma_0=2.1\ 10^{-17}$ J, this is perhaps still within the limits of error.
For all the specimens the ratio of the time t_1 to the rupture time t_r is between 0.34 to o.38.
In column 6 is to see that a material, especial concrete, is able to build up higher energy barriers when the creep propagation is slower or when the sustained load is lower.
For the tested concrete with an average strength of $\sigma_b=35.4\ N/mm^2$ two relations can be derived:

$$^{10}\log(t_1) = 39.51 - 50.9.\sigma_0/\sigma_b \tag{17 a}$$

$$^{10}\log(t_r) = 39.95 - 50.9.\sigma_0/\sigma_b \tag{17 b}$$

By solving (4) for load controlled experiments it was possible to fit data form σ,ε curves with a correlation coefficient better than 0.99, see figure 9 and 10.
The correspondent parameters such as the activation energy and the activation volume are in the same range as the values obtained from the creep curves. There is found a relation between the activation volume and the stress for stresses > $0.3.\sigma_b$.

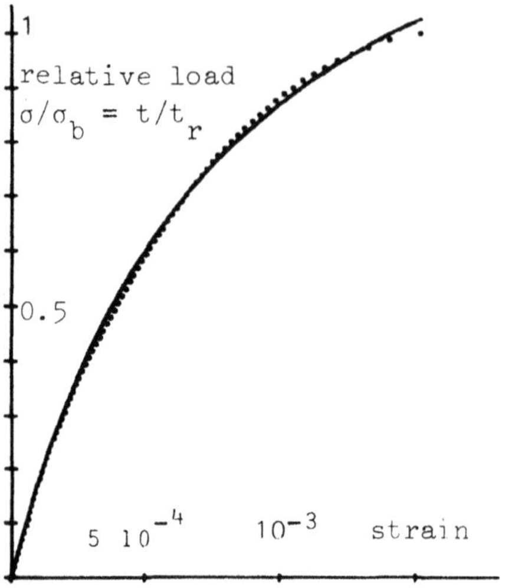

fig 9 σ,ε diagram

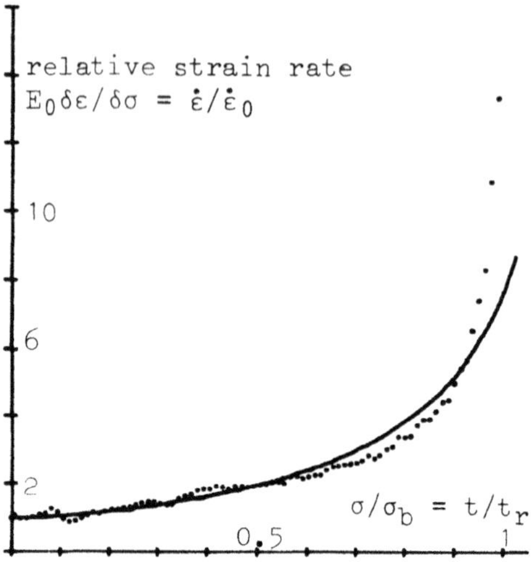

fig 10 strain rate versus time

CONCLUSIONS

The above can describe the total creep curve with correlation coefficient better then 0.997. The calculated parameters depend strongly on the fitting procedure and on the place of the ε,t points on the curve. In spite of (15) is not confirmed, the above can be useful to predict the creep deformation and the rupture time and therefore important for practical use.

REFERENCES

1. A.S. Krausz and H. Eyring, "Deformation Kinetics", John Wiley & Sons, New York (1975)

2. J.C. Pollet and S.J. Burns, "thermally activated crack propagation - Theory", Int. of Fracture. vol.13,no.5(oct 1977) 667-679.

3. R.L. Day and B.R. Gamble, "The use of activation spectrum theory to describe creep in concrete", This proceedings.

4. A.S. Krausz, "Time dependent crack propagation in linear-elastic solids", J. Appl. Phys. 49 (7) July 1978, p 3774-3778.

5. P. Klug and F.H. Wittmann, "Activation energy of creep of hardened cement paste", Materieux et Constructios. 2, 7,(1969) p 11-16

SHRINKAGE AND SWELLING OF TWO HYDRATED ALITE PASTES

L. J. Parrott and J. F. Young
Cement and Concrete Association, England and
University of Illinois at Urbana-Champaign USA

ABSTRACT

Measurements of length changes during drying and rewetting were correlated with weight changes and microstructural data for two hydrated alite pastes. It is concluded that the irrecoverable shrinkage that developed during first drying probably resulted from a rearrangement of the hydrate particles under the action of capillary tension stresses. The irrecoverable shrinkage exhibited a maximum at a relative humidity of about 0.35 which was also reflected in the total shrinkage. Recoverable shrinkage is thought to result from changes in surface energy of the solid phases and capillary tension effects. There was evidence that the change in microstructure during first drying can be partially recovered by the development of a disjoining pressure between hydrate particles during resaturation.

RESUME

Pour deux pâtes d'alite hydraté on a correlé les mesures des variations de longueur pendant le séchage et le mouillage, avec les variations de poids et avec les données sur la microstructure. On en a conclu que le retrait irréversible qui est apparu lors du premier séchage, est dû, probablement, à un réarrangement des particules hydratées sous l'action des tensions capillaires. Le retrait irréversible présente un maximum pour une humidité relative de 0.35 environ. Ce maximum est aussi présent dans le retrait total. Le retrait irréversible est dû aux variations de l'énergie de surface des phases solides et des effets de la tension capillaire. On remarque que les changements dans la microstructure pendant le premier séchage peuvent être partiellement rétablis par l'apparition d'une pression de disjonction entre les particules hydratées pendant le mouillage.

Introduction

In this study the length changes of two hydrated pastes with water/alite ratios of 0.4 and 0.6 by weight are examined as a function of hygrometric conditioning. The specimens were dried directly to a selected relative humidity to minimize the duration of testing and consequent aging effects. This procedure also simplifies the interpretation of drying behavior. It contrasts with many earlier studies (e.g. References 1 and 2) where specimens were exposed to a sequence of relative humidities and either periods of equilibration were limited or tests were carried out over prolonged periods. It has been shown that significant structural changes occur in cementitious pastes during first drying (1, 3-5). Thus interpretation of length change results should include a consideration of this changing structure. Previous shrinkage work (5) was restricted to one relative humidity and w/c ratio, while the present work considers drying over a wide range of relative humidities (0.97-0.07) and at two w/c ratios. In addition, new approaches to estimating structural changes are used.

The experiments presented here included length and weight change measurements on specimens of hydrated alite paste that had an effective drying thickness of 3 millimeters. A specimen was dried at a selected relative humidity for 78 days, rewet for 21 days and then dried for a second time. Interpretation of structural changes is thought to be more easily accomplished with the present experimental procedure than with sequential sorption steps. Measurements were carried out on companion specimens to establish the microstructure of the pastes. The results that relate to pore structure have already been reported (4) and it has been shown that

i) The drying process develops at a rate that is dependent upon relative humidity and is particularly slow at relative humidities in the range 0.3 to 0.8.

ii) The alite hydrates do not form a rigid gel. The hydrate particles take up new positions during first drying and, although substantial recovery was noted during rewetting, some micropores apparently become isolated and are subsequently inaccessible to water and other fluids.

Analysis of the silicate structure during drying, which was studied previously (5), was also considered in the present work and results will be reported elsewhere. Silicate polymerization does not seem to alter greatly when mature pastes are dried in the absence of any applied load.

Experimental

The experimental techniques were similar to those reported earlier (4) so only the essential details will be given here. Two slabs of alite paste were cast; one with a water/alite ratio of 0.4 and the other with a ratio of 0.6 by weight. The slabs were sawn to produce prisms with an effective drying thickness of 3 millimeters as illustrated in Figure 1. After they were 85% hydrated (5 to 6 weeks in limewater at 25°C) the prisms were dried at a selected relative humidity in the range 0.07 to 0.97. Drying was accomplished with the aid of vacuum desiccators that contained an appropriate saturated salt solution (see Table 1). Weight changes during drying were measured to ± 0.2 mg (ie 0.001% of the specimen weight) using a Mettler Type H35 balance. Strain changes were measured with a dial gauge comparator. The readings were reproducible to ± 5 x 10^{-6} (ie 0.06% of the average shrinkage) and periodic readings on brass and Invar reference specimens indicated that long-term drift was no greater than 10 x 10^{-6}.

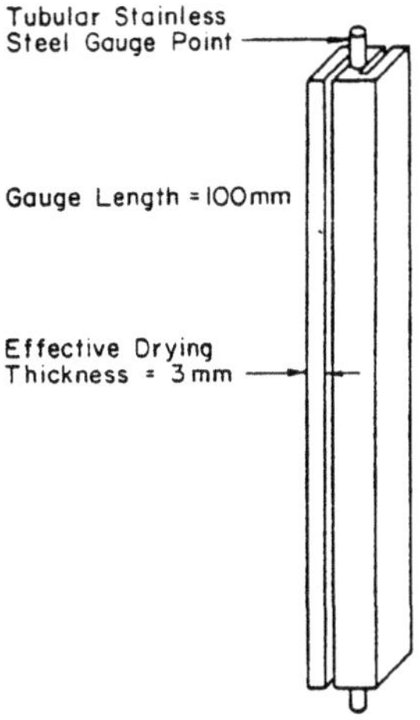

FIG. 1. Alite Paste Specimen

TABLE 1

Saturated Salt Solutions for Relative Humidity Control

Salt	K_2SO_4	$BaCl_2$	$(NH_4)_2SO_4$	$CuCl_2$	NaBr	KNO_2	NaI	$MgCl_2$	CHCOOK	LiCl	NaOH
Relative humidity at 25°C	0.97	0.90	0.80	0.68	0.58	0.48	0.39	0.33	0.22	0.11	0.07

After 78 days of drying the prisms were rewet in two stages: The first stage involved a transfer to the desiccator with the next highest relative humidity (see Table 1) where the prism was retained for 7 days. The second stage involved resaturation under limewater for 14 days. Weight and length changes during a second period of drying were also determined for comparison with the first drying data.

Results

Kinetics of Shrinkage

Examples of length change data are illustrated in Figures 2 and 3 for two hydrated alite specimens. The changes are plotted against the square root of time. The extent to which equilibrium was achieved during first drying is shown in

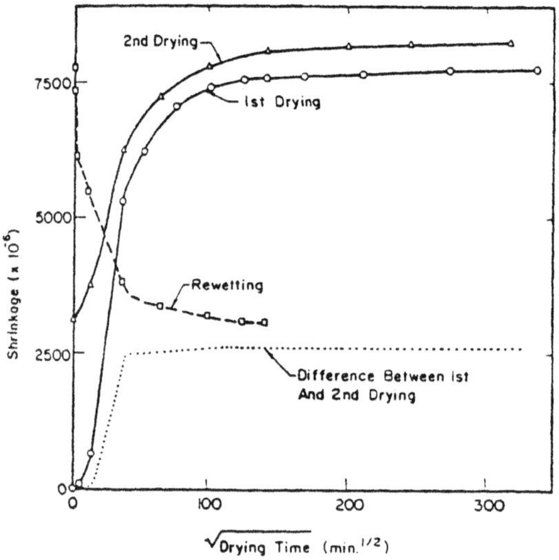

FIG. 2. Shrinkage Versus Square Root of Time for 1st and 2nd Drying at a Relative Humidity of 0.07 (water/alite = 0.4)

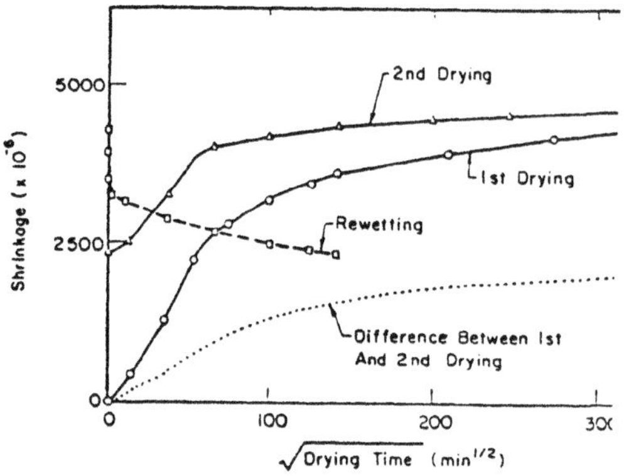

FIG. 3. Shrinkage Versus Square Root of Time for 1st and 2nd Drying at a Relative Humidity of 0.58 (water/alite = 0.4)

Figure 4. It can be seen that even after 78 days of drying no specimens are at weight and length equilibrium. Lack of equilibrium is particularly noticeable at relative humidities in the range 0.3 to 0.9.

Irrecoverable Shrinkage

Rewetting did not lead to full recovery of length as may be seen in Figures 2 and 3. The irrecoverable and recoverable components of shrinkage were greatly dependent upon the relative humidity of drying (see Figure 5). It can be seen that there is a maximum in the irrecoverable shrinkage which is also reflected in the total shrinkage. The rate at which the irrecoverable shrinkage developed

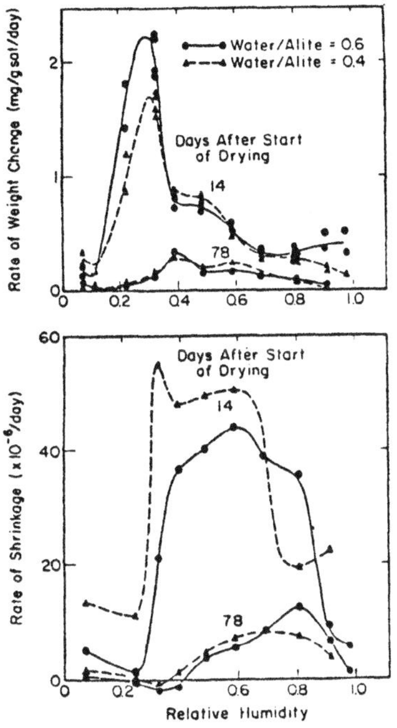

FIG. 4. Rates of Weight Change and Shrinkage Versus Relative Humidity

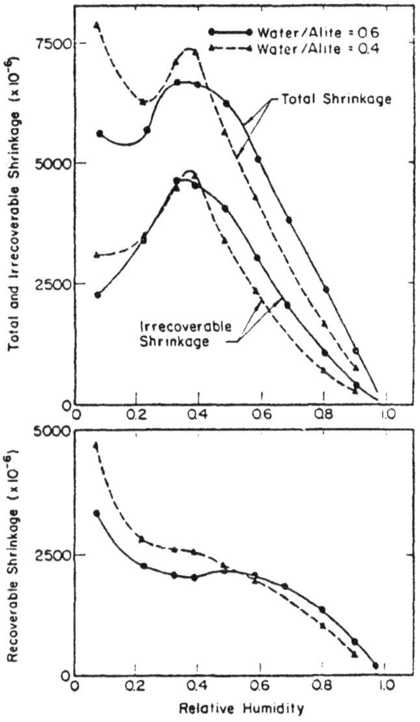

FIG. 5. Total, Irrecoverable and Recoverable Shrinkage Versus Relative Humidity (Dried for 78 Days; Rewet for 21 days)

with time after the start of first drying was studied in two ways. Figure 6 shows typical results from the direct approach of rewetting prisms after selected periods of drying. It can be seen that irrecoverable shrinkage develops mainly between 1 and 14 days after the start of drying. This conclusion was also true for specimens with a water/alite ratio of 0.6 dried at a relative humidity of 0.33. The irrecoverable component of shrinkage developed more slowly when specimens were dried at a relative humidity of 0.8. The method of rewetting specimens after selected drying times would have required large numbers of prisms for a complete investigation of irrecoverable shrinkage development. An alternative approach is demonstrated in Figures 2 and 3 where it can be seen that most of the irrecoverable shrinkage develops during first drying. Thus an indication of the rate of irrecoverable shrinkage development can be obtained by differencing the first and second drying shrinkage curves. Figure 2 shows that this strain difference develops rapidly and then stabilizes

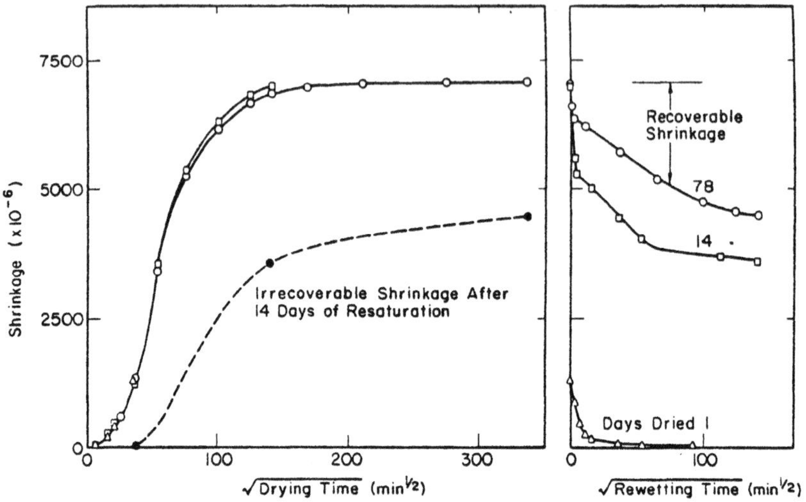

FIG. 6. Shrinkage and Recovery After 1, 14 and 78 days of Drying at a Relative Humidity of 0.33 (water/alite = 0.4)

abruptly one day after the start of drying at a relative humidity of 0.07. However, Figure 3 shows that the strain difference develops gradually over a prolonged period for drying at a relative humidity of 0.58.

Shrinkage-Weight Change Relationships

In addition to relationships between length change and relative humidity or time of drying it is instructive to examine the length change versus weight change relationships for the alite specimens: An example is shown in Figure 7 for drying at a relative humidity of 0.33. The length change was proportionally much less recoverable than the weight change. Also resaturation involved a transition from a region where small expansions were accompanied by large weight gains to a region where moderate expansions were associated with virtually no weight changes. The second drying curve was quite different from the first and again it seems that the irrecoverable shrinkage on first drying develops at an intermediate stage of drying. The shrinkage versus weight loss data for 0.4 water/alite pastes dried at different relative humidities are shown in Figure 8. Although, shrinkage is not a unique function of weight loss this could be an acceptable approximation for some purposes if drying was restricted to relative humidities above 0.3.

An examination of irrecoverable shrinkage, as indicated by the difference between first and second shrinkage (see Figures 2 and 3), is plotted against the weight loss during first drying in Figures 9 and 10. It can be seen

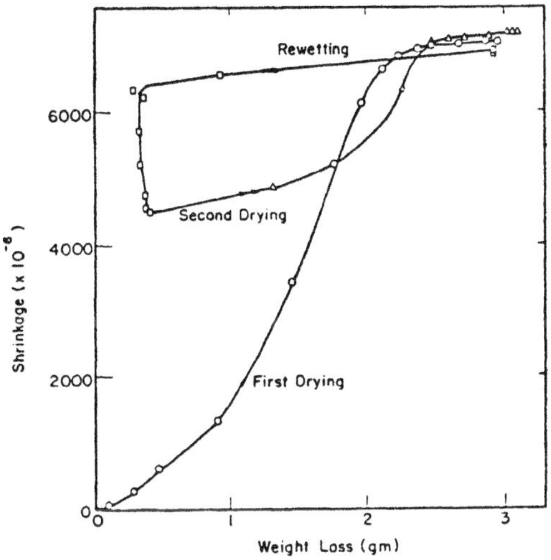

FIG. 7. Shrinkage Versus Weight Loss (water/alite = 0.4, Relative Humidity = 0.33)

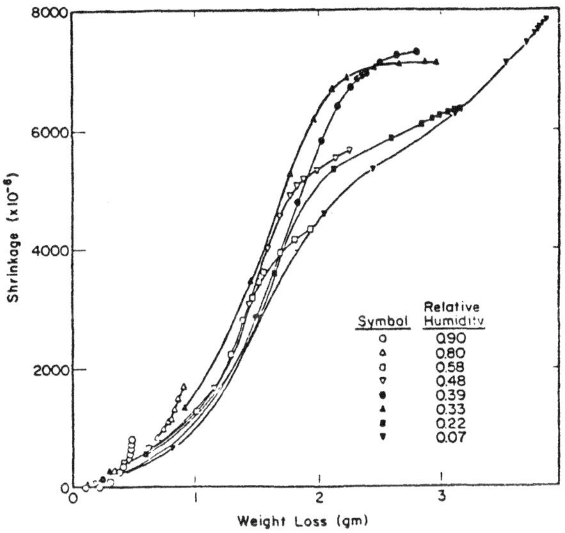

FIG. 8. Shrinkage Versus Weight Loss During First Drying (water/alite = 0.4)

that this measure of irrecoverable shrinkage develops in a fairly restricted range of intermediate water content; Also the water content range is dependent upon the water/alite ratio. It is particularly noticeable at the higher water/alite ratio (Figure 10), that a substantial amount of water can be lost without much irrecoverable shrinkage occurring.

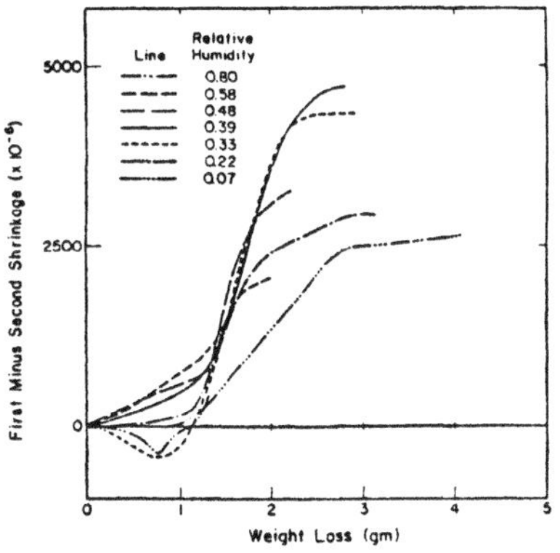

FIG. 9. 1st minus 2nd Shrinkage Versus First Drying Weight Loss (water/alite = 0.4)

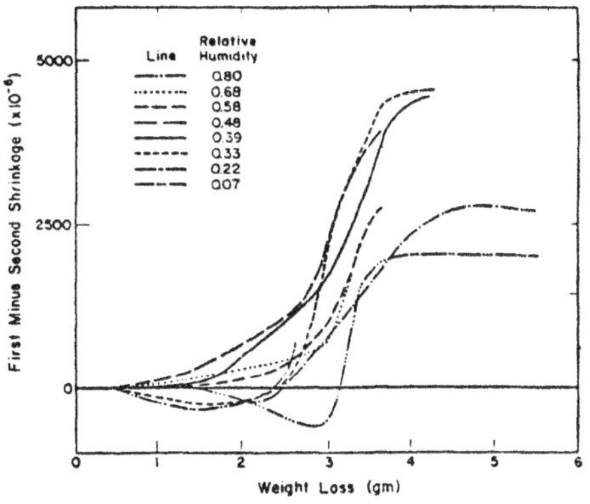

FIG. 10. 1st minus 2nd Shrinkage Versus First Drying Weight Loss (water/alite = 0.6)

Discussion

Effect of Relative Humidity

The response of a paste to loss of moisture depends on the magnitude of the relative humidity to which it is exposed. The results for first drying at a relative humidity of 0.07 exhibit the following features

a) The period in which irrecoverable shrinkage develops is limited to about one day (Fig. 2)

b) The period in which the specimen contains an intermediate amount of water is limited to about one day (see Figures 9 and 10 and ref. 4)

c) Irrecoverable shrinkage is about one third of the total shrinkage (Fig. 5)

d) Weight and length change approach equilibrium relatively quickly (Figures 2 and 4 and ref. 4)

First drying at an intermediate relative humidity, 0.58 for example, yields features which contrast strongly with those given above

e) The period in which irrecoverable shrinkage develops is extended and presumably would exceed the 78 day period of first drying (Fig. 3)

f) The specimen contains an intermediate moisture content (Figures 9 and 10 and ref. 4) for long periods

g) Irrecoverable shrinkage is about two thirds of the total shrinkage (Fig. 5)

h) Weight and length changes do not reach equilibrium during the 78-day period of first drying (Figures 3 and 4 and ref. 4).

Possible Origins of Shrinkage

A simple mechanism that can account for these contrasting observations must be able to explain both recoverable and irrecoverable shrinkage. It must also be able to explain why irrecoverable shrinkage develops only at a specified range of moisture contents and is greatest at intermediate relative humidities. This behavior can be best explained by considering a creep response to a stress developed in the skeletal structure of the paste during drying. The origin of this stress is thus of foremost interest.

Capillary Tension: One possibility that fits the data is that stress develops as a result of capillary tension in the pore liquid. This could explain why the maximum in the irrecoverable and total shrinkage occurs at a relative humidity of about 0.35 (Fig. 5). Furthermore, it can explain why irrecoverable shrinkage terminates abruptly at a low relative humidity (Fig. 2) and develops gradually at intermediate relative humidities (Fig. 3). As the relative humidity is lowered both the maximum stress, and the time of imposition of the stress, will increase until a relative humidity of about 0.4 when the stress is relieved as pores empty. This view could also explain why a narrow range of water loss is so important, why there is a small rate of expansion at relative humidities in the range 0.3 to 0.4 when the water loss is still positive (Fig. 4), and why the large weight loss is accompanied by only a small amount of shrinkage during the later stages of drying at 0.33 relative humidity (Fig. 7).

Some other relevant data concerning the swelling and weight gains during the adsorption step after first drying are illustrated in Figure 11. To normalize the data the swelling has been divided by the weight gain. The sensitivity of the swelling results to relative humidity is consistent with stresses in the paste being controlled by capillary tension effects at intermediate relative humidities. Figure 11 also shows the calculated surface area of the liquid/vapor interface. This area was obtained by dividing the weight of water sorbed by the expected increase in the thickness of the adsorbed water film (6). It should be emphasized that the calculated area is not that of the solid/liquid interface and that there is no correction included for capillary condensation effects. A significant feature of Figure 11 is that unit adsorption on a fairly large surface at low relative humidities causes only a minor expansion relative to that associated with the uptake of a similar quantity of water at intermediate relative humidities.

Disjoining Pressure

Although capillary tension effects seem helpful in explaining the shrinkage data it is possible to explain many of the results in terms of a model in which the shrinkage stress originates from a reduction in disjoining pressure (7) in narrow pores. There are, however, two results which suggest that shrinkage does not simply result from a reduction in disjoining pressure. First, the small expansion rate that was shown in Figure 4 and was discussed above, would not be expected. Second, exchange of water for methanol followed by exchange of the methanol for n-pentane did not cause any significant shrinkage. In spite of the completeness of the fluid exchange process (see ref. 4) the length changes were extremely small. According to Derjaguin (6) a considerable change in disjoining pressure should result from the replacement of a polar molecule (water or methanol) with a non-polar molecule (n-pentane). The results suggest that the solid hydrate particles in the alite paste were not subject to a

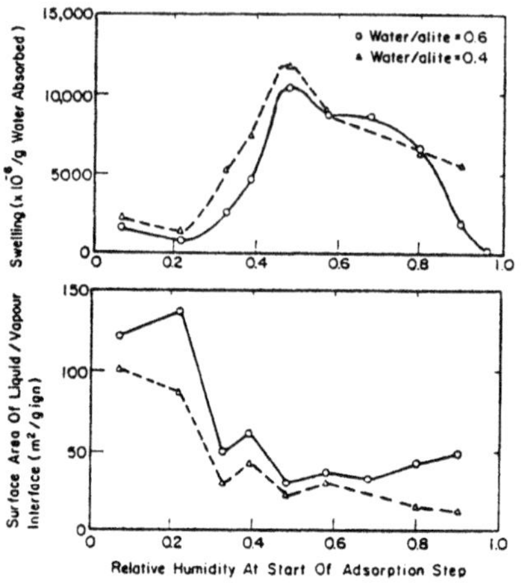

FIG. 11. Swelling and Surface Area of the Liquid/Vapor Interface Accessible During Adsorption Steps

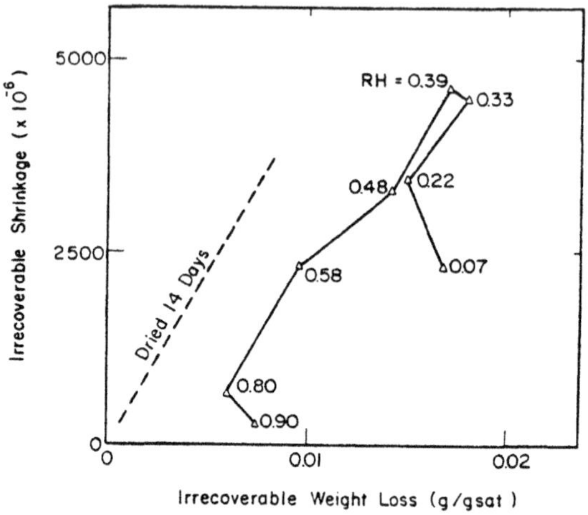

FIG. 12. Irrecoverable Shrinkage Versus Irrecoverable Weight Loss (water/alite = 0.4, dried 78 days)

disjoining pressure in the original saturated state. Paradoxically the results for resaturation (eg Fig. 7) imply that rewetting does involve the generation of a disjoining pressure since the later stages of swelling are not accompanied by any significant weight gain (ie the swelling is apparently associated with a redistribution of pore water). Perhaps this paradox could be resolved by assuming that at the start of drying the spacing of hydrate particles is sufficient for menisci to form between them, thereby causing a reduction of interparticle spacing. This change would be partially recoverable under the action of a disjoining pressure developed between hydrate particles during resaturation.

Irrecoverable Shrinkage

It can be seen from Figure 7 that the irrecoverable shrinkage is accompanied by an irrecoverable weight loss. A plot of the irrecoverable weight loss against relative humidity of drying (4) had many features in common with the irrecoverable shrinkage plot in Figure 5. This similarity is explored in more detail in Figure 12. The figure shows a correlation between irrecoverable weight loss and irrecoverable shrinkage although water may be irrecoverably lost at high relative humidities without much irrecoverable shrinkage developing. Some limited data obtained after 14 days of drying supports the view that the irrecoverable weight loss has two components of which one dominates irrecoverable shrinkage. Helmuth and Turk (1) have examined two simple models for predicting the ratio of the irrecoverable volumetric shrinkage to the volume of water irrecoverably lost.

The models were based upon a) irrecoverable shrinkage resulting directly from a pore collapse (ratio = 1) and b) irrecoverable shrinkage developing at a constant pore volume to total volume ratio (ratio > 2). The models did not adequately predict the results (ratio $\sim$ 0.5 which was obtained by Helmuth and Turk and also found in Figure 12) and a third possibility can be introduced by way of explanation. The present authors, like Helmuth and Turk, favor model a) above, but would impose the condition that due to rearrangement of the pore structure repenetration with water after drying is incomplete.

The present ideas relating to rearrangement of the hydrate particles during first drying and the effect of this upon irrecoverable shrinkage are consistent with the ideas of Pickett (8) and of Feldman and Swenson (9). They also concluded that irrecoverable shrinkage developed at relative humidities above 0.5 and was related to how the original dispersion of hydrate particles could allow subsequent rearrangement during first drying. Furthermore they demonstrated that certain admixtures can affect the dispersion of hydrates and cause a high irrecoverable shrinkage without influencing the recoverable shrinkage. Bentur, et al. (10) reached similar conclusions after observing that pastes containing $CaCl_2$ had large total and large irrecoverable shrinkages. These were attributed to the effects of $CaCl_2$ upon the "pore component."

It should be noted that in this study a correlation between the surface area measured by nitrogen adsorption and irrecoverable shrinkage was not observed, since the surface area is mainly recoverable if the pastes are resaturated before drying by solvent replacement (4). Although this is apparently in contradiction to earlier results (5), it should be noted that if the pastes were not resaturated before solvent replacement then there is an apparent reduction in surface area. This latter approach is closer to the drying conditions used earlier and indicates that the relationship between surface area reduction and irrecoverable shrinkage is not a fundamental one.

Recoverable Shrinkage

The recoverable shrinkage at high humidities would be due to the elastic response of the material to the imposed stress, as well as recoverable volume changes in the pore system. At low relative humidities it seems that changes in surface energy of the solid best account for the increase in recoverable shrinkages in the region where capillary stresses no longer operate.

A feature in Figures 2, 3 and 6 is the discontinuity in the swelling behavior where a transition occurs from fast swelling to slow swelling about

10 minutes after immersion. This points marks the time at which swelling continues without further weight gains. This suggests that the slow component results from a redistribution of water within the material which causes dispersion of the hydrate particles. The fast component shows a sensitivity to relative humidity that would be expected if it were due to changes in the capillary tension and surface energy during the filling of the larger pores.

Silicate Polymerization

Earlier studies (5) suggested that silicate polymerization during drying might be the cause of irrecoverable shrinkage in well-hydrated C_3S pastes. This conclusion was not substantiated in the present work using gel permeation chromatographic analysis of the trimethylsilyl derivatives of the paste. The main effect of silicate structure upon shrinkage is thought to be via its control of the response of the skeletal part of the paste to the stresses induced by drying.

Conclusions

The results from a study of the length changes, weight changes and pore structure changes of two hardened alite pastes during drying and rewetting led to the following main conclusions:

1. The shrinkage resulting from first drying was only partially recoverable with rewetting. This was due to an irrecoverable shrinkage component that developed at an intermediate stage of drying.
2. The irrecoverable component of shrinkage was associated with an irrecoverable loss of water. This water loss was greater in volume than the irrecoverable volumetric shrinkage and suggested that first drying caused isolation of some spaces.
3. Irrecoverable shrinkage was attributed to capillary tension forces in the liquid phase that stressed and rearranged the hydrated particles.
4. A maximum in the irrecoverable shrinkage, that was reflected in the total shrinkage, was observed at about 0.35 relative humidity. The reduction in the irrecoverable shrinkage with relative humidity below 0.35 was thought to be due to the limited period of exposure of the paste to capillary tension stresses.
5. Recoverable shrinkage was thought to be due to changes in surface energy of the hydrate particles and to capillary tension effects, depending on the relative humidity.

6. Swelling during resaturation exhibited two distinct stages wherein the capillary tension effects are first negated and then redistribution of water occurs. Since the later stages of swelling are not accompanied by any significant weight gain it appears that disjoining pressure is generated that partially, but not fully, redisperses the hydrated alite particles.

References

1. R. A. Helmuth and D. H. Turk, "The Reversible and Irreversible Drying Shrinkage of Hardened Portland Cement and Tricalcium Silicates Pastes," J. Portland Cement Assoc., R&D Labs., 8-21 (May 1967).
2. R. F. Feldman, "Sorption and Length Change Scanning Isotherms of Methanol and Water on Hydrated Portland Cement," Proc. Fifth Int. Congr. Chem. Cement, Tokyo, Vol. III, 36-44 (1968).
3. C. M. Hunt, L. A. Tomes and R. L. Blaine, "Some Effects of Aging on the Surface Area of Portland Cement Paste," J. Res. Nat. Bureau Standards, 64A, 2, 163-169 (1960).
4. L. J. Parrott, W. Hansen and R. L. Berger, "Effect of Drying and Rewetting upon the Pore Structure of Hydrated Alite Paste," to be published.
5. A. Bentur, R. L. Berger, F. V. Lawrence, Jr., N. B. Milestone, S. Mindess and J. F. Young, "Creep and Drying Shrinkage of Calcium Silicates Pastes, III. A Hypothesis of Irreversible Strains," Cem. Concr. Res., 9, 83-96 (1979).
6. J. Hagymassy, Jr., S. Brunauer and R. Sh. Mikhail, "Pore Structure Analysis by Water Vapor Adsorption, I. t-Curves for Water Vapor," J. Colloid Interface Sci., 29, 3, 485-491 (1969).
7. B. V. Derjaguin and N. V. Churaev, "Structural Component of Disjoining Pressure," J. Colloid Interface Sci., 49, 2, 249-257 (1974).
8. G. Pickett, "Effect of Aggregate on Shrinkage of Concrete and Hypothesis Concerning Shrinkage," J. Amer. Concr. Inst., 27, 5, 581-590 (1956).
9. R. F. Feldman and E. G. Swenson, "Volume Changes on First Drying of Hydrated Portland Cement with and without Admixtures," Cem. Concr. Res., 5, 1, 25-35 (1975).
10. A. Bentur, N. B. Milestone, J. F. Young and S. Mindess, "Creep and Drying Shrinkage of Calcium Silicates Pastes, IV. Effects of Accelerated Curing," Cem. Concr. Res., 9, 2, 161-170 (1979).

Acknowledgements

The first author is indebted to the Cement and Concrete Association, U.K. for leave of absence to undertake this study. Stimulating discussions with W. Hansen, Prof. R. L. Berger and Prof. F. V. Lawrence are gratefully acknowledged. This work was supported by National Science Foundation Grant Number ENG 78-09529.

CREEP AND MICROSTRUCTURAL CHANGES IN CONCRETE

A K Mullick

Cement Research Institute of India New Delhi

ABSTRACT

A hypothesis relating to microstructural changes in the cement paste following creep is presented. The changes are such as to render the microstructure of cement paste in a more stable configuration and thereby, to impart 'pseudo' strain-hardening to the composite. A part of the irreversible creep in maturing concrete is caused by such microstructural changes which can be taken as a measure of the stabilisation effects. The creep response of concrete under subsequent stress applications is reduced because of the microstructural changes and is a function of the stabilisation caused by the previous stress history. Phenomenological evidence in support of the hypothesis is provided by experimental data on creep under constant and variable stress regime.

On a présenté une hypothese sur les changements de microstructure se produisant dans la pâte de ciment à la suite du fluage. Ces changements donnent une configuration plus stable à la microstructure de la pâte de ciment et communiquent ainsi au mélange un pseudodurcissement sous déformation. Une partie du fluage irréversible du béton en prise est dûe à des changements de microstructure que l'on peut considérer comme une mesure des effects de stabilisation. La réaction au fluage du béton soumis à des contraintes ultérieures est diminuée en raison des changements de microstructure et elle est fonction de la stabilisation produite par les contraintes antérieures. L'évidence phénoménologique soutenant cette hypothèse est fournie par les données expérimentales sur le fluage dans des conditions de contraintes constantes et variables.

INTRODUCTION

Although the various theories of creep of concrete differ in regard to the mechanisms at work, there is a near unanimity in the pivotal role ascribed to the solid C-S-H component, the pore component and water in the cement paste in the time-dependent deformations of concrete (1, 2). It is conceivable that macroscopic deformation of concrete caused by various stress, temperature and humidity regimes is a result of the response of the cement paste system at the micro level - ie, individual responses of these random components and possible interactions between them. Within the broad framework of classical 'seepage' and 'viscous' theories of creep of concrete, various mechanisms like solidification and recrystallisation of the solid components, formation of new bonds between gel particles, reduction in inter and intra-crystalline spacing and permanent displacement of hydrate layer in relation to each other have been postulated (1, 2, 3). All these processes would amount to changes in the thermodynamic state of the system (4).

While it can be debated, whether one or all these mechanisms are indeed operative during creep of concrete, it is important to conceive that the microstructure of the cement paste, being in a metastable state, is capable of such changes which lends the hydrate system to a more permanent and stable configuration (5). The new states would have lower entropy and could be of higher activation energies (4, 6). The significance of lower entropy and higher activation energy are that the microstructural changes taking place in the hydrate would be partly irreversible and subsequent mechanical responses of the composite at a macroscopic scale would be reduced. This may be called 'pseudo' strain-hardening of concrete.

HYPOTHESIS AND ITS TEST

If such microstructural changes do take place in the hydrate system during creep of concrete, the pertinent question will be whether the imprint of such changes can be measured. Some attempts have been made in the past to monitor concommitant changes in the pore structure as well as the solid components of the cement paste following time-dependent deformation. Mills held that changes took place in the pore geometry in the cement paste and found that the reduction in the pore spaces not accessable to large-molecular fluids, ie, zones of restricted adsorption, were related to the volumetric strains caused by creep and/or shrinkage (7). Other subsequent evidences include decrease in the surface area of the components (as measured by nitrogen adsorption) and increase in polymerisation bonding of C-S-H components (8). Changes in the inter-layer volume and density of solid phase have also been reported by others (9).

It is significant to note that all these experiments involved drying of the specimens. Such evidence in cases of application of mechanical stress alone, under saturated conditions, have not been reported. In fact, it has been doubted whether increased polymerisation bonding and changes in the pore geometry and dimensions may be caused by mechanical stress (5). Perhaps structural changes caused by mechanical stress are of sub-microscopic dimensions not amenable to direct measurements and, more importantly, are of random nature at the microscopic scale. An attempt was, therefore, made in the present investigation, to provide phenomenological evidence at macroscopic scale in support of the following hypotheses:

i/ Microstructural changes take place during creep of concrete, which render the hydrate phase more stable and rigid and, therefore, are partly responsible for the irreversible component of creep.

ii/ As a sequel to such microstructural changes, the creep responses of such strain-hardened concrete during subsequent strain applications will be smaller than that of a 'virgin' specimen of identical state of hydration and subjected to sustained stress for the first time.

iii/ Such reduction in the creep response of preloaded specimens is related to the irreversible creep generated during the prior loading, the latter thus being an 'index' of microstructural changes.

Suitable experimental programme was devised for testing the above hypotheses. Creep tests on 100 mm x 300 mm cylinderical specimens of maturing concrete under saturated conditions and at 21 C were carried out, both under constant and time-varying compressive stress regimes using a hydraulic loading system. Details of the experimental procedure are given elsewhere (10).

Experimental Results and Discussions

Schematic description of the creep tests, showing the ages of application and removal of the constant and variable stresses and the respective stress-strength ratio (R) are shown in Table 1. The constant stress tests of Series A and B were employed to study the influence of age and level of stress at the first loading and its duration on creep and subsequent creep recovery. Tests in Series C, D and E were designed to study the influence of previous stress history commencing at the age of 7 days involving different

TABLE 1

Description of Creep Tests

Series	Scheme	Variables
A	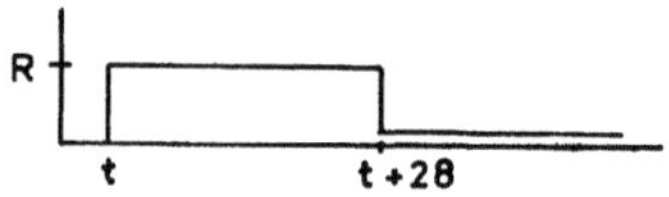	t – 3, 7, 10, 14, 28, 35, 42 and 49 days R – 0.195 to 0.34
B	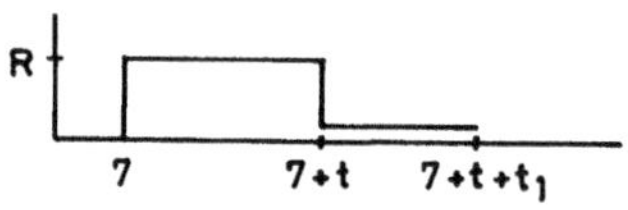	t – 3, 7, 21, 28 and 35 days t_1 – min 14 days, usually 28 days R – 0.26 to 0.40
C	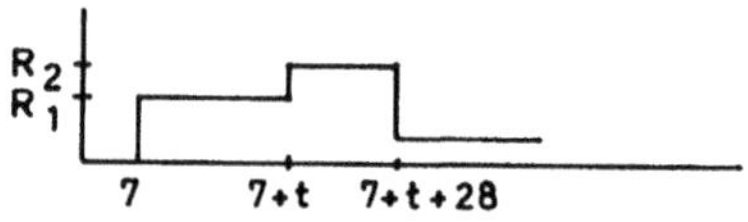	t – 3, 7, 21, 28 and 35 days R_1 – 0.12 to 0.295 R_2 – 0.23 to 0.38
D	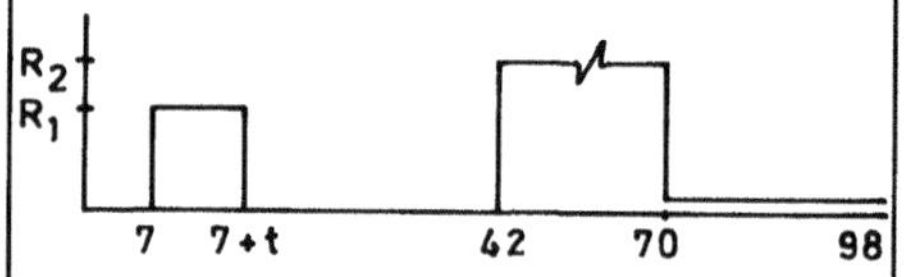	t – 7, 21, 28 and 35 days R_1 – 0.17 to 0.38 R_2 – 0.09 to 0.36
E	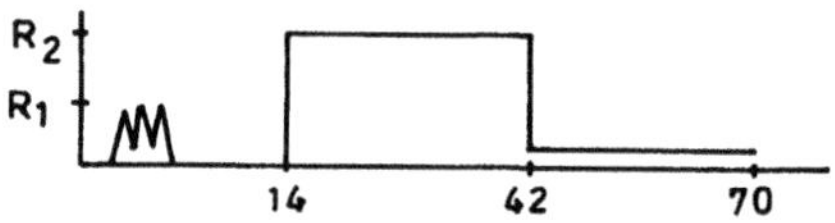	R_1 – 0.4 (cyclic loading) R_2 – 0.14 to 0.42

levels and duration of stress on creep response on subsequent stress application. In Series C, the stress was suddenly increased at different ages of 10, 14, 28, 35 or 42 days. In Series D, a stress either equal, larger or smaller than the previous, was applied at the age of 42 days after a gap between first and second loadings. The previous stress history in Series E comprised cyclic loads to R = 0.4 at 7, 8 or 9 days by means of an MTS electrohydraulic servo-controlled testing machine. Stress pulse was applied in a triangular mode, the time between successive peaks being 2 seconds. The total number of cycles was about 42500.

The time taken for the application of the desired stress was 18 seconds and the first measurement of strain could be completed within 150 seconds of the commencement of the test. The magnitude of stress during a test could be varied at will by using rams of different diameters. The instantaneous strain was reckoned from the dynamic modulus of elasticity of concrete at the respective ages.

Creep-Stress Relation

The relationship between creep and stress-strength ratio (R) was found to be non-linear (10). The relationship was linearised in the manner described by Gamble and Thomass (11) such that isochronous creep was linearly related to a stress factor F(R). For short durations (eg upto 12 hours) after commencement of test, characteristic creep (ie, creep ÷ F(R) was equal for different ages of loading. For longer durations under load, it was lower, later the age of loading (12).

Creep Recovery and Irreversible Creep

Most of the creep recovery took place within 14 days and it tended to reach a limiting value within 28 days. It was found that creep recovery was also non-linearly related to the stress-strength ratio and the same linearising factor derived from creep tests was valid for creep recovery, except that its magnitude was smaller than that of creep. Since no significant microcracking was anticipated in the ranges of stress involved, it would suggest that the processes involved in creep and creep recovery (eg mass transfer) could themselves be non-linearly dependent upon stress. For creep tests started at different ages and unloaded 28 days thereafter, the recoverable proportion of creep increased with the age of loading. Thus, more mature the concrete was at the time of loading the higher was the creep recovery. For creep tests started at 7 days and unloaded after various durations the recoverable proportion of creep decreased as duration under load increased.

The irreversible creep was obtained by subtracting recoverable creep from the total. While the recoverable proportion of creep tended to reach the limiting value within a short time the irreversible component increased gradually with time at a declining rate (Fig 1). The magnitude of irreversible creep was different for different ages and durations of loading, being greater earlier the age of loading or duration under load.

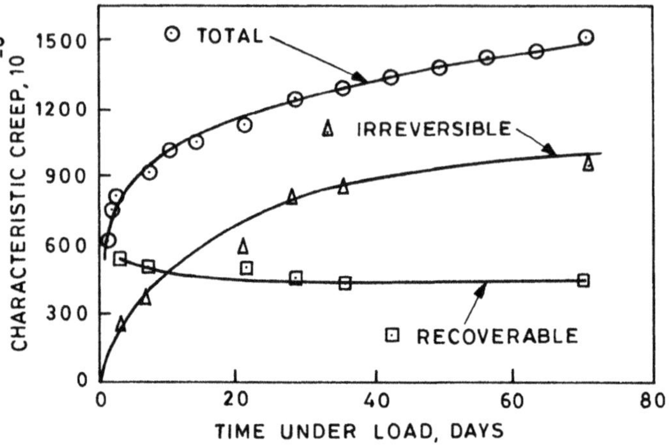

FIG 1

Components of Creep

These phenomena can be partly explained in terms of continuing hydration while under load. In a maturing concrete, more volume of hydration products are formed after the commencement of loading. Because of transfer of load to the newly formed gel, larger volume of hydration products are operative in restraining the recovery of strain upon removal of causitive stress, thereby preventing complete recovery of strain. Continued hydration yet, could not be the only cause of irreversible creep. If that were so, concrete loaded at sufficiently later age would not exhibit any irreversible creep. As such, the contribution of stress itself on the irreversible creep must be recognised. The proportion of irreversible creep during different durations of load and for different ages of loading is compared with the accompanying relative gain in strength in Fig 2; the latter representing qualitatively the relative gain in hydration while under load. The data points in Fig 2 are arranged in two groups: for tests started at 7 days age and unloaded after different duration of loading (shown by solid line), and tests commenced at different ages of loading and unloaded after 28 days (shown by dotted lines).

Both the groups of data show the same trend in that lower the ratio of strength at loading and unloading, the higher the proportion of irreversible creep. Significantly however, the two curves do not coincide, especially when the strength ratio is near to unity. A small change in strength after loading (say 5 per cent) is achieved within a short time for an young concrete but may be achieved by mature concrete over a longer period of time. The microstructural changes that are caused by application of stress cannot be expected to be similar in both the cases.

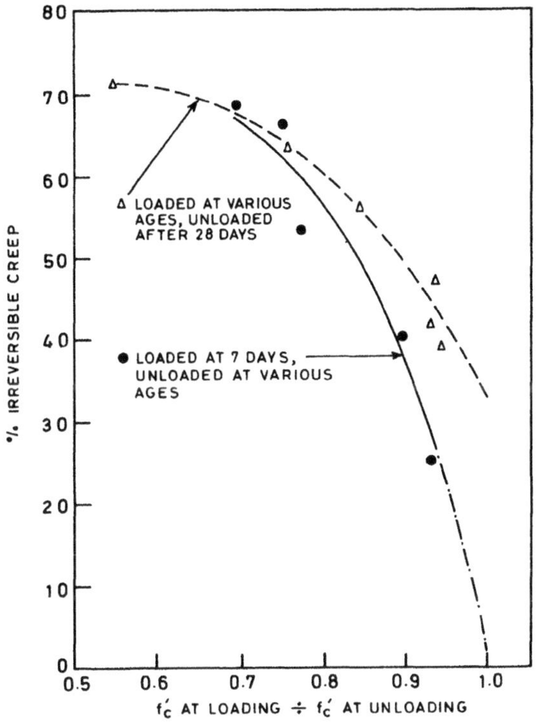

FIG 2

Influence of Continuing Hydration on Irreversible Creep

Consequently, the irreversible creep arising out of structural changes due to load are different in the two cases and that is why the two curves do not coincide. These results show that irreversible creep is not the result of continuing hydration alone and micro-structural changes must be another reason.

Influence of Stress-History on Creep

Results of tests in series C, D and E are summarised in Table 2. The additional creep response under subsequent loading was always lower than the 'reference' creep of virgin specimens subjected to the same stress-strength ratio at the corresponding age. Similar trend has indeed been observed by many others in the past (13). The reduction in irreversible creep during reloading was, however, more pronounced than in total creep.

TABLE 2

Components of Creep in Variable Stress Tests
(All Strains in 10^{-6})

Scheme	Test #	Previous Strain History		During Re-loading		Reference 'virgin' specimen	
		Total creep	Irr. creep	Total creep	Irr. creep	Total creep	Irr. creep
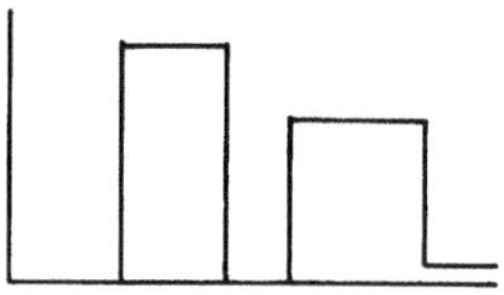	D 7	490	213	153	44	211	99
	D 8	492	267	115	43	178	86
	D 9	379	236	49	12	97	46
	D 10	648	462	-41	9	169	84
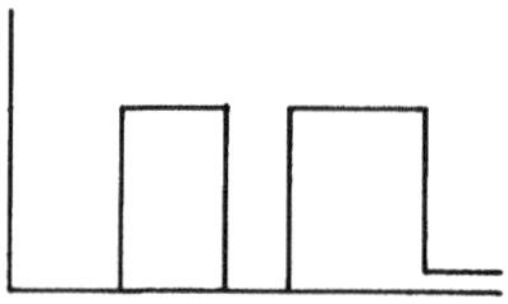	B 1	439	112	322	-	385	-
	D 4	356	159	189	66	203	91
	D 5	374	203	161	59	220	106
	D 6	398	248	157	26	235	110
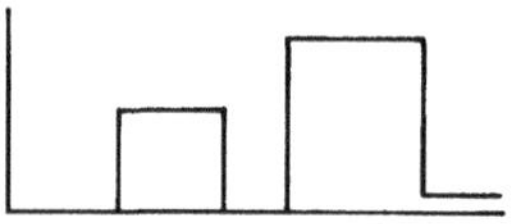	D 1	185	66	368	147	522	245
	D 2	335	182	289	106	464	218
	D 3	216	135	300	112	420	197
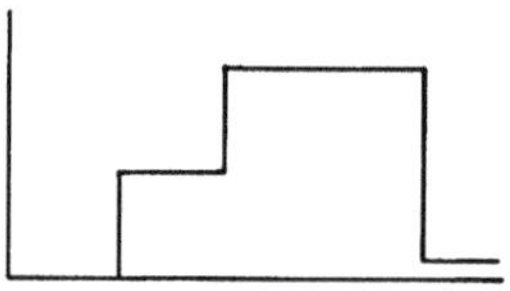	C 1	71	19	196	-	271	-
	C 2	167	68	268	178	427	230
	C 3	331	180	165	113	331	140
	C 4	272	170	312	-	569	-
	C 5	428	287	130	76	551	272
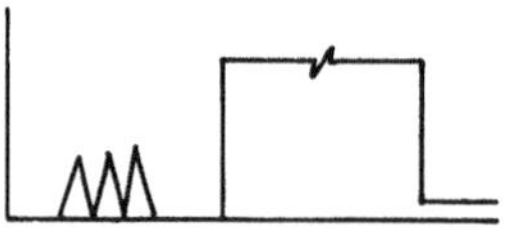	E 1	-	66	547	306	722	416
	E 2	-	66	366	169	383	219
	E 3	-	66	158	66	159	88

The reduction in creep response increased with increase in duration of first loading. For the same duration under first loading, the reduction in creep was more, greater the magnitude of previous stress. Another significant result was that creep-R relationship during second loading was more linear than on first loading (10). This may suggest that creep of concrete is due to a number of mechanisms having different activation energies, some of which may be alleviated by previous stress-history. Thus the effects of previous stress history can be qualitatively summed up as; it reduced creep under subsequent loading, which was more linear with respect to stress and more recoverable in nature.

The effects of stress-history were summed up in terms of previous irreversible creep. Fig 3 show that the reduction in creep response as well as in its irreversible component were dependent upon the previous irreversible creep. Considering that the data in Fig 3 comprised different durations and levels of stress and interactions of continuing hydration in different measures, the trend is considered significant.

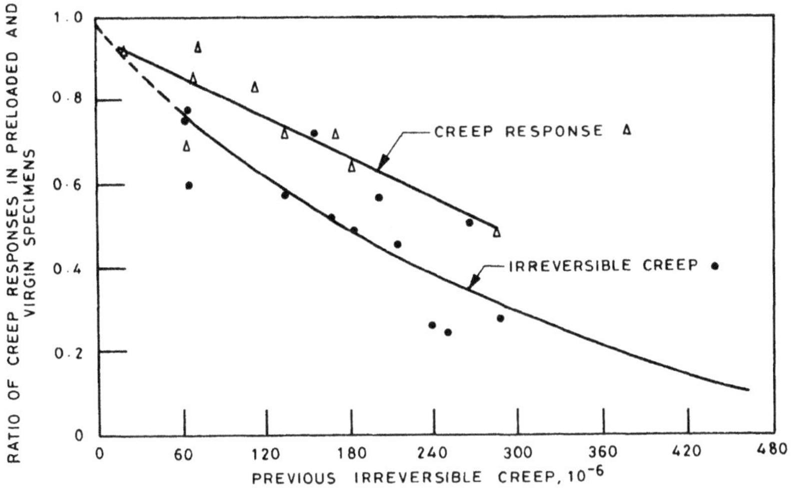

FIG 3

Dependance of Creep Responses
on Previous Irreversible Creep

The magnitude of total creep after second loading to a higher stress was remarkably close to the 'reference' creep of a virgin specimen subjected to the same stress at the later age. In such cases where the second stress (strictly speaking, stress-strength ratio) was lower than in earlier age, total creep was greater than the reference creep, mainly because of accumulation of irreversible creep from previous loading.

STRAIN-HARDENING OF CONCRETE

Fig 4 presents a schematic summary of the results. For a virgin specimen, the non-linear relationship between creep and stress-strength ratio is shown by the curve OAB. For a preloaded specimen with previous irreversible creep = OC, the relation between creep and stress strength ratio is almost linear along the path CD till it meets the virgin curve at point D. For a stress-strength ratio equal to R_1 or greater, the total creep is the same as in a virgin specimen. For reloading to a lower stress-strength ratio $R_2 < R_1$, the total creep EF is greater than AF in the virgin specimen. A family of similar curves for different amounts of irreversible creep from previous loading obtained in the present investigation is shown in Fig 5.

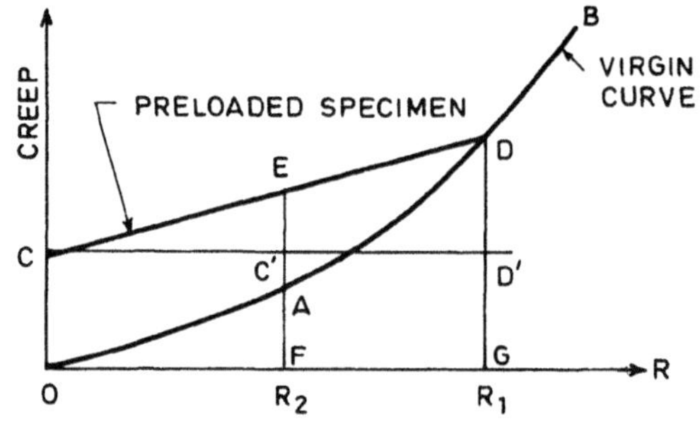

FIG 4

Schematic Summary

The results presented herein will suggest that concrete during creep undergoes some sort of strain-hardening. Reduction in creep due to increase in hydration is manifested by its dependance on age of loading. Enhanced maturing by heat curing also reduces creep and makes it more recoverable in nature (14).

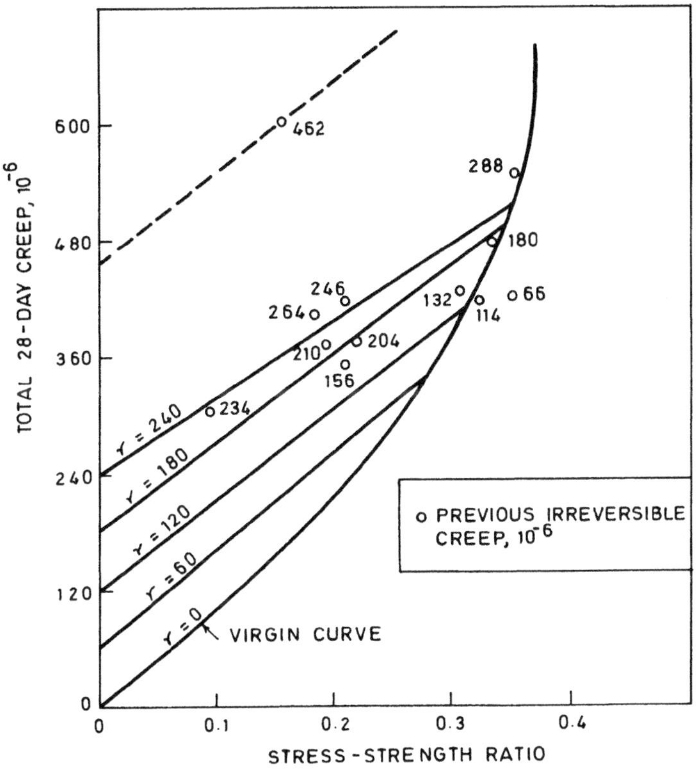

FIG 5

Total Creep in Preloaded Specimens, Series C, D

Compared to that, the additional effects of microstructural changes can be called 'stress-induced' maturity, because of which, simple 'age-hardening' models always over-estimate creep and creep recovery under variable stress (10, 13).

CONCLUSION

It is possible that creep of concrete is caused by a number of process involving the solid hydrate and water bound in it with different degrees of firmness. Some mechanisms may involve purely reversible mass transfer and the response to stress is governed by the state of hydration in the paste. There could be other mechanisms which bring about a permanent change in the gel structure, rendering it into a more stable configuration. Without reference to any particular mechanism, the results provide phenomenological evidence of such microstructural changes taking place, a hypothesis which have subsequently been advanced by others also.

ACKNOWLEDGEMENTS

This investigation was carried out by the author for his doctoral thesis at the University of Calgary Canada under the supervision of Dr B R Gamble. Financial assistance was provided by the University and NRC Canada. The paper is published with the permission of Director-General Cement Research Institute of India.

REFERENCES

1. T C Powers, 'Some Observations on the Interpretations of Creep Data', Bulletin RILEM, 33 (1966) 381-391.

2. R F Feldman and P J Sareda, 'A Model for Hydrated Portland Cement Paste as Deduced from Sorption-length change and Mechanical Properties', Materials and Structures Bulletin RILEM, 6 (1968) 509-520.

3. F Slate and B L Meyers, 'Some Physical Processes Involved in Creep of Concrete', International Conference on 'Structure, Solid Mechanics and Engineering Design in Civil Engineering Materials', Southampton, (1969) 769-773.

4. Z P Bazant, 'Constitutive Equation for Concrete Creep and Shrinkage based on Thermodynamics of Multi-phase Systems', Materials and Structures Bulletin RILEM 3, (1970) 3-36.

5. F H Wittman, 'Properties of Hardened Cement Paste', Principal Report, Sub-Theme VI-2, 7th International Congress on the Chemistry of Cement, Paris (1980).

6. B R Gamble and J M Illston, 'Rate of Deformation of Cement Paste and Concrete During Regimes of Variable Stress, Moisture, Contact and Temperature' Conference on 'Hydraulic Cement Pastes: Their Structure and Properties', Sheffield (1976).

7. R H Mills, 'Molecular Sieve Effect in Concrete', Proceedings 5th International Symposium on the Chemistry of Cement, Tokyo, Paper III-46 (1968) 74-85.

8. A Bentur, R L Berger, F V Lawrence, Jr., N B Milestone, S Mindess and J F Young, 'Creep and Drying Shrinkage of Calcium Silicate Pastes III.A Hypothesis of Irreversible Strains, Cement and Concrete Research, 9 (1979) 83–96.

9. B B Hope and N H Brown, 'A Model for the Creep of Concrete' Cement and Concrete Research, 5 (1975) 577–586.

10. A K Mullick, 'Effects of Stress History on the Microstructure and Creep Properties of Maturing Concrete', PhD Thesis, The University of Calgary Canada (1972).

11. B R Gamble and L H Thomass, 'The Creep of Maturing Concrete Subjected to Time-Varying Stress', Proceedings 2nd Aust. Conference on the 'Mechanics of Structures and Materials' Adelaide (1969).

12. A K Mullick and B R Gamble, Discussion on 'Maturing Creep of Portland Cement Paste', Journal American Concrete Institute, 69 (1972) 361–363.

13. A D Ross, 'Creep of Concrete under Variable Stress', Journal American Concrete Institute, 54 (1958) 739–758.

14. L J Parrott, 'Recoverable and Irrecoverable Deformation of Heat-Cured Cement Paste', Magazine Concrete Research 29 No 98, (1977) 26–30.

ISOTHERMAL DRYING SHRINKAGE AND WETTING SWELLING OF HARDENED CEMENT PASTE

S Sabri and J M Illston

The Hatfield Polytechnic, Hatfield, Herts, UK.

ABSTRACT

The results are reported of shrinkage and swelling tests on tubular specimens of hardened cement paste with a 1mm wall thickness. Attention is concentrated on the first cycle of humidity from saturation to complete dryness. The resulting hysteresis and irreversible shrinkage is examined in the light of parallel measurements of moisture movement. For this purpose two states of water are identified by thermogravimetric methods and the two states identified as A - free and adsorbed water and B - water associated with the structure of ettringite and C-S-H gel. Reversible shrinkage is attributed to the movement of type B water, while irreversible shrinkage is thought to derive from the movement of type A water leading to the permanent closing of spaces between hydrate particles; this in turn leads to the formation of additional spaces accommodating type B water.

هذه المقاله تعرض نتائج قياسات لاختبارات الانكماش والتمدد لعينات رقيقه من عجينه الاسمنت المتصلده بسماكة ١ مم . انصب الاهتمام على الدوره الاولى من تغيير الرطوبه من حالة بلل كلى الى حالة جفاف كامل. لختبر الانكماش الغير عكوس على ضوء قياسات لحركة المياه داخل عجينة الاسمنت. لهذا الغرض امكن التعرف على نوعين من المياه

ا- مياه حره و مياه ممتزه طبيعيا .

ب- مياه اماهه فى مادة "الاترنجيت" و "كالسيوم سيليكات هيدرات".

الانكماش المعكوس يحدث نتيجه حركه المياه من نوع "ب" و الانكماش الغير عكوس يحدث نتيجة حركه المياه من نوع "ا" والتى تودى الى حالة انقباض للاماكن الموجوده بين جزيئات عجينه الاسمنت و بالتالى تودى الى زيادة الاماكن المحتله بالمياه من نوع "ب" .

INTRODUCTION

It has long been recognised that the shrinkage of concrete is largely attributable to loss of water from the hardened cement paste (hcp). The desorption isotherm shows a continuous loss of water over the full range of humidity from saturation to complete dryness; this observation carries the implication that the water in hcp is held in a variety of states in pores with a spectrum of sizes, and that several different physical causes of shrinkage are in operation. It is apparent that these various shrinkage mechanisms overlap, and it has not proved possible to separate the states of water or the mechanisms in a quantitative manner. The problem is exacerbated by the slowness of moisture movement through the material, which in most experiments has led to the existence of such large moisture gradients and such long periods to reach equilibrium as to introduce considerable distortions in the measurements.

The semi-isothermal method of derivative thermogravimetry (1) offers a means of making a quantitative differentiation between states of water in hcp, and it is the purpose of this paper to reiterate findings on the water distribution with changing humidity and to relate these findings to concomitant shrinkage. The irreversible behaviour observed on first drying is of particular interest, and attention is concentrated on the first humidity cycle. To eliminate the worst effects of the presence of moisture gradients thin specimens and controlled rates of drying were employed.

EXPERIMENTAL : SHRINKAGE MEASUREMENTS

Specimen Preparation

A type I cement was mixed at a water cement ration (W/C) of 0.47, cast as hollow cylindrical tubes of wall thickness 1 mm, external diameter 32 mm, and length 100 mm, in specially designed perspex moulds (2), stored at 20°C for one day in 100% R H chamber, demoulded and cured at 20°C in deionized water for 27 days before the start of drying.

Apparatus and Experimental Conditions

Length changes were measured using linear variable differential transformer (LVDT) transducers. The measurements were performed in invar frames containing the samples which were placed with the transducers inside a standard vacuum desiccator which was stored in a cabinet at a constant temperature of 20°C.

The relative humidity was controlled by creating alternative sources and sinks of moisture while continuous determination of the actual relative humidity inside the desiccator was found by means of electric hygrometry. All measurements were performed under high vacuum conditions and this eliminated any possible carbonation of the hcp samples during the experiments.

Experimental Procedures

Shrinkage was determined by progressively drying three identical samples simultaneously in small steps. The measurements were made over the full range of humidity from 0 to 100% R H, and the rates of shrinkage were observed at each of several constant humidities. The size of each drying step was about 10% change in R H.

The samples were kept at each level of R H for 100 hours except during the last step between 11 and 0% R H when they were kept for more than 40 days. Drying below 11% R H resulted in a large and relatively slow shrinkage and 100 hours of drying were not sufficient to reach near-equilibrium.

At the end of the drying period at 0% R H the relative humidity was increased in steps until complete saturation was again reached. About 200 hours was allowed for each wetting step.

Results

For each drying step, about 80% of the shrinkage after 100 hours occurred during the first 9 hours of drying. Shrinkage continued at a much slower rate thereafter. As previously indicated the one exception was the step from 11 to 0% R H when there was no indication of a marked change in shrinkage rate during the early hours of drying.

The development of wetting swelling with time was somewhat different to that of drying shrinkage. Except during the first step of wetting from 0 to 11% R H, the swelling versus time curves showed no initial rapid rate, and deformation was found to be more time dependent as the relative humidity was increased. During the first step of wetting, however, swelling was fast and the samples reached constant length and weight during the first day of exposure to water vapour.

A complete first drying shrinkage-wetting swelling isotherm at 20°C is shown in Figure 1. It can be seen that:

1. Drying shrinkage is large between 100 and 40% R H.

2. Very little shrinkage is observed between 40 and 20% R H. The marked break in the drying shrinkage curve at about 40% R H is not observed on the wetting swelling curve.

3. Nearly half of the total shrinkage occurs below 11% R H.

4. A large hysteresis loop extending over the full humidity range is present.

5. About 40% of the first drying shirnkage is irreversible or complete resaturation.

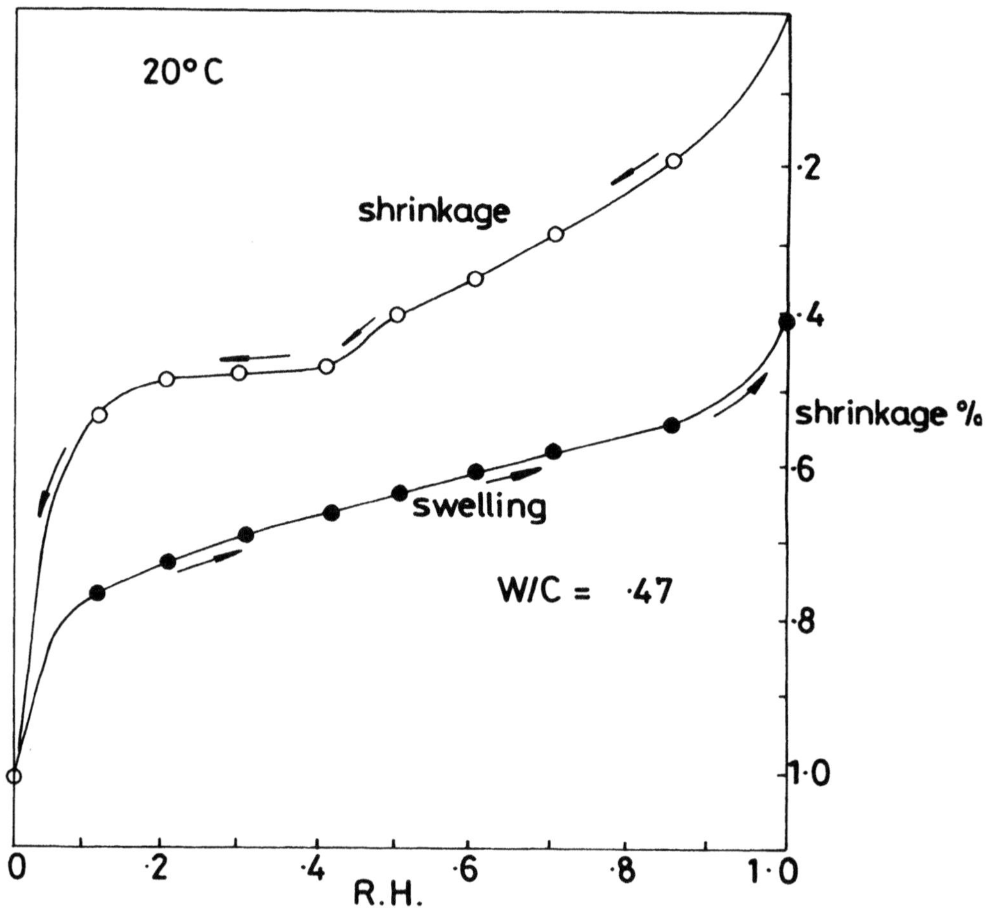

FIG. 1

Drying shrinkage - wetting swelling isotherm at 20°C for thin samples of hcp cured for 28 days before the start of drying.

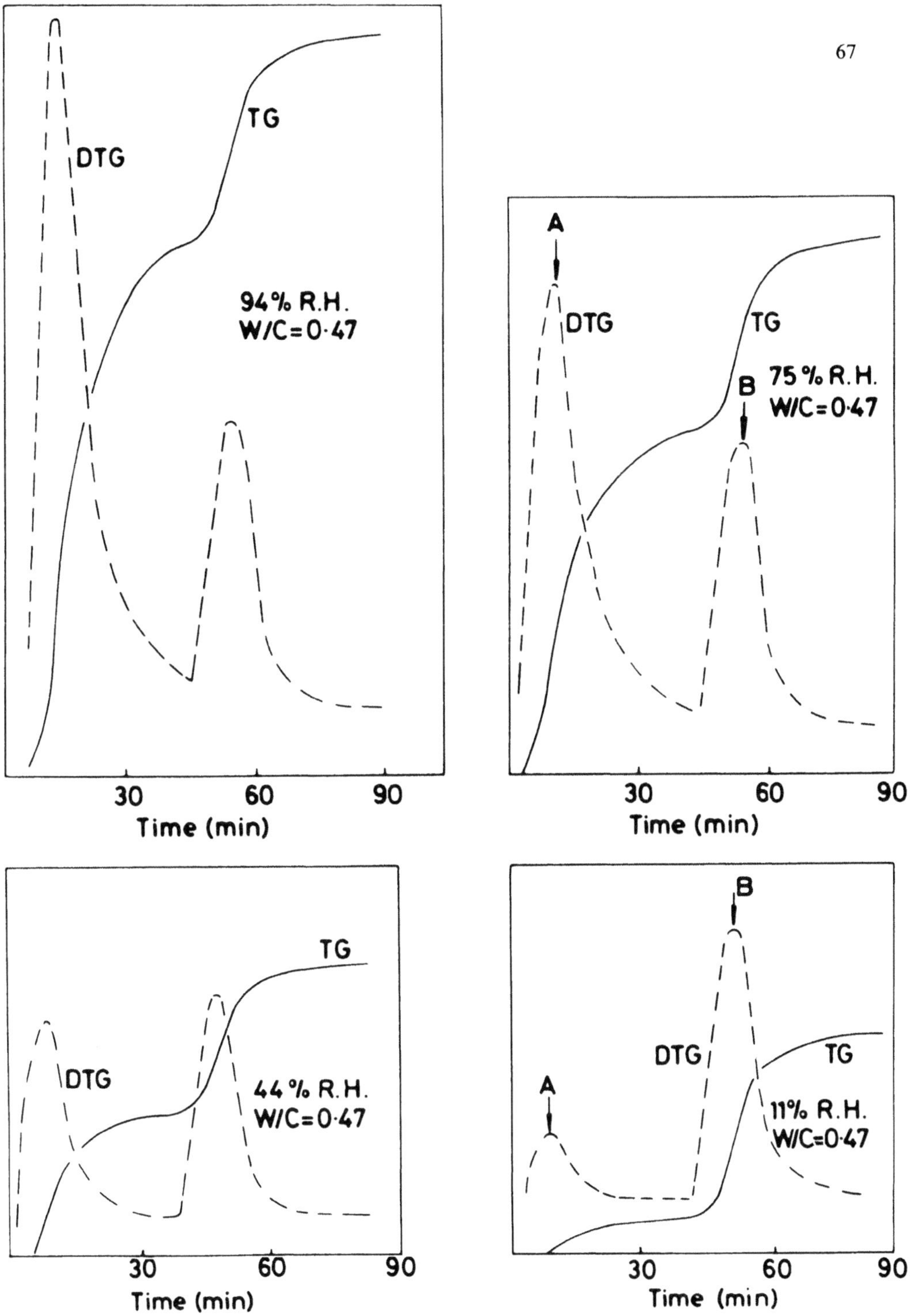

FIG. 2

Thermogravimetric and differential thermogravimetric curves, using the semi-isothermal method, for samples of hcp conditioned to various levels of relative humidity.

Thermogravimetric Analysis

The distribution of evaporable water in hcp was studied using a simultaneous semi-isothermal method of thermogravimetry and derivative thermogravimetry. The results were presented in an earlier paper (3).

Figure 2 shows typical semi-isothermal TG/DTG curves for samples of hcp conditioned to different levels of relative humidity. The semi-isothermal TG/DTG method allows the separation of TWO distinct peaks on the DTG curve in the temperature range 20 to 100°C. The two types of water in the states corresponding to the lower and the higher temperature peaks are designated water type 'A' and water type 'B' respectively. Water type 'A' is taken to include all the free and physically adsorbed water present in the paste and water type 'B' is regarded as a structural component of some hydrates, such as C-S-H gel and ettringite, that are not stable with regard to changes in moisture content.

Through the study of the distribution of evaporable water in hcp, a picture of the microstructure of hcp has been developed (2). It is thought to be composed of aggregations of some basic structural units which might vary considerably in shape and dimension. Each aggregation of units forms a 'gel particle' which might be composed of one or many more structural units. The basic units within each particle in the initially saturated state are separated by layers of water about one molecule deep (water type 'B') while water type 'A' exists in spaces between gel particles and aggregations of gel particles. This picture of the microstructure of hcp is very similar to that of Feldman (4), if the water type 'B' is regarded as equivalent to 'interlayer water'.

Relation Between Water State and Shrinkage

A separation of the effect of water type 'A' and water type 'B' on first drying shrinkage from saturation to an intermediate level of R H was not possible because both types of water are lost simultaneously from the paste even during the early hours of drying. Attempts to determine the role of each type of water on drying shrinkage through the study of the changes in shrinkage rates during drying was also unsuccessful (2). However, by examining the desorption-adsorption isotherms of water type 'A' and water type 'B', shown in Figures 3 and 4 respectively, it can be seen that most of the water that is lost below 11% R H is of type 'B'. It should be noted that the isotherm for water type 'B' is plotted non-dimensionally to include a correction for the hydration that occurs during the test. Water type 'A' is not greatly affected by hydration and for comparison the content of water type 'B' in the saturated state before drying started was 0.064. It follows that the large shrinkage that is observed below 11% R H must be mainly due to the loss of the hydrate water type 'B'. Water in ettringite is thought not to contribute significantly because similar shrinkage was also observed in this region when drying C_3S pastes (5) which contain no ettringite. Hence the large shrinkage of hcp occurring below 11% R H can be attributed to the loss of the hydrate water in the C-S-H gel component of the paste.

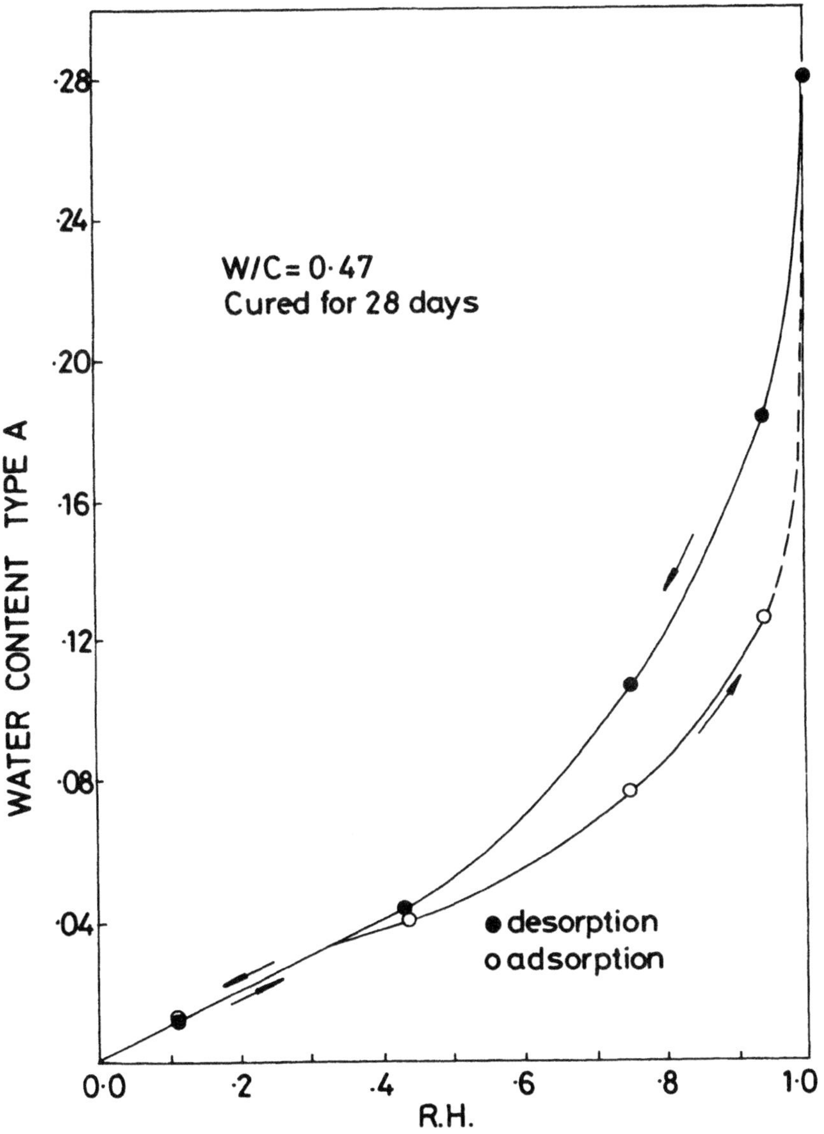

FIG. 3

Adsorption - desorption isotherm for water type 'A'

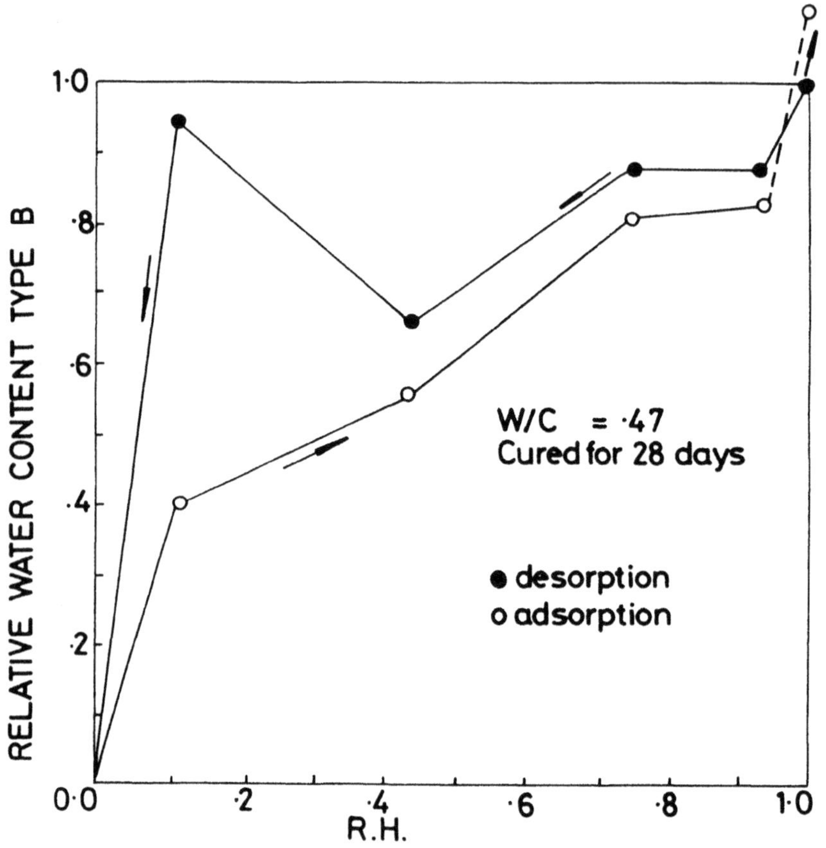

FIG. 4

Adsorption - desorption isotherm for water type 'B'
Individual samples dried directly from saturation to each relative humidity.

After the first day of exposing dry paste to water vapour, only water type 'B' is absorbed by the paste; the amount of water type 'A' present in the paste actually decreased after reaching a maximum during the first day exposure (2). The samples, however, continued to expand after the first day of exposure to water vapour. It can be concluded that the swelling of dry paste on exposure to water vapour is mainly due to the absorption of water type 'B'. The rate of water absorption by ettringite is low below 70% R H, where most of the wetting swelling occurs, and it follows that swelling of hcp is mainly due to the absorption of the hydrate water if C-S-H gel.

The assumption that loss or gain of type 'B' water is directly proportional to shrinkage allows further deductions to be made. Thus while it is evident that type 'B' water is associated with all the shrinkage below 11% R H, it is responsible for only a part of that occurring at higher humidities. The loss of water type 'A' thus also causes shrinkage and no doubt mechanisms such as capillary tension are operative on drying.

The amounts of both types of water are fully recovered on resaturation, but the recovery of type 'B' alone can account for virtually all of the recoverable shrinkage. The irrecoverable shrinkage must therefore be related to a structural change occurring during the process of drying, but not resulting in any large redistribution of water. Other observations not reported here in detail (2) confirm the feature shown in Figure 4 of an increase of water type 'B' on resaturation. A variety of drying regimes all resulted in significant increases, even after allowance was made for hydration during the humidity cycle. The structural change occurring during drying can then be identified as the formation of additional spaces containing type 'B' water, presumably at the expense of space containing type 'A' water.

The view is taken here that new spaces of type 'B' can be created by a mechanism in which neighbouring structural units which are partially or totally separated by water type 'A' in the initially saturated state, approach each other creating more spaces of type 'B'. This packing of the basic structural units leads to an increase in the average size and to a permanent decrease in the external surface area of particles. Most of the surface-to-surface movement that results in an externally measured deformation occurs during the first few hours of drying. This movement is reversible unless enough time is allowed for chemical bonds to develop between surfaces, so that even though shrinkage itself is not appreciably increasing after the first few hours of drying, first drying shrinkage becomes more irreversible with the drying time. This has been observed (6). The increase in the volume of spaces type 'B' with drying-wetting cycles should also lead to an increase in shrinkage below 11% R H with drying and wetting cycles. This has also been confirmed by experiment (7).

Conclusions

1. Drying shrinkage and wetting expansion can be related to the loss and gain of two types of water which are identified as:

   A. Free water and physically adsorbed water.

   B. Water forming a structural component of hydrates including C-S-H gel and ettringite.

2. Reversible shrinkage is attributed largely to the movement of type 'B' water.

3. Irreversible shrinkage is associated with the formation during the drying and wetting cycle of additional spaces containing type 'B' water.

References

1. B El-Jazairi and J M Illston, "A Simultaneous Semi-Isothermal Method of Thermogravimetry and its Application to Cement Pastes", Cem.Con.Res., Vol. 7, 247 (1977).

2. S Sabri, "The Response of Hardened Cement Paste to Changes in Ambient Relative Humidity and Temperature", a PhD thesis submitted to the University of London (1980).

3. S Sabri and J M Illston, "The Distribution of Evaporable Water in Hardened Cement Paste", a paper presented to the 7th International Congress on the Chemistry of Cement, Paris (1980).

4. R F Feldman and P J Sereda, "A Model for Hydrated Portland Cement Paste as Deduced from Sorption-Length Change and Mechanical Properties", Materiaux et Constructions, Vol. I, No. 6, 509-520 (1968).

5. R A Helmuth, "Dimensional Changes and Water Vapour Sorption of Hydrated Tricalcium Silicate", MSc Thesis, Graduate School of Illinois Institute of Technology, Jan. (1965).

6. R A Helmuth and D H Turk, "The Reversible and Irreversible Drying Shrinkage of Hardened Portland Cement and Tricalcium Silicate Pastes", J. of P.C.A. Res. and Devel. Lab., Vol. 9, 8-21 (1967).

7. K G Krasilnikov and N N Skoblinskaya, "The Change of Cement Stone Porous Structure in the Process of Expansion and Shrinkage at Adsorption of Water", Proc. of Int. Symp. Rilem, Part IV, Prague, E-51-69 (1973).

Part II
Stochastic Nature of Creep and Shrinkage

EXPERIMENTAL STUDY OF THE VARIABILITY OF SHRINKAGE IN CONCRETE

F. Alou and F.H. Wittmann

Swiss Federal Institute of Technology Lausanne
Laboratory for Building Materials Science
Lausanne - Switzerland

ABSTRACT

Cylindrical specimens between 100 and 500 mm in diameter have been used to investigate shrinkage of two different concrete mixes. In addition the influence of varying the composition of the cement paste on shrinkage of concrete was studied using cylinders 160 mm in diameter. Six different paste compositions were tested. Shrinkage was followed for four months under controlled climatic conditions of 18°C and 65% R.H. the final shrinkage being determined by extrapolation. During the first few days variation in the measured shrinkage was influenced by the precision of the measuring technique. However, after a drying period of about one month, the scatter of the materials properties became dominant. The values of shrinkage obtained indicate that the predictions given by the CEB-FIP and various other standards considerably underestimate the deformations caused by shrinkage.

RESUME

On a mesuré le retrait sur des éprouvettes de diamètre variable d'une part et sur des éprouvettes de dimensions constantes d'autre part. Pour les éprouvettes de diamètre variable, 10 à 50 cm, on a utilisé deux bétons et pour les échantillons de même géométrie (cylindres 16/32), on a considéré six bétons dont la qualité de la pâte était variable (essentiellement définie par le rapport E/C).

Les mesures ont été effectuées pendant 4 mois sous des conditions constantes : 18°C et 65% HR. Par le calcul, nous avons déterminé les retraits finals. La variabilité trouvée est la somme de plusieurs variabilités. La variabilité causée par la technique des mesures est prédominante au début et la variabilité intrinsèque du matériau joue le rôle le plus important après quelques semaines. Les valeurs que nous avons obtenues montrent que les prévisions CEB-FIP et celles préconisées par différentes normes sont nettement insuffisantes pour estimer les déformations dues au retrait.

INTRODUCTION

At present the compressive strength of concrete is considered as a statistical variable in most of the national standards. On the contrary the variability of deformations is rarely taken into account if the behaviour of different materials is to be compared (1, 2). One of the aims of this contribution is to present some experimental results and to draw attention to this problem. In this way a basis for the comparison between theoretical predictions is provided (3).

In addition the influence of specimen geometry has been investigated and models for the drying and shrinkage of porous materials, such as hardened cement paste and concrete, have been evaluated (4, 5).

In the Laboratory for Building Materials at Swiss Federal Institute of Technology Lausanne a comprehensive study on the influence of the composition of concrete on shrinkage and creep has been carried out. First results of this work will be presented in this report.

In order to investigate the effect of varying the specimen diameter (between 100 and 500 mm) only two types of concrete have been used, one of medium quality and one of high quality. The samples used to measure the amount of shrinkage on drying were stored at 18°C and 65% R.H. and up to the present time have reached an age of about one hundred and fifty days. The measurements of creep, with and without exchange of water from the exterior, have only been in progress for about thirty days. For these tests three stress levels were used, these were 10, 20 and 30% of the ultimate strength.

The measurements of shrinkage of concrete of varying composition started about four months ago. In the first set of experiments the sieve composition and the volume of cement paste were held constant and only the quality of the paste was varied that is the proportion of cement and water. The specimens, which were cylinders of diameter 160 mm and height 320 mm, were also stored at 18°C and 65% R.H.

For these reasons this contribution is limited, from the point of view of variability, to the measurement of shrinkage of specimens of variable diameter and volume of paste on one hand and on the other hand to differing quality without taking into account the measurement of creep. Corresponding creep values will be presented and discussed in a later report.

PREPARATION OF CONCRETE SPECIMENS

Composition of Concrete

For the fabrication of different types of concrete only one type of aggregate and one cement have been used.

The aggregates were supplied by the Gravière Cantonale of Bioley-Orjulaz and were separated into five classes : 0/1; 0/3; 3/8; 8/15 and 15/30. They were essentially calcareous (fine grain of siliceous limestone) and the majority (about 80% by weight) rounded.

The quantity of soft fragment of rocks in the aggregates was defined by the swiss standard SIA 162 and was below 6%. For this reason they were suitable for the fabrication of high resistance concrete.

In Figure 1 the sieve analysis of the five constituents is shown.

The cement used was a normal Portland cement CPN conforming to the swiss standard SIA 215 concerning cements (this is similar to that of cement type I described in the ASTM standards).

Sieve Composition

Two sieve compositions of continuous type have been used which are represented in Figure 2.

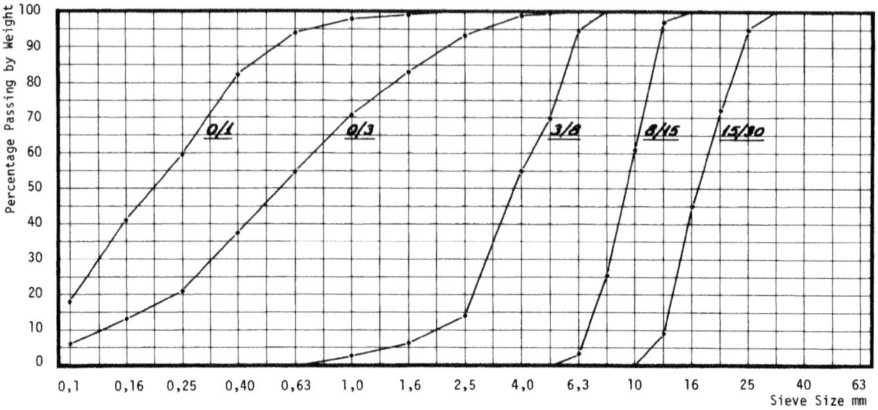

Figure 1

Sieve Analysis of Aggregates

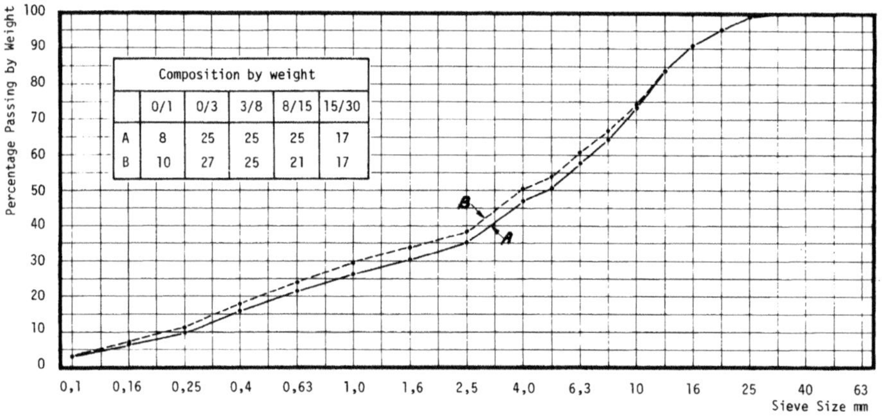

Figure 2

Grading curves of aggregates for high quality (A) and medium quality (B) concrete

Composition A, having less fine elements than composition B, was adopted for the concrete of high quality to be employed for the fabrication of specimens of varying geometry. This grading curve (A) was also used for the concrete where the quality of the cement paste was altered.

Composition B was used only for the concrete of medium quality where the surplus of fine elements allowed the production of concrete with a low void content.

Cement Content

For the concrete used for studying the influence of specimen geometry on shrinkage two cement contents were used, 332 kg/m^3 for high quality concrete and 227 kg/m^3 for medium quality concrete with grading curves A and B respectively.

Concerning the concretes of varying quality of paste the cement content was between 265 and 405 kg/m^3. For these tests only grading curve A was used.

Mixing Water

The quantity of water used in the concretes for tests on variable geometry (concretes A and B) was determined by obtaining about the same workability. This corresponded to concretes which were slightly plastic, easy to handle and to compact by vibration (slump 10 to 15 mm; VB 6 to 8).

For the concretes of variable paste quality (concretes I to VI) the quantity of water was determined a priori from the relationship W/C, taking into account the workability necessary to obtain concretes as free of voids as possible considering the means of pouring used. For this reason it was not possible to compact the concrete perfectly with for example a level of cement of 400 kg/m^3 and a relationship of $W/C = 0.25$.

Admixtures

For certain concretes with paste of variable quality (concretes IV to VI) an admixture of a superfluidising type based on lignosulphonate was used to reduce the level of water. This was necessary because the preordained quantity of water did not lead to workable concrete. Even with this additive it was not possible to pour correctly the concrete with a cement content of 400 kg/m^3 in 100 litres of water ($W/C = 0.25$).

Mixing Procedure, Specimen Geometry, and Storage

All the concretes were blended with a mechanical mixer. A blending time of two minutes was allowed from the time when all the components (aggregates, ciment, water and the admixture if used were introduced).

The specimens of concretes type I to IV (concretes of variable paste quality) were cylinders with ϕ = 160 mm and h = 320 mm.

For concretes 1 and 2 moulds of rigid PVC tubes were used, the diameters being 100, 150, 200, 250, 300, 400 and 500 mm.

The height of all the specimens made with concretes 1 and 2 was 1500 mm.

All the specimens were stored in the same manner that is :
- until 7 days : in the mould (watertight) at 18°C
- after 7 days : in air at 18°C and 65% R.H.

EXPERIMENTAL PROCEDURE

Compressive Strength

The compressive strength was measured using cylinders 160 mm in diameter and 320 mm high. These samples were also stored as described above that is 0 to 7 days in the mould and then in air at 18°C and 65% R.H. After 7 days they were straightened in water in order to rectify any lack of planarity.

Tables I and II give the pertinent characteristics of fresh concrete and also the values of compressive strength obtained.

Concrete No	Composition of Concrete: Cement kg/m^3	Water kg/m^3	Air l/m^3	Aggregates kg/m^3	Water/Cement by weight	Workability: Slump cm	°VB sec	Compressive Strength: 7d	28d
1	332	160	13	1938	0,48	1	7÷8	33,3	42,5
2	227	164	12	2019	0,72	1,5	6÷7	20,7	30,7

Remarks : a. Concrete without admixtures
b. The compressive strength has been determined by means of cilinders ϕ = 160 mm, h = 320 mm

Table I

Composition and properties of concrete mixes 1 and 2 used to study the influence of the geometry on shrinkage

Concrete No	Composition of Concrete: Cement kg/m³	Water kg/m³	Air l/m³	Aggregates kg/m³	Admixture % c.w.	Water/Cement by weight	Workability: Slump cm	OVB sec	Compressive strength 28 days N/mm²
I	266	174	11	1960	-	0,65	3	3÷4	26,0
II	296	162	16	1952	-	0,55	1	7÷8	36,6
III	332	150	23	1938	-	0,45	-	12	46,1
IV	379	139	18	1928	1	0,37	-	20	52,3
V	378	132	25	1922	2	0,35	1	6÷7	54,1
VI	404	137	18	1901	2	0,34	1	7÷8	56,3
Remark : The compressive strength has been determined by means of cilinders ϕ = 160 mm, h = 320 mm									

Table II

Composition and properties of concrete mixes I to VI used to study the influence of the quality of cement paste on shrinkage

Shrinkage

The measurements of shrinkage were carried out on all the various types of concrete that is mix numbers 1 and 2 and I to VI. Drying was begun after seven days and the climatic conditions were $18^{\circ} \pm 1^{\circ}C$ and $65 \pm 5\%$ R.H.

A Huggenberger extensometer was used with a measuring range of 250 mm (1 division : 0.001 mm being 0.004 o/oo).

The measuring points, which were made from stainless steel, were glued to the concrete surface using a special adhesive. In order to ensure good adhesion the layer of grout in the region to be bonded was removed.

RESULTS OF MEASUREMENTS OF SHRINKAGE

Samples of Varying Diameter

The results of the measurements of shrinkage are summarized in Tables III and IV, the mean shrinkage is given together with the standard deviation and the coefficient of variation.

Specimen		Time of Drying in Days						
		2	5	11	30	47	85	109
∅ 10 cm	$\bar{x}$	9	79	154	230	307	322	339
	s	12,8	18,3	34,6	39,5	39,8	45,5	43,0
	c.v.	1,42	0,23	0,22	0,17	0,13	0,14	0,13
∅ 15 cm	$\bar{x}$	1	50	97	158	217	246	263
	s	6,8	19,2	6,8	13,3	18,0	23,1	20,0
	c.v.	6,83	0,38	0,07	0,08	0,08	0,10	0,08
∅ 20 cm	$\bar{x}$	3	35	74	126	180	209	231
	s	8,3	15,5	10,6	12,4	19,9	23,0	23,4
	c.v.	2,75	0,44	0,14	0,10	0,11	0,11	0,10
∅ 25 cm	$\bar{x}$	4	22	64	105	145	175	191
	s	12,7	12,0	33,2	28,5	25,2	21,0	20,8
	c.v.	3,16	0,55	0,52	0,27	0,17	0,12	0,11
∅ 30 cm	$\bar{x}$	5	17	36	65	94	118	138
	s	21,7	5,0	3,3	14,4	10,1	5,2	13,7
	c.v.	4,25	0,30	0,09	0,22	0,11	0,04	0,10
∅ 40 cm	$\bar{x}$	0	22	32	58	83	106	120
	s	-	11,6	19,9	21,8	22,5	28,0	22,9
	c.v.	-	0,52	0,62	0,38	0,27	0,26	0,19
∅ 50 cm	$\bar{x}$	6	18	17	30	44	64	79
	s	16,5	16,2	13,2	12,0	12,7	18,5	16,1
	c.v.	2,75	0,90	0,78	0,40	0,29	0,29	0,20

Remarks

1. The first row of values of each specimen is the average shrinkage in μm/m, the second row is the standard deviation and the last row the variation coefficient.
2. There have been 6 measurements for each type of specimen.
3. The measurements started at t = 7 days.

Table III

Measured shrinkage of specimens having different diameters
Concrete 1 : ($\sigma_{f\ 28\ days}$ = 42.5 N/mm2)

Specimen		Time of Drying in Days						
		2	5	16	23	40	78	102
∅ 10 cm	$\bar{x}$	-9	32	105	174	226	264	287
	s	10,5	10,8	12,4	17,7	15,5	16,3	17,4
	c.v.	1,17	0,34	0,12	0,10	0,07	0,06	0,06
∅ 15 cm	$\bar{x}$	-32	-4	57	89	118	181	202
	s	7,3	7,3	8,9	8,3	7,7	25,6	28,6
	c.v.	0,23	1,83	0,16	0,09	0,06	0,14	0,14
∅ 20 cm	$\bar{x}$	-16	10	17	54	105	148	169
	s	6,5	8,3	16,1	23,7	38,7	50,3	62,7
	c.v.	0,41	0,83	0,95	0,44	0,37	0,34	0,37
∅ 25 cm	$\bar{x}$	-14	2	29	55	82	120	140
	s	5,2	8,3	14,4	13,6	18,6	17,0	21,4
	c.v.	0,37	4,16	0,50	0,25	0,23	0,14	0,15
∅ 30 cm	$\bar{x}$	-24	-16	17	32	45	85	107
	s	17,6	14,2	21,3	18,2	15,8	19,2	22,7
	c.v.	0,73	0,89	1,25	0,57	0,35	0,23	0,21
∅ 40 cm	$\bar{x}$	-23	-12	-1	11	25	58	79
	s	14,4	13,9	11,9	11,9	12,8	14,1	13,6
	c.v.	0,63	1,15	11,9	1,09	0,51	0,24	0,17
∅ 50 cm	$\bar{x}$	-15	-4	25	30	36	65	82
	s	10,0	8,6	16,1	15,5	13,5	20,2	16,8
	c.v.	0,67	2,16	0,64	0,52	0,37	0,31	0,21

Remarks

1. The first row of values of each specimen is the average shrinkage in μm/m, the second row is the standard deviation and the last row the variation coefficient.
2. There have been 6 measurements for each type of specimen.
3. The measurements started at t = 7 days.

Table IV

Measured shrinkage of specimens having different diameters
Concrete 2 : ($\sigma_{f\ 28\ days}$ = 42.5 N/mm²)

In Figures 3 and 4 the variation in shrinkage is plotted as a function of duration of drying for specimens of differing diameter and for concretes type 1 and 2.

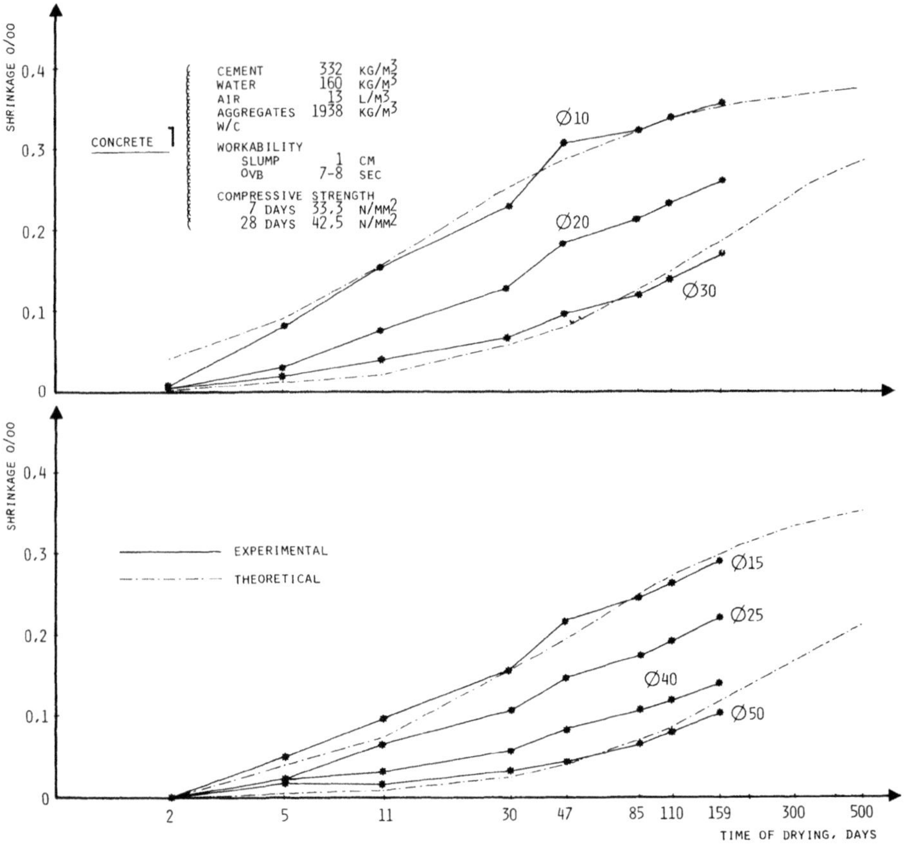

Figure 3

Comparison between experimental values and calculated shrinkage for specimens having different diameters. Concrete mix number 1.

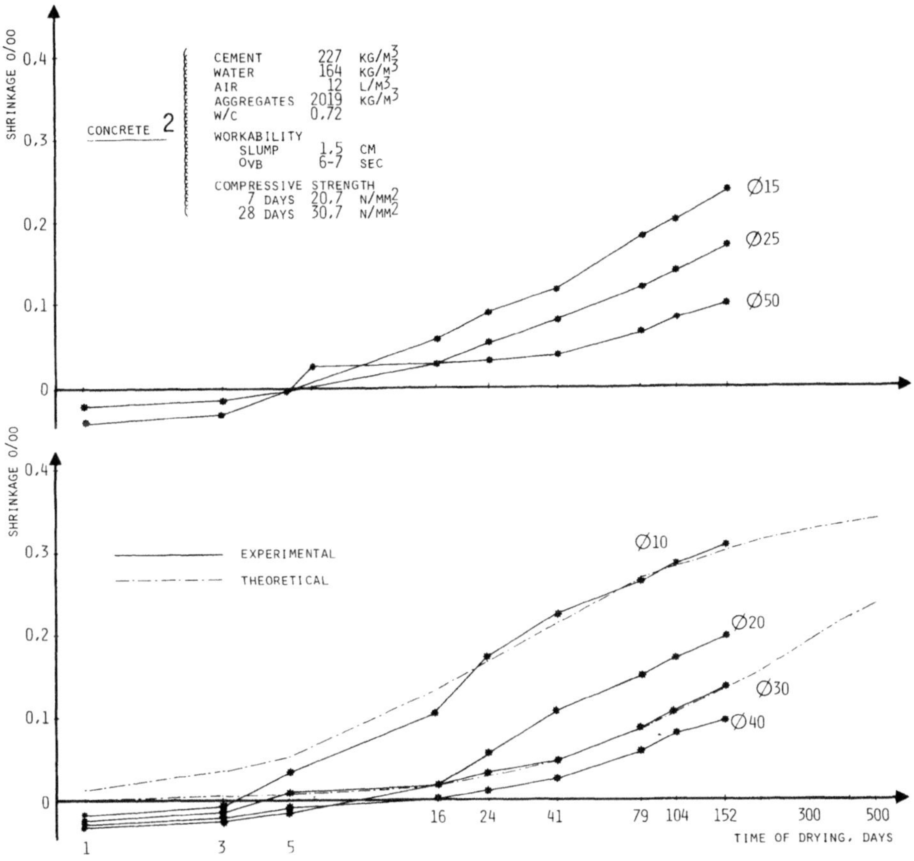

Figure 4

Comparison between experimental values and calculated shrinkage for specimens having different diameters. Concrete mix number 2.

For concrete type 2 (medium quality) swelling was observed during the first few days of drying. This is probably due to expansion arising from the temperature rise during the hydration of cement. Contrary to expectation this phenomenon was not observed for concrete 1 which is of higher quality and had a larger cement content than concrete 2. A possible reason for this behaviour is that the higher tensile strength of concrete type 1 counteracts surface cracking.

Theoretical curves of shrinkage for specimens of 100 mm diameter were calculated from the relationship $\varepsilon = at/b+t$, this has given the following values for a and b :

Concrete 1 : $\bar{a} = 0{,}388$; $\bar{b} = 16$

Concrete 2 : $\bar{a} = 0{,}361$; $\bar{b} = 27$

For specimens greater than 100 mm in diameter it has been assumed that the value of a is independent of the diameter. This permits the calculation of $b = f$ (radius) $= b_o\, r^2$ and these results are given in Table V.

Figure 5 show the variation of $\bar{b}$ with r, the radius of the samples in cm for the concretes 1 and 2.

		Diameter in cm 10	15	20	25	30	40	50	Average
Concrete 1	$\bar{b}$	16	46	66	94	175	204	396	
	s	3,0	6,6	9,8	16,7	26,0	38,0	39,0	
	b_o	0,64	0,82	0,66	0,60	0,78	0,51	0,63	0,66
Concrete 2	$\bar{b}$	27	80	117	149	260	525	337	
	s	3,2	4,8	15,1	14,8	19,5	177,0	49,0	
	b_o	1,08	1,42	17,0	0,95	1,16	1,31	0,54	1,09

Table V

Values of $\bar{b}$, s, and b_o for different diameters

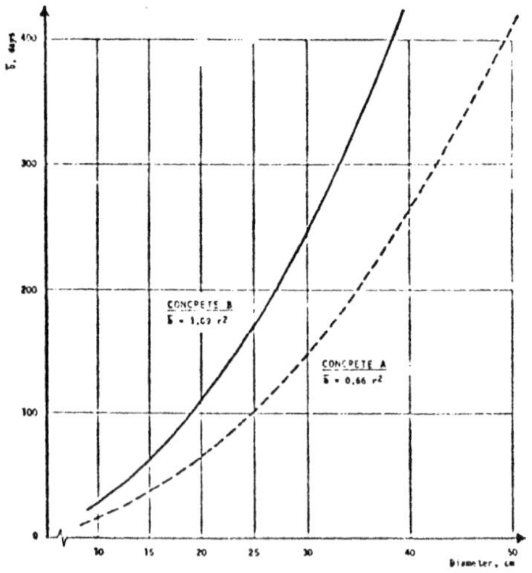

Figure 5

Variation of $\bar{b}$ with the radius

<u>Samples with different quality of cement paste</u>

The results from concretes of types I to VI are compiled in Table VI.

Concrete No	Cement kg/m^3	Water kg/m^3	Admixture % c.w.	Water/Cement by weight		Time of Drying in Days 1	4	7	22	29	69	90	117
I	266	174	n/a	0,65	$\bar{x}$	8	31	47	105	125	194	225	244
					s	9,1	13,4	12,1	16,8	21,1	22,9	25,3	28,0
					c.v.	1,14	0,43	0,26	0,16	0,17	0,12	0,11	0,11
II	296	162	n/a	0,55	$\bar{x}$	9	37	59	126	149	215	246	263
					s	7,1	14,6	8,4	15,2	21,0	17,1	19,1	21,6
					c.v.	0,78	0,39	0,14	0,12	0,14	0,08	0,08	0,08
III	332	150	n/a	0,45	$\bar{x}$	8	44	71	145	163	219	245	260
					s	4,8	6,7	10,0	12,6	16,5	17,2	23,4	21,4
					c.v.	0,60	0,15	0,14	0,09	0,10	0,08	0,10	0,08
IV	379	139	1	0,37	$\bar{x}$	16	65	99	186	207	265	293	305
					s	4,5	10,1	14,2	19,6	19,4	25,6	29,2	30,1
					c.v.	0,28	0,16	0,14	0,11	0,09	0,10	0,10	0,10
V	378	132	2	0,35	$\bar{x}$	16	64	100	190	213	285	319	336
					s	4,0	7,9	10,4	18,1	18,4	26,3	28,0	30,6
					c.v.	0,27	0,12	0,10	0,10	0,09	0,09	0,09	0,09
VI	404	137	2	0,34	$\bar{x}$	19	61	99	189	210	280	312	326
					s	5,7	16,2	20,1	28,3	29,0	35,5	41,0	41,4
					c.v.	0,30	0,27	0,20	0,15	0,14	0,13	0,13	0,13

<u>Remarks</u>

1. The first row of values of each specimen is the average shrinkage in μm/m, the second row is the standard deviation and the last row the variation coefficient.
2. There have been 12 measurements for each type of specimen.
3. The measurements started at t = 7 days.

Table VI

Mean value and variability of measured shrinkage of specimens with different quality of cement paste as function of duration of drying

Figures 6 and 7 indicate the variation in shrinkage as a function of drying time for concretes I to VI. In these diagrams the theoretical values of shrinkage calculated from the expression $\varepsilon = at/b+t$ are shown together with the statistical distribution of values of a which correspond to the final shrinkage ($t = \infty$).

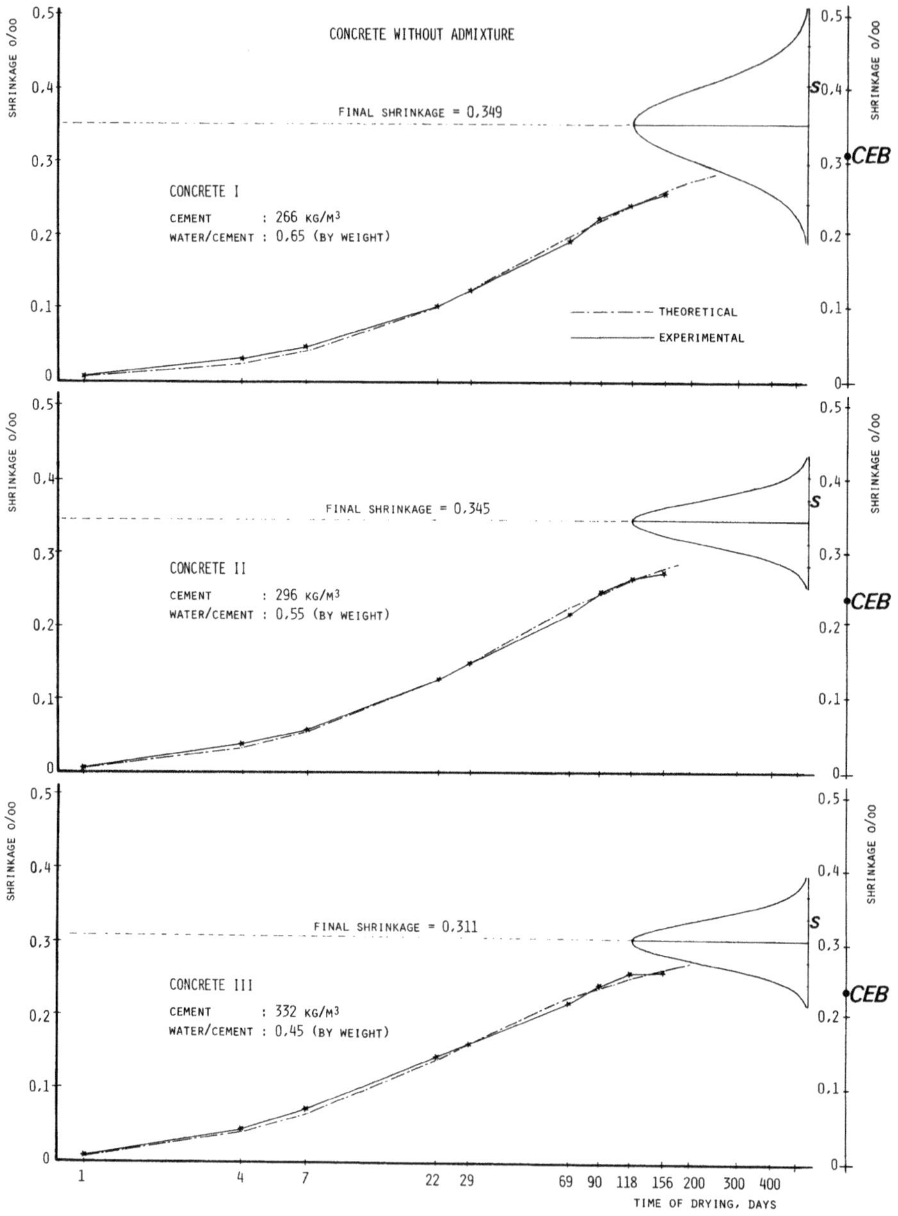

Figure 6

Comparison between experimental and calculated shrinkage and the spreading of the final shrinkage. Concrete without admixtures.

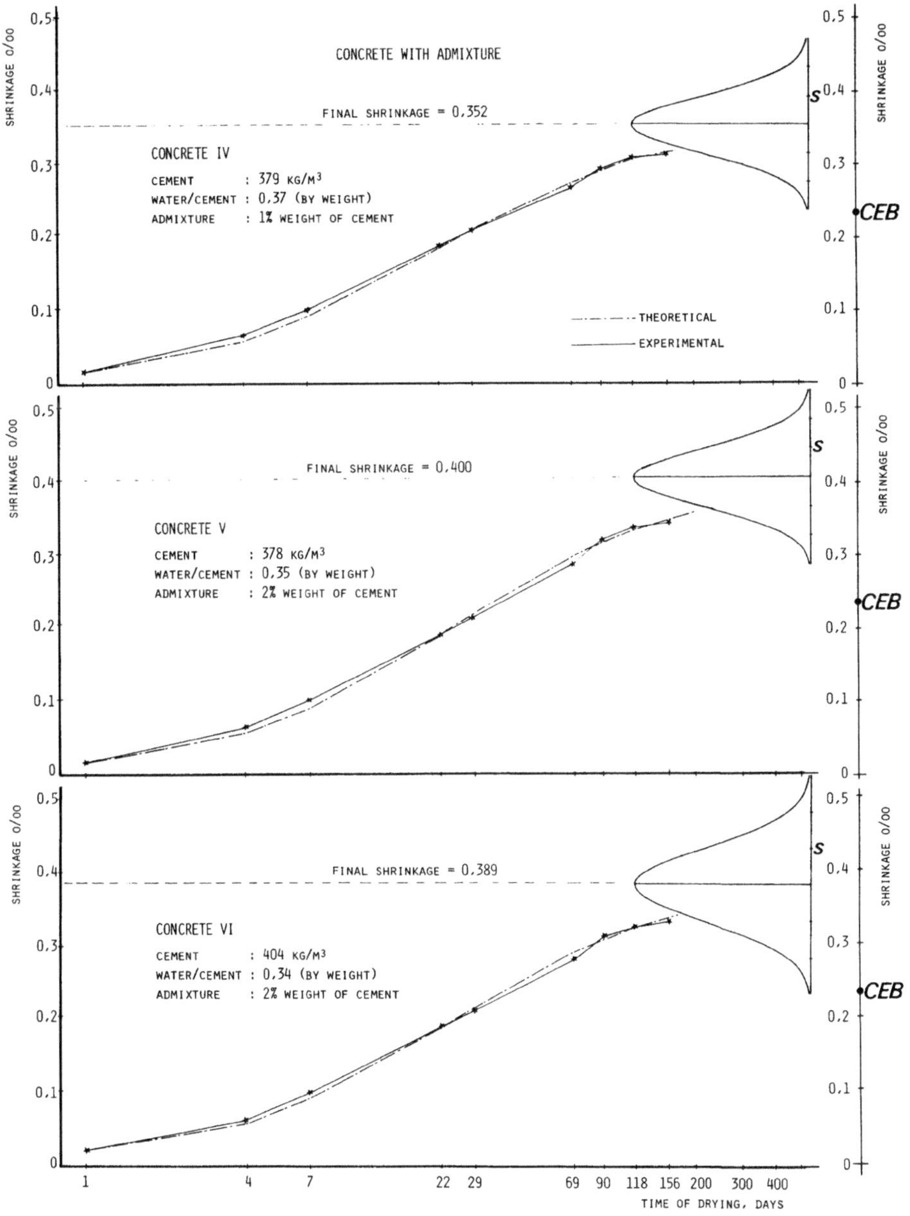

Figure 7

Comparison between experimental and calculated shrinkage and the spreadings of the final shrinkage. Concrete with admixtures.

Table VII summarizes the values obtained for concretes I to VI with the final shrinkage calculated from the CEB-FIP.

Concrete No	Cement kg/m3	W/C	$\bar{a}$ (o/oo) s	$\bar{b}$ (days) s	Final Shrinkage in o/oo according to the CEB-FIP ①	②	③	④	⑤	⑥	⑦	Remarks
I	266	0,65	0,349 0,052	56 20	0,311	0,311	0,373	0,373	0,393	0,434	76,73%	Concrete without admixtures
II	296	0,55	0,345 0,030	40 7,4	0,311	0,233	0,323	0,242	0,340	0,394	99,99%	
III	332	0,45	0,311 0,029	28 3,8	0,311	0,233	0,258	0,194	0,273	0,359	99,64%	
IV	379	0,37	0,352 0,038	21 3,0	0,311	0,233	0,200	0,150	0,212	0,414	99,92%	Concrete with admixtures
V	378	0,35	0,400 0,039	26 2,5	0,311	0,233	0,176	0,132	0,186	0,464	99,99%	
VI	404	0,34	0,389 0,048	25 3,8	0,311	0,233	0,176	0,132	0,186	0,468	99,94%	

① workability not taken into account
② workability taken into account
③ derived values from ① by taking into account the composition of the concrete (coefficient K_b of the "Recommandations internationales pour le calcul et l'exécution des ouvrages" CEB-FIP 1970, Prague)
④ derived values from ② by taking into account coefficient K_b
⑤ according to "Recommandations internationales pour le calcul et l'exécution des ouvrages" CEB-FIP 1970, Prague
⑥ deformation according to fractile of 5%
⑦ fractile determined from column ②

Table VII

Comparison of the obtained values with the CEB-recommendations

It is evident that the values of final shrinkage deduced from our tests are always greater than those obtained from the present recommendations of the CEB-FIP. It is very interesting to note that the estimates CEB-FIP 1970 are in better agreement with the measured values especially for concrete without admixture.

DISCUSSION OF RESULTS AND CONCLUSIONS

The results that have been obtained thus far are preliminary and for this reason the conclusions which may be drawn are tentative. Firstly, the drying time of about four months is too short to deduce with sufficient accuracy the final shrinkage. Secondly, the number of results obtained is somewhat limited at present. In addition, to our knowledge, there is little literature on this subject to allow comparison with a solid foundation.

The study of the variation of different deformations is indispensable in practice since the probability of cracking depends essentially on this value. The loss in watertightness or the reduction in the level of prestressing in modern concrete structures are just some relevant examples. On this subject it must be noted that one extreme value suffices to result in cracking and as a consequence produces the loss in watertightness.

Il all cases it is essential to determine the probability of finding extreme values.

It is interesting to compare the values obtained with those recommended by the CEB-FIP/2 (Table VII).

If our values are confirmed by supplementary experiments which are in progress, this indicates that the values given by CEB-FIP (78) ar markedly too low. In effect, if a value of 5% excess is assumed and the mean value of shrinkage is 0.349 o/oo (concrete I) a value of shrinkage of 0.434 o/oo is obtained whereas the value indicated from the CEB-FIP is only 0.311 o/oo. Thus the value recommended by the CEB-FIP should be increased by about 40%. For the other types of concrete without additive, which have been tested, the increase in the values of CEB-FIP is still higher (concrete type II 69% and type III 54%).

The table below gives the values of final shrinkage which appear in various standards for the types of concrete without additive that have been used in this work.

Standard	Country	Final Shrinkage o/oo
CEB-FIP	--	0,23 à 0,311
SIA 162	Switzerland	0,35
DIN 4227	Germany	0,25
NEN 3861	The Netherlands	0,33

The variation that has been observed is due to at least two factors, that caused by the measuring technique and that resulting from the material. However, for example in Tables III, IV and VI, it can be estimated that after about 30 days of drying the variation in material properties predominate.

These results are in agreement with those obtained by Reinhardt, Pat and Wittmann (6).

It must be underlined once more that we have only examined the variation in shrinkage under controlled and constant climatic conditions. In practice the regional climate must always be taken into consideration where the temperature, wind speed and humidity are stochastic variables and consequently further increase the variability (7). Therefore the variability found under well controlled laboratory conditions must be looked upon as being a lower bound.

REFERENCES

1. Norme SIA 162.

2. CEB-FIP, "Code modèle pour les structures en béton", Avril 1978, Secrétariat permanent : 6, rue Lauriston, Paris.

3. Cinlar E., Z.P. Bazant and E. Osman, "Stochastic process for extrapolating concrete creep", Journal Eng. Mech. Div., ASCE 103, pp. 1069-1088 (1977).

4. Roelfstra P.E. and F.H. Wittmann, "Een rekenmodel ter bepaling van de tijdsafhankelijke vervorming van uitdrogend beton", Cement 32, pp. 348-351, (1980).

5. Wittmann F.H. and P.E. Roelfstra, "Total deformation of loaded drying concrete", Cement and Concrete Research 10, pp. 601-610 (1980).

6. Reinhardt H.W., M.G.M. Pat and F.H. Wittmann, "Variability of creep and shrinkage of concrete", Contribution au colloque sur le retrait et le fluage du béton, Lausanne (1980).

7. Cornelissen H.A.W., "Kruip van beton - Een stochastische grootheid", Thèse de doctorat, TH Eindhoven, Pays-Bas (1979).

VARIABILITY OF CREEP AND SHRINKAGE OF CONCRETE

H.W. Reinhardt[1)], M.G.M. Pat[1)] and F.H. Wittmann[2)]

1) Delft University of Technology, Stevin Laboratory Delft, The Netherlands
2). Swiss Federal Institute of Technology, Laboratory for Building Materials, Lausanne, Switzerland

ABSTRACT
So far only few reports on the variability of creep and shrinkage have been published. In many practical cases, however, it is necessary to know the probability of occurence of extreme values. Therefore four different test series, each consisting of indentical concrete specimens with a given mix proportion, have been prepared. Shrinkage and creep have been measured for 300 days. For the creep experiments a load of 20% and 40% of the corresponding prism strength has been applied. In each test series between four and eight identical specimens have been tested. Experimental results have been evaluated statistically. For shrinkage a coefficient of variation in the range of 0,07 and 0,14 was found. For the total time-dependent deformation (creep plus shrinkage) the coefficient of variation after 300 days of loading was found to be between 0,05 and 0,10. Creep recovery was measured for 28 days.

ZUSAMMENFASSUNG
Bisher erschienen nur wenige Veröffentlichungen zur Streuung von Kriechen und Schwinden. In vielen Fällen ist es jedoch notwendig die Auftretenswahrscheinlichkeit extremer Verformungswerte zu kennen.
Versuche mit vier unterschiedlichen Betonmischungen wurden ausgeführt. Kriechen und Schwinden wurden 300 tagelang messend verfolgt. Bei den Kriechversuchen wurde eine Last von 20% bzw. 40% der Prismenfestigkeit aufgebracht. Die einzelnen Testserien umfassten zwischen vier und acht Prüfkörper. Die Messergebnisse wurden statistisch ausgewertet. Der Variationskoeffizient der Schwindverformung liegt zwischen 0,07 und 0,14, während für den Variations-

koeffizienten der gesamten zeitabhängigen Verformung Werte zwischen 0,05 und 0,10 gefunden wurden. Das Rückkriechen wurde 28 tagelang verfolgt.

1 INTRODUCTION

Over decades, the analysis of civil engineering structures has been done in a deterministic way where material properties and loads were assumed to be constants. More recently, the analysis has changed to a semi-probabilistic one - en example is the CEB-FIP model code (1) - where the material strength is assumed to be statistically distributed and the load to be constant. In several national codes the wind load is also treated as a stochastic process.

The statistical distribution of strength of a material can be determined by experiments. Then a certain value can be taken as a rather safe lower limit - for instance 5% failure probability - which is used in the statistical analysis.

So far, this concept has proved to be successful and consequently it should be extended to cover in addition to strength other material properties. The statistical distribution function of creep and shrinkage of concrete so far has got only little attention. With the aim to contribute to fill this gap experiments on gravel concrete have been carried out. Results will be described and analysed in the following chapters.

2 EXPERIMENTS

2.1 Scope of Experiments and Materials Used

Four mixes of gravel concrete were casted and investigated during one year. Shrinkage and creep at two stress levels and at constant temperature and relative humidity were measured. The number of equally treated specimens was between four and eight.

The concrete composition is given in Table I. All gradings are within the prefered zone 3 of the CEB-FIP (1) grading curves. The cement used was a Portland B according to Dutch Standards. The workability of all mixes was almost the same. The air content varied between 1,3 and 2%. All specimens were cured at the same conditions: 7 days at 20°C and 99% R.H. and then at 20°C and 65% R.H.

The compressive strength of the concrete was determined by using cylinders with a diameter of 150 mm and a height of 400 mm.

The values of the 28-day cylinder strength are given in Table II, as average values of three tests. These values were taken as a reference for the chosen stresses in the creep tests which were 20% and 40%, of the cylinder strength respectively (see Table II). The stress was applied at an age of 28 days and kept constant during 300 days. The surrounding controlled climate during the tests was 20°C and 65% R.H. After unloading, the deformation of the specimens was measured for another 28 days.

Shrinkage measurements were carried out on separate specimens under the same climate conditions during 300 days.

TABLE I: Concrete Composition

Mix no	Cement content kg/m³	Aggregate content kg/m³	Grading of aggregate perc. passing, sieves in mm							fine material < 250 µm m³/m³	water cement ratio	compaction index (Walz)
			0,25	0,5	1	2	4	8	16			
1	243	1938	7	13	21	32	48	70	100	0,21	0,60	1,13
2	247	1938	7	13	21	32	48	70	100	0,21	0,60	1,11
3	300	1904	5	11	20	31	48	70	100	0,19	0,55	1,08
4	349	1831	3	9	18	29	46	69	100	0,16	0,50	1,08

TABLE II: Cylinder Strength at 28 Days and Stresses Applied in Creep Tests

Mix no	Cylinder strength N/mm²	Applied stresses in creep tests N/mm²	N/mm²
1	23,9	4,8	9,6
2	22,9	4,6	9,2
3	22,6	4,5	9,0
4	29,6	5,9	11,8

2.2 Test Set-up and Experimental Procedure

2.2.1 Specimens. The specimens for the creep and shrinkage measurements were prisms 150x150x600 m^3 which were cast in a horizontal position.

The upper and lower faces of the shrinkage specimens were sealed with an aluminium sheet in order to prevent evaporation. Because of the use of stiff steel moulds, the faces of the creep specimens were sufficiently plane and parallel and did not need any processing. On two opposite sides of the specimens, dial gauges were mounted with a measuring length of 300 mm and a scale division of 0,01 mm.

2.2.2 Creep rig. The creep rig consists of a closed frame of steel plates and tension rods which are stressed by a hydraulic jack. The jack is mounted between the upper loading plate and the upper specimen. The force is measured by an electric load cell and centered by a ball hinge, as indicated in Fig. 1. Primarily, the pressure is supplied by a hydraulic pump equilibrating the pre-sure in the nitrogen pressure tank. When the desired load is reached the pump is removed and the non-return valve prevents decompression of the loading equipment. The nitrogen pressure tank has sufficient capacity to compensate oil lack and creep deformation. Nevertheless, the hydraulic pressure was checked at regular time intervals and adjusted to the nominal pressure if the load had decreased.

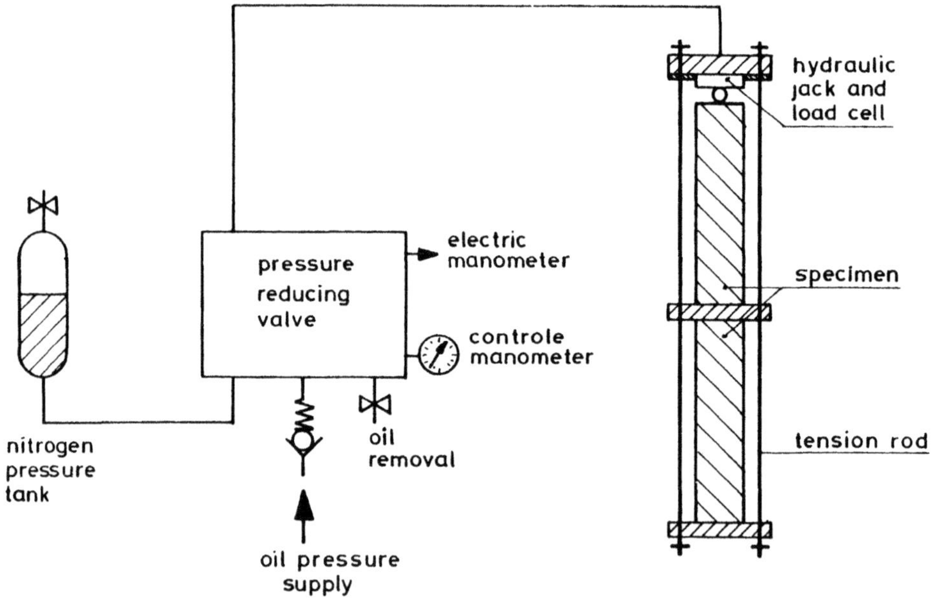

Fig. 1. Creep rig, schematically.

2.2.3 Measurements. Shrinkage measurements started on the 7th day after casting and were continued for 321 days. Before the first measurement and after the last measurement the specimens were weighed on a precision beam--scale.
Creep measurements have been taken at approximately logarithmic intervals, i.e. 1, 2, 5, 10, 30 minutes, 1, 2, 6 hours, 1, 2, 7, 14, 30, 60, 100, 150, 300 days after application of load.

After the creep and shrinkage tests, all specimens were loaded to failure. The stress-strain relation and the elastic modulus were determined.

3 RESULTS

3.1 Shrinkage

In Fig. 2 the shrinkage measurements are plotted versus time for all 5 or 6 specimens of the four tested mixes. Mix 1 and 2 were almost the same with a comparatively low cement content, whereas mix 3 and 4 had a lower water cement ratio and higher cement content.

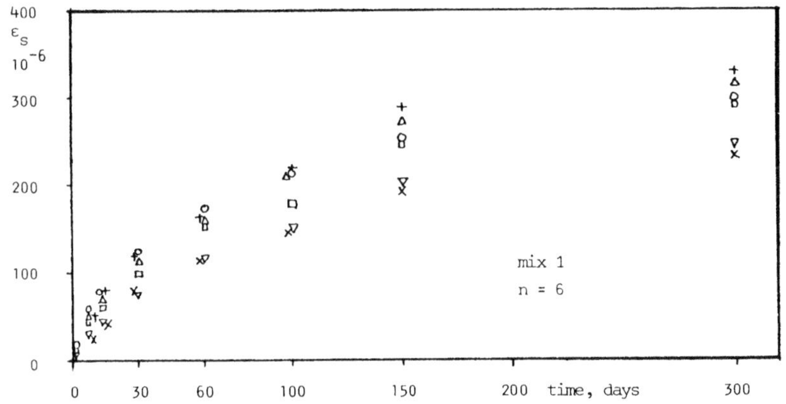

Fig. 2. Shrinkage vs. time of drying of series of identical specimens of the four mixes.

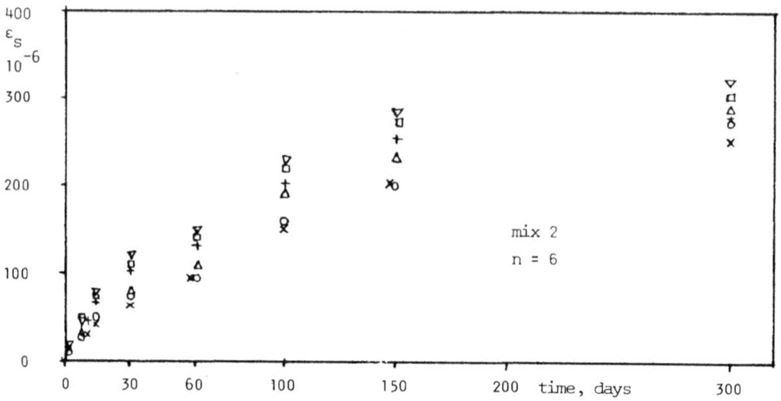
400
εs
10-6
300
200
100
0
0
30
60
100
150
200
time, days
300
mix 2
n = 6

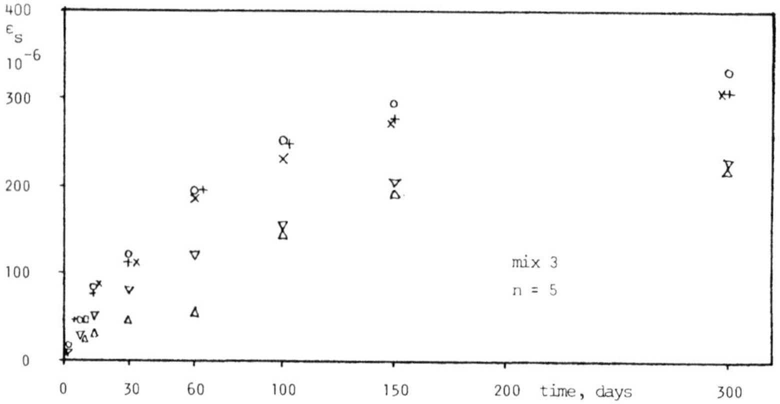
400
εs
10-6
300
200
100
0
0
30
60
100
150
200
time, days
300
mix 3
n = 5

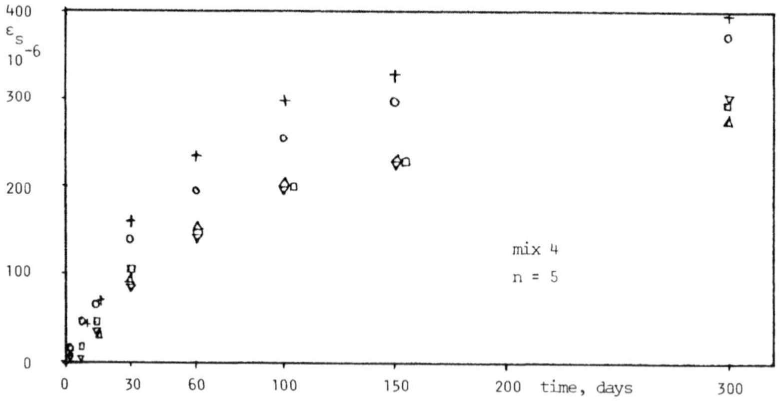
400
εs
10-6
300
200
100
0
0
30
60
100
150
200
time, days
300
mix 4
n = 5

Water content of the fresh concrete and water loss during shrinkage is shown in Table III. After the shrinkage measurements, the specimens were dried in an oven at 105 °C. The water loss during drying is also given in Table III.

TABLE III: Water Loss During Drying

Mix no	Water in fresh concrete $1/m^3$	Water loss during shrinkage		Water loss during drying	
		$\bar{w}$ $1/m^3$	v.c. %	$\bar{w}$ $1/m^3$	v.c. %
1	146	47,8	2 6	58,9	2,9
2	148	38,2	3,3	59,9	1,6
3	165	34,8	4,3	71,4	3,0
4	175	42,8	6,0	87,1	3,4

$\bar{w}$ = average, v.c. = coefficient of variation

3.2 Creep and shrinkage

The sum of creep and shrinkage deformation of the loaded specimens is plotted as function of time in Fig. 3.

All experimental results are described in detail in a report of the Stevin Laboratory (2).

In the following paragraph these results will be evaluated statistically.

4 ANALYSIS OF THE RESULTS

4.1 General Remarks

All results will be treated in the same way: first, the average of the experimental results of equal specimens which are measured at a given time, will be calculated, second, the standard deviation of these results will be calculated together with the 90% confidence limits, third the results of all specimens of one mix and one stress level will be fitted by regression analysis and the 90% confidence limits of the parameters of the fitted curve will be given.

So far, no difference has been made between various contributions to the overall scatter of the results such as measuring

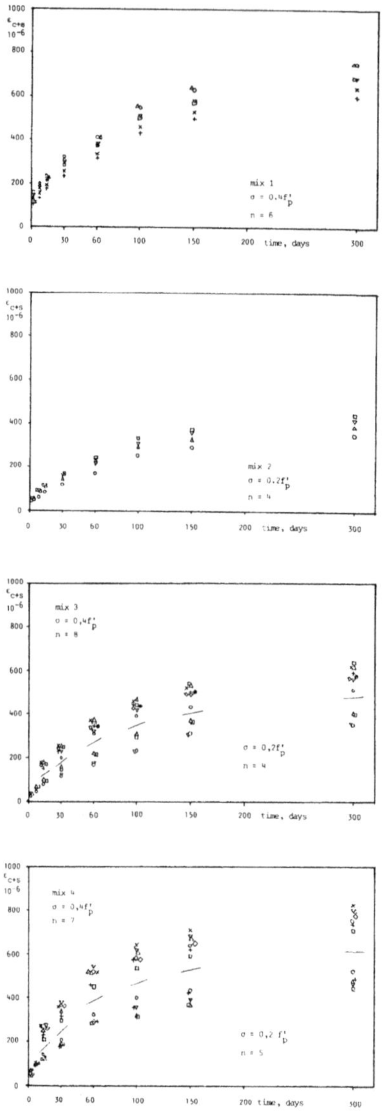

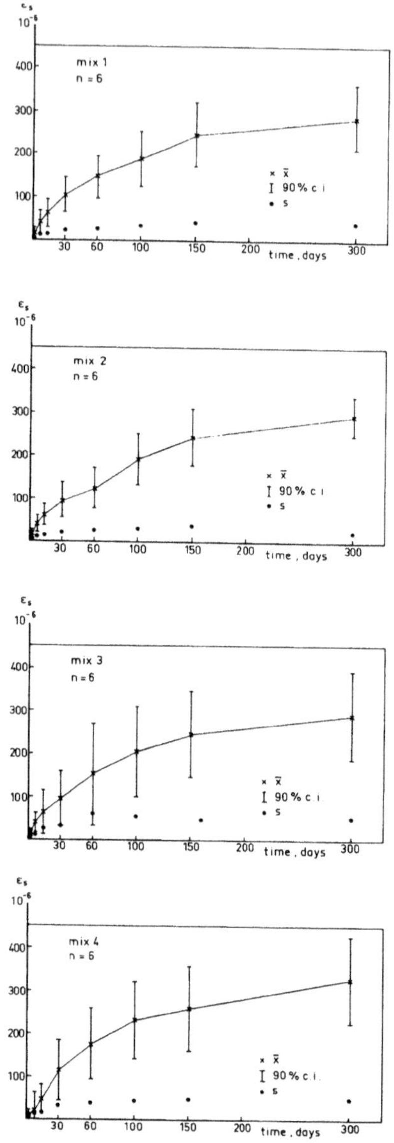

Fig. 3. Creep and shrinkage vs. time of identical specimens of four mixes.

Fig. 4. Shrinkage mean, standard deviation and 90% confidence limits for mix 1 through 4.

inaccuracy, mix deviation, load and climate control, material variability. These influences are treated as random phenomena. Measurements which could be identified as being wrong, were eliminated. Therefore the number of observations is not always the same.

4.2 Shrinkage

The results of the shrinkage measurements are summarized in Table IV for all mixes. Standard deviation and coefficient of variation are given as well. The same results plotted on a linear scale (Fig. 4) show the relatively large scatter quite clear. In the beginning of the measurements the measuring accuracy is supposed to have a remarkable influence. The dial gauge division being 0,01 mm and the measuring length being 300 mm a standard deviation of about $15 \cdot 10^{-6}$ can be expected. That means that up to about 14 days a considerable part of the scatter of the results can be attributed to this effect. Lateron, the material scatter is going to be predominant and the standard deviation reaches values as high as $50 \cdot 10^{-6}$ which is equivalent to a coefficient of variation of about 15%. This is even higher than the scatter of the standard strength tests where the coefficient of variation was found to be about 8%.

For practical applications the experimental results should be extrapolated in time in order to get the asymptotic value of shrinkage. A hyperbolic function is convenient for this purpose. Ross among other authors has used the following formula:

$$\varepsilon_s = \frac{at}{b + t} \tag{1}$$

with the asymptotic value of a. The results obtained from each individual specimen, measurements of all specimens of one mix, and the mean values of the four mixes (see Table IV) were evaluated on an elementary statistical basis separately (least square method). Table V shows the calculated results, i.e. the coefficient of the regression line and the square of the coefficient of correlation. It turns out that the absolute value of a does hardly vary within one mix, irrespective of the manner of calculation. From the individual results it can be seen that the coefficient of variation is of the same order of magnitude as it is for the experimental results at 300 days (see Table IV).

TABLE IV: Mean Value $\bar{x}$, Standard Deviations and Coefficient of Variation v.c. of Shrinkage Tests, in 10^{-6}

Mix no number of spec		time, days 2	7	14	30	60	100	150	300
1	$\bar{x}$	14	43	64	102	147	188	242	286
n = 6	s	7	15	17	22	25	34	40	39
	v.c.	0,51	0,34	0,26	0,21	0,17	0,18	0,16	0,14
2	$\bar{x}$	15	39	61	92	121	193	243	286
n = 6	s	6	11	14	22	25	31	35	24
	v.c.	0,43	0,28	0,22	0,24	0,20	0,16	0,14	0,08
3	$\bar{x}$	15	40	64	95	151	207	247	292
n = 6	s	5	12	26	32	60	53	49	51
	v.c.	0,34	0,29	0,40	0,34	0,40	126	0,20	0,18
4	$\bar{x}$	6	22	49	117	175	230	260	329
n = 6	s	7	21	18	35	41	45	50	51
	v.c.	1,17	0,93	0,38	0,30	0,24	0,20	0,19	0,15

TABLE V: Statistical Evaluation of Shrinkage Results by Regression Analysis

Mix no	Data treated		a 10^{-6}	r^2
1	individual	$\bar{x}$	344	0,98
		s	38	
		v.c.	0,109	
	all		334	0,78
	mean		337	0,99
2	individual	$\bar{x}$	359	0,95
		s	24	
		v.c.	0,068	
	all		358	0,78
	mean		355	0,95
3	individual	$\bar{x}$	396	
		s	56	0,98
		v.c.	0,143	
	all		388	0,48
	mean		387	0,99
4	individual	$\bar{x}$	392	
		s	55	0,97
		v.c.	0,141	
	all		397	0,75
	mean		391	0,99

TABLE VI: Mean Value, Standard Deviation, and Coefficient of Variation of Total Time-Dependent Deformation (creep and shrinkage without elastic strain), in 10^{-6}

Mix no stress level no of spec		Time, days 2	7	14	30	60	100	150	300
1	$\bar{x}$	133	172	212	284	374	504	576	687
$\sigma = 0{,}4\ f'_p$	s	21	24	25	34	41	50	55	62
n = 6	v.c.	0,16	0,14	0,12	0,12	0,11	0,10	0,10	0,09
2	$\bar{x}$	54	82	106	151	214	297	341	402
$\sigma = 0{,}2\ f'_p$	s	7	11	14	24	30	32	35	41
n = 4	v.c.	0,12	0,14	0,13	0,16	0,14	0,11	0,10	0,10
3	$\bar{x}$	35	60	87	136	200	269	340	382
$\sigma = 0{,}2\ f'_p$	s	7	9	11	22	32	39	30	27
n = 4	v.c.	0,21	0,16	0,12	0,16	0,16	0,15	0,09	0,07
3	$\bar{x}$	73	119	163	239	346	437	504	589
$\sigma = 0{,}4\ f'_p$	s	20	24	29	19	22	23	34	38
n = 8	v.c.	0,28	0,20	0,18	0,08	0,06	0,05	0,07	0,06
4	$\bar{x}$	59	96	129	187	296	349	399	483
$\sigma = 0{,}2\ f'_p$	s	9	6	7	13	16	35	32	28
n = 4	v.c.	0,15	0,06	0,06	0,07	0,05	0,10	0,08	0,06
4	$\bar{x}$	112	187	255	347	505	593	655	774
$\sigma = 0{,}4\ f'_p$	s	15	27	25	32	37	36	39	42
n = 7	v.c.	0,13	0,14	0,10	0,09	0,07	0,06	0,06	0,05

4.3 Creep and Shrinkage

The same procedure as just described for shrinkage deformation only, has been applied to evaluate the sum of creep and shrinkage measurements (total time-dependent deformation). Table VI contains the results of the four mixes, subdivided into two categories of two different stress levels. In comparison with the shrinkage results, the scatter of total time-dependent deformation is considerably smaller. After a duration of load of 300 days the coefficient of variation varies between 5 and 10%. Figs. 5 and 6 make this feature evident. If the principle of superposition were valid without any restriction, it couldn't happen that the scatter of the result of the sum of two independent events i.e. creep and shrinkage, was less than the scatter of one single event i.e. shrinkage only.

The limited accuracy of the dial gauges used in this experimental study may explain the comparatively high scatter at small deformations.

Creep and shrinkage are not independent and therefore the creep deformation cannot be calculated by subtracting the mean shrinkage strain which was determined on separate specimens (3,4). Therefore no attempt has been made here to analyse creep and shrinkage from measured total time-dependent deformation separately.

It can be anticipated that the elastic deformation of concrete shows a certain scatter too. In Table VII the elastic deformation of two of the test series is shown as an example. The coefficient of variation is found to be 0,08 and 0,03 respectively. For comparison the total time-dependent and the total deformation are also given in Table VII. It is noteworthy that all three strains have about the same variability for a given test series.

4.4 Creep Recovery

For 28 days after unloading the strain measurements were continued. Table VIII shows the mean value, the standard deviation and the coefficient of variation of the sum of creep recovery and shrinkage deformation of the four mixes. The shrinkage deformation, however, may be assumed to be small. The scatter is rather large, with a coefficient of variation after 28 days of 0,10 to 0,25. Because of the small strains the scatter is probably caused to a great extend by the inaccuracy of the measuring device.

The elastic deformation after unloading is indicated in Table IX. For comparison the total time-dependent (creep recovery plus shrinkage) and the total deformation as measured 28 days

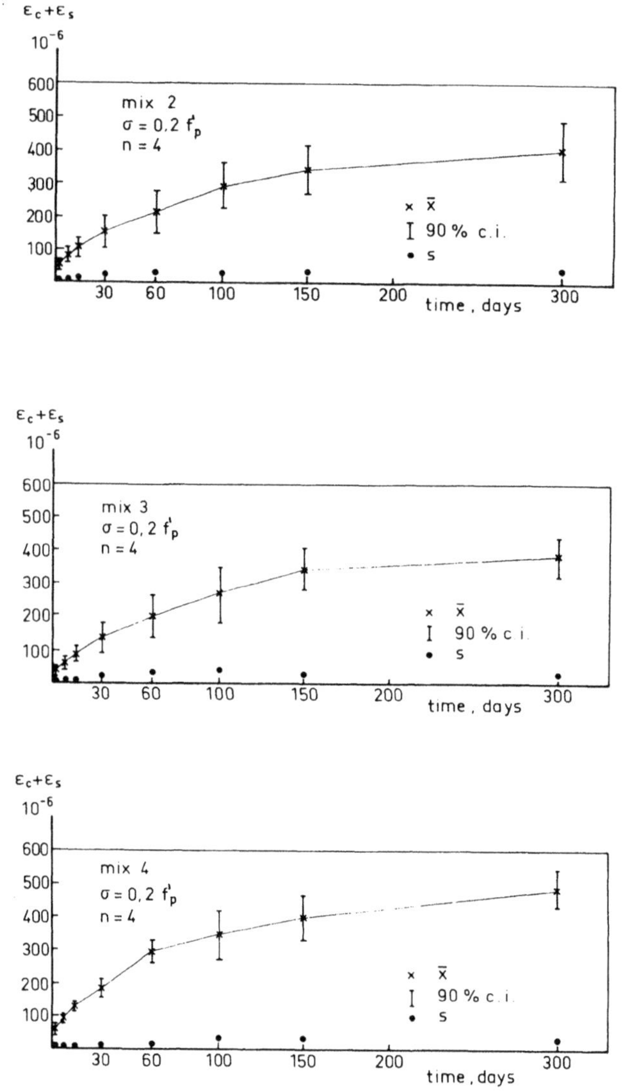

Fig. 5. Creep + shrinkage for $\sigma = 0{,}2\ f'_p$

mean, standard deviation and 90% confidence interval.

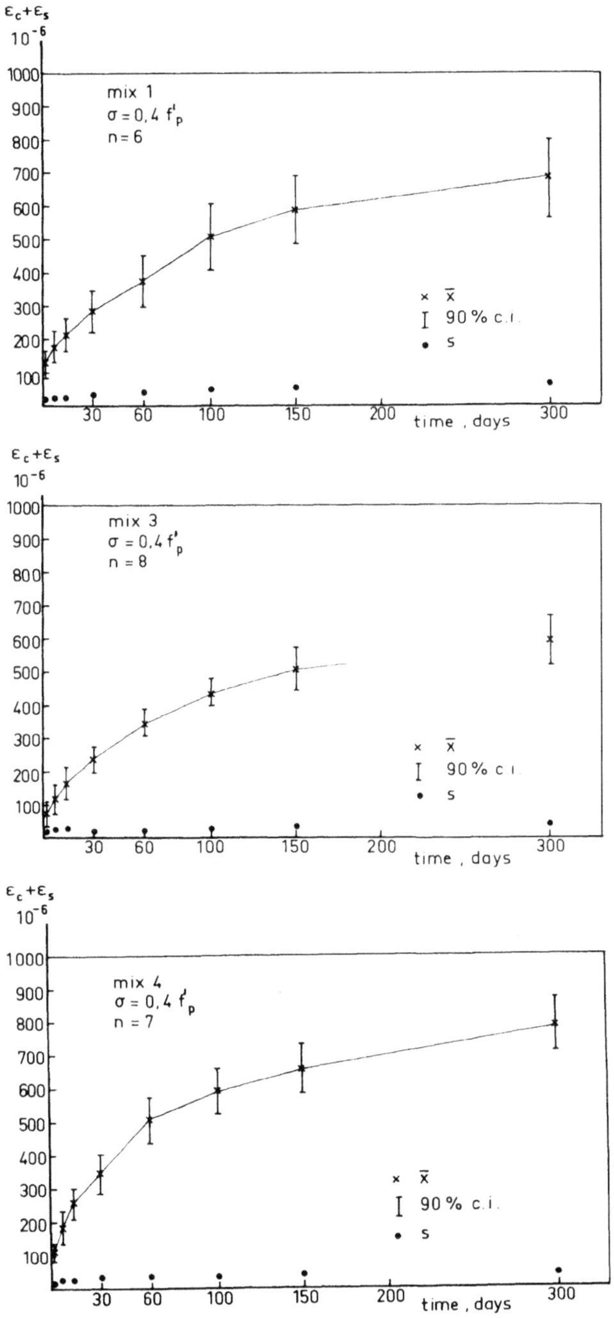

Fig. 6. Creep + shrinkage for $\sigma = 0,4\ f'_p$

mean, standard deviation and 90% confidence interval.

TABLE VII: Elastic Deformation ε_{el}, Total Time-Dependent Deformation ε_{cs} and Total Deformation ε_{tot} after a Duration of Load of 300 Days for two Different Series

Mix no	Spec no	ε_{el}	ε_{cs} at 300 d	ε_{tot} at 300 d
1	1461	274	757	1031
$\sigma = 0{,}4\ f'_p$	1462	276	693	969
	1463	301	757	1058
	1464	285	671	956
	1465	243	601	844
	1466	246	643	889
	$\bar{x}$	271	687	958
	s	23	62	82
	v.c.	0,08	0,09	0,09
3	3461	281	520	801
$\sigma = 0{,}4\ f'_p$	3462	279	566	845
	3463	271	618	889
	3465	261	598	859
	3466	253	620	873
	3467	263	570	833
	3468	268	580	848
	$\bar{x}$	268	589	857
	s	9	38	33
	v.c.	0,03	0,06	0,04

TABLE VIII: Mean Value, Standard Deviation and Coefficient of Variation of Reversible Strain (creep + shrinkage) during the first 28 Days after Unloading

mix no stress level no of spec		Time after unloading, days 1	2	7	14	28
mix 1 $\sigma = 0{,}4\ f'_p$ n = 6	$\bar{x}$ s v.c.	27 5 0,20	24 5 0,21	32 6 0,19	36 6 0,18	38 6 0,17
mix 2 $\sigma = 0{,}2\ f'_p$ n = 4	$\bar{x}$ s v.c.	12 2 0,17	12 4 0,30	16 3 0,17	18 3 0,17	19 5 0,25
mix 3 $\sigma = 0{,}2\ f'_p$ n = 4	$\bar{x}$ s v.c.	15 5 0,35	17 5 0,31	18 4 0,20	19 2 0,11	18 4 0,22
mix 3 $\sigma = 0{,}4\ f'_p$ n = 8	$\bar{x}$ s v.c.	24 7 0,27	26 7 0,28	30 8 0,28	35 9 0,26	35 8 0,24
mix 4 $\sigma = 0{,}2\ f'_p$ n = 5	$\bar{x}$ s v.c.	17 3 0,15	17 3 0,17	20 3 0,15	27 5 0,20	29 6 0,20
mix 4 $\sigma = 0{,}4\ f'_p$ n = 6	$\bar{x}$ s v.c.	33 2 0,07	38 4 0,10	44 4 0,09	51 6 0,12	53 5 0,10

after unloading are also entered in Table IX. The comparatively high coefficient of variation of the total time-dependent deformation which is essentially creep recovery may be at least partly due to the limited accuracy of the measurements.

5 DISCUSSION AND CONCLUSIONS

A stochastic model for describing creep deflection of reinforced concrete beams has been published by Benjamin, Cornell and Gabrielson (5). These authors used a Poisson process. Cinlar and co-workers (6) more recently treated the stochastic nature of creep theoretically. They modeled creep by means of a non-stationary stochastic process and assumed creep to be a local gamma process. Jordaan (7) questions whether macroscopic creep really is a stochastic process; but his argument that because of the huge number of gel particles taking part in the creep process stochastic variation cannot be observed is certainly an unjustified simplification of complex materials behaviour.

Cornelissen (8) studied different influences of creep of concrete such as water-cement ratio, cement content or temperature. He concludes that in reality creep of concrete turns out to be a stochastic quantity because of the interaction of all the different unpredictable influences.

In the present study all influences from outside have been carefully kept constant. If we neglect the contribution of a given measuring technique for the moment there remain two basically different components of the variability of creep and shrinkage of concrete: (I) materials inherent inhomogeneities and (II) environmental influences.

In a given concrete specimen, the amount of aggregate, the geometrical arrangement, the water-cement ratio and properties of cement vary locally in a statistical way. That leads to a threedimensional field of largely varying creep properties. The distribution of creep centres is superimposed by the threedimensional stress field within the concrete specimen. It is well known that high stress concentrations can be observed in the inhomogeneous structure of concrete. The interaction of the stress field with the spatial distribution of creep centres leads to the stochastic nature of creep under carefully controlled environmental conditions.

In real structures the overall variability will be dominated by external stochastic influencing factors such as temperature and humidity. And in addition to this the variability of parameters of concrete technology has a significant influence.

TABLE IX: Elastic Deformation after Unloading ε'_{el}, Total Time-Dependent Deformation ε'_{cs} (creep recovery and shrinkage) and Total Deformation ε'_{tot} as Measured 28 Days after Unloading

mix no	Spec no	ε'_{el}	ε'_{cs} at 28 d	ε'_{tot} at 28 d
	3461	210	37	247
	3462	223	28	251
	3463	253	37	290
	3464	220	17	237
3	3465	233	40	273
	3466	208	43	251
$\sigma = 0{,}4\ f'_p$	3467	198	37	235
	3468	218	38	256
	$\bar{x}$	220	34,60	255
	s	17	8	18
	v.c.	0,08	0,24	0,07

REFERENCES

1. CEB-FIP Model code for concrete structures (1978).

2. Pat, M.G.M. Kruipspreiding, deel 1: Proefresultaten. Stevinrapport 5-80-1 (Delft, 1980).

3. Roelfstra, P.E. and F.H. Wittmann. "Een rekenmodel ter bepaling van de tijdsafhankelijke vervorming van uitdrogend beton". Cement (The Netherlands) 22(1980) 348-351.

4. Wittmann, F.H. and P.E. Roelfstra. "Total deformation of loaded drying concrete". Cem. Concr. Res. 10(1980).

5. Benjamin, J.R., C.A. Cornell and B.I. Gabrielson. "A stochastic model for the creep deflection of reinforced concrete beams". Proc. Int. Symp. Flexural Mechanics of Reinforced Concrete, American Concrete Institute, Detroit, SP 12 (1965) 557-580.

6. Cinlar, E., Z.P. Bazant and E. Osman. "Stochastic process for extrapolating concrete creep". J. Eng. Mech. Div. ASCE 103(1977) 1069-1088.

7. Jordaan, I.J. "Models for creep of concrete with special emphasis on probabilistic aspects". Mat. and Struct. 13(1980) 29-40.

8. Cornelissen, H. Creep of concrete - a stochastic quantity. PH.D. Thesis, TH Eindhoven, The Netherlands (1979).

CREEP OF CONCRETE - A STOCHASTIC QUANTITY

H.A.W. Cornelissen

Delft University of Technology,
dep. of Civil Engineering, Delft, The Netherlands

ABSTRACT

The magnitude of creep is regarded as a stochastic quantity, because of random variations of important influential factors.
A test-program on a statistical basis has been carried out to investigate the influence of actual variations of the concrete composition (wcr;acr), on creep. The 28-day creep tests (under compression) were performed at 25°C; 50% RH as well as at two ambient conditions with alternating temperature and relative humidity, derived from the real conditions in The Netherlands. Portland and Portland blast-furnace cement was used in the test program. It is concluded that extensive variations of the creep deformations will be caused by actual variations of the water-cement ratio's, and by variations of the temperature even in the normal range.

SAMENVATTING

Kruip wordt beschouwd als een stochastische grootheid ten gevolge van toevallige variaties van belangrijke invloedsfactoren. Een proevenprogramma op statistische basis werd uitgevoerd om de invloed van praktijkspreidingen in de betonsamenstelling na te gaan. Drukkruipproeven (28 dagen belast) werden uitgevoerd bij 25°C; 50% RV en bij omgevingscondities met wisselende temperatuur en relatieve vochtigheid, afgeleid van de werkelijke condities in Nederland. Voor de proeven werden portland en hoogoven cement gebruikt. Geconcludeerd wordt dat kruip aan spreiding onderhevig is ten gevolge van de praktijkvariaties van de water-cement factor en variaties van de temperatuur zelfs in de normale range.

1. INTRODUCTION

The approach made to the variability of the magnitude of the creep deformations deals especially with either the stochastic character of the fundamental creep process {1,2}, or with deviations from creep formulas founds in the experimental results of different investigators {3}. These deviations can be attributed to the different experimental conditions and the differences between the types of cement and aggregates.
In practice however variations of the creep deformations including deviations from the design values, occur as a result of actual variations of the influential factors. Because of random variations in some important influential factors, the resulting creep is regarded as a stochastic quantity.
It can be mentioned that the variability of the compressive strength of concrete has been approached in the same way {4}.

This research program deals with the influence on creep of variations in the concrete composition occuring at the building site. With reference to the Dutch temperature and relative humidity conditions, the creep experiments are performed at two artificial alternating T and RH conditions, (day-night cycle), reflecting the real conditions during a typical summer and winter month. In addition the creep is investigated at constant T and RH conditions ("laboratory-conditions"). Because of the importance of creep in prestressed concrete structures, the choice of the types of cement involved was based on this application. The types of cement were: extra-rapid hardening portland cement (in Dutch description pc C), rapid hardening portland cement (pc B) and rapid hardening portland blast-furnace cement (hoc B). With respect to the last mentioned type of cement only a few experimental results are available. In {5} two contradicting publications on creep of hoc B are indicated. It is expected that in The Netherlands the use of hoc B will increase because of the lower energy demand for manufacture and the relatively low demand for the scarce raw material marl.
The selected influential factors are:
- the concrete composition (actual variations)
- T and RH conditions (real and laboratory conditions)
- the type of cement.

The most important factors that are held constant, are:
- the type of aggregate and the grading curve
- size and shape of the test specimens
- the curing conditions
- the load and the age at loading
- the loading period.

2. EXPERIMENTS

- Concrete compositions.

A mix intended to have a water-cement ratio (wcr) of 0,45 and an aggregate cement ratio (acr) of 5,0 showed after analysis of the fresh concrete (600m^3), that the realized mixes at the building site had water-cement and aggregate-cement ratio's respectively in the range of 0,40 to 0,50 and 4,5 to 5,5. The results are used as an indication to determine the concrete composition involved in the experiments (see FIG 1 and TABLE 1).

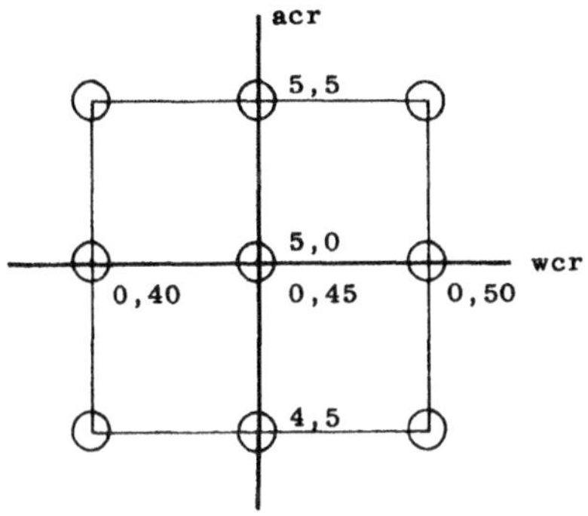

FIG.1. The concrete compositions in terms of wcr and acr.

wcr	acr	cement (kg/m^3)	water (kg/m^3)	aggregate (kg/m^3)
0,40	4,5	407,8	163,1	1835,2
0,40	5,0	378,0	151,0	1891,0
0,40	5,5	352,8	141,1	1940,3
0,45	4,5	399,6	179,8	1798,2
0,45	5,0	371,0	166,9	1855,0
0,45	5,5	346,6	156,0	1906,0
0,50	4,5	391,7	195,8	1762,6
0,50	5,0	364,0	182,2	1822,0
0,50	5,5	340,6	170,3	1873,3

TABLE 1. The concrete compositions.

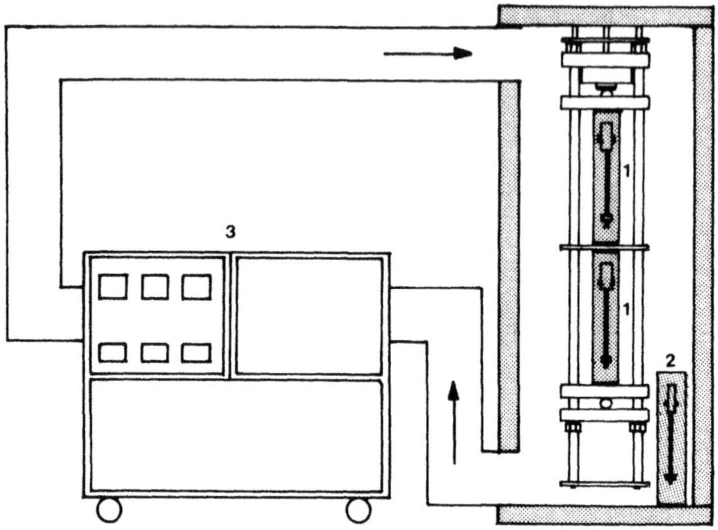

FIG 2. Test set up for creep measurements
(1=creep, 2=shrinkage, 3=T and RH conditioner)

- Curing and loading.

After two weeks curing in water (20°C), followed by one week at 25°C and 50% RH, the prisms (100x100x500mm^3) were loaded uniaxially for four weeks. The load on each prism was the same, resulting, because of their different compositions, in different stress-strength ratio's per prism (30-40%). The cube strengths are given in TABLE A-1.
The deformations were measured at two opposite sides parallel to the direction of the load (see FIG 2).

- Artificial temperature and relative humidity conditions.

In the Netherlands during the period 1960 to 1970, the year 1961 showed average conditions in relation to temperature and relative humidity. The data concerning a typical winter and summer month in that year, were translated in an artificial summer and winter climate with alternating temperature and relative humidity (every 12 hours).
The translated conditions are shown in FIG 3.

- Experimental scheme on a statistical basis.

The set up of the experiments was based on a 3^4factorial design (4 parameters each at 3 levels).
Due to the capacity of the loading frame, at the two alternating T and RH conditions, one third of the possible combinations was performed in these conditions. The experimental scheme is shown in TABLE 2.

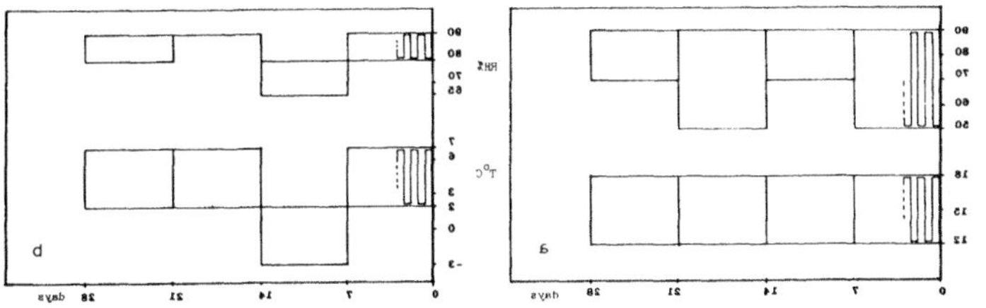

FIG 3. Artificial "summer" conditions (a) and "winter" conditions (b).

wcr	acr	constant (T=25°C; RH=50%)			artificial summer climate			artificial winter climate		
		* pc C	pc B	hoc B	pc C	pc B	hoc B	pc C	pc B	hoc B
0,40	4,5	xx	xx	xx	x					x
0,40	5,0	xx	xx	xx			x		x	
0,40	5,5	xx	xx	xx		x		x		
0,45	4,5	xx	xx	xx		x		x		
0,45	5,0	xx	xx	xx	x					x
0,45	5,5	xx	xx	xx			x		x	
0,50	4,5	xx	xx	xx			x		x	
0,50	5,0	xx	xx	xx		x		x		
0,50	5,5	xx	xx	xx	x					x

* Chemical and physical properties of the cements are given in TABLE A-2.

TABLE 2. The experimental scheme.

3. ANALYSIS OF THE EXPERIMENTAL RESULTS

The experimental creep results were transformed to specific creep data being: $\varepsilon_{csp}(t)=(\varepsilon_t(t) - \varepsilon_e(t_o) - \varepsilon_s(t))/\sigma$

$\varepsilon_t(t)$ is the total deformation, $\varepsilon_e(t_o)$ is the elastic deformation and $\varepsilon_s(t)$ is the deformation of the separate shrinkage specimen, σ is the acting compressive stress. As a function of time the specific creep is described by a power function:

$$\varepsilon_{csp}(t) = a_o t^n$$

The calculated values of a_o and n are given in TABLE A-3 and A-4. Linear equations were fitted to the 7, 14, 21 and 28 days specific creep data. These equations were of the type:

$$\underline{\varepsilon}_{csp}(t_1) = \beta_o x_o + \beta_1 x_1 + \beta_2 x_2 + \dots + \beta_k x_k + \underline{e}$$

The dependent variables $x_o - x_k$ represent the levels of the influential factors, $\underline{e}$ is the random error. The parameters $\beta_o - \beta_k$ are estimated and in addition the variances of these parameters. The models only consist of significant terms.

4. RESULTS AND CONCLUSIONS

Some important conclusions, valid within the limitations proposed are given below:
1. At T is 25°C and RH is 50% an estimation of the specific creep (for instance after 28 days under load) as a function of the wcr, the acr and type of cement is given by:

$$\hat{\varepsilon}_{csp}(28\text{ d}) = b_o + b_1 W + b_2 W2 + b_3 T + b_4 C + b_5 C2 + b_6 WT + b_7 WC + b_8 WC2 + b_9 W2C + b_{10} W2C2 + b_{11} TC + b_{12} TC2$$

and the variance:

$$\hat{var}\underline{\varepsilon}_{csp}(28\text{ d}) = \hat{var}\underline{b}_o + W^2 \hat{var}\underline{b}_1 + \dots + TC2^2 \hat{var}\underline{b}_{12}$$

W, T and C represent the coded levels of the wcr, acr and type of cement (TABLE A-5)

The calculated values of b_i and $\hat{var}\underline{b}_i$ are shown in TABLE A-6.

Using one type of cement the model is as follows

$$\hat{\varepsilon}_{csp}(28\text{ d}) = b'_o + b'_1 W + b'_2 W2 + b'_3 T + b'_4 WT$$

The relations, obtained with this formula are shown in FIG A-1, A-2 and A-3)

2. The standard deviation of the average specific creep as a result of the experimental procedure, proves to be 0,07 - 0,09 (10^{-5} per N/mm^2).

3. Variations in the concrete composition (especially the wcr) result in extensive variations of the creep deformations. In an actual situation the intention to produce a mix with a wcr of 0,45 and an acr of 5,0, can result in water-cement ratio's and the aggregate cement ratio's being in the range of 0,40 - 0,50 and 4,5 - 5,5 respectively. The specific creep (in 10^{-5} per N/mm^2) will then vary from 1,8 to 4,6 using pc C, from 2,5 to 5,4 using pc B and from 1,7 - 2,6 using hoc B (see FIG A-1, A-2 and A-3).

It is noted that the differences between the required and realized values of the wcr, as measured in ready mixed concrete plants in The Netherlands, showed a variation of + 0,06 (i.e 1,96 x standard deviation). This is shown in FIG 4 {6}. The value 0,06 is very close to the variation of the wcr (being + 0,05), used in our research program.

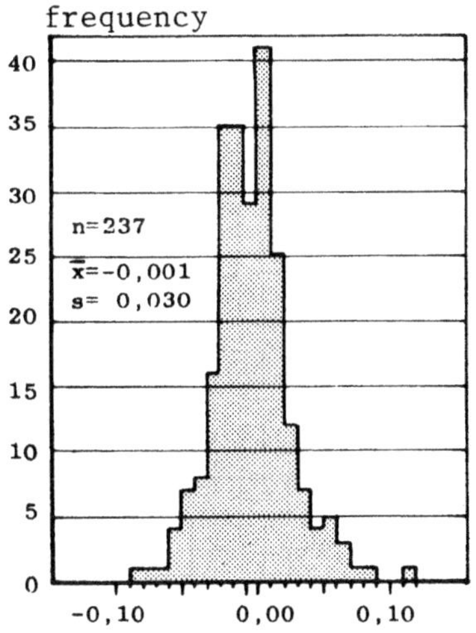

FIG 4. Differences between required and realized values of the wcr, as measured in ready-mixed concrete plants in The Netherlands {6}.

4. Because of the square relation between creep and the important influential factor wcr, a normal distribution of the water-cement ratio's will result in a skew distribution of the creep (see FIG 5). Therefore there will be more creep deformation than calculated, when the average required concrete composition is considered.

5. Rapid hardening portland blast-furnace cement concretes in general showed less creep than both the involved portland cement concretes. Moreover there was a smaller effect of fluctuations of the concrete composition on creep variations, when portland blast-furnace cement was used.

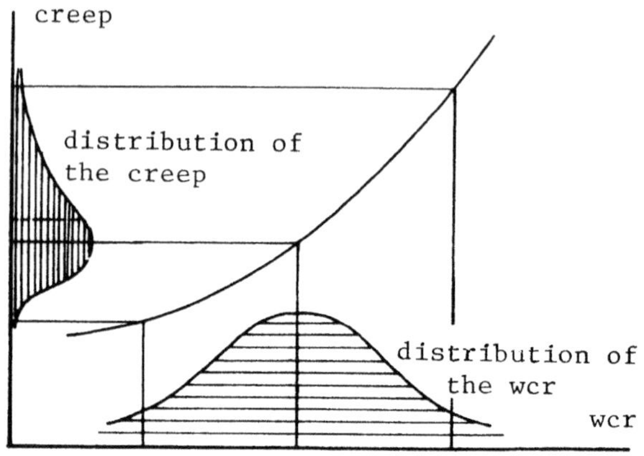

FIG 5. A normal distribution of the wcr results in a skew distribution of the magnitude of creep.

6. The effect of the different T and RH conditions on creep was mainly caused by the differences in the average values of the temperature. In the performed experimental program the alternations of the T and RH were observed to have no influence. In FIG 6 the effect is shown of the T and RH conditions on the specific creep. The creep was calculated from linear models, fitted to the experimental data.

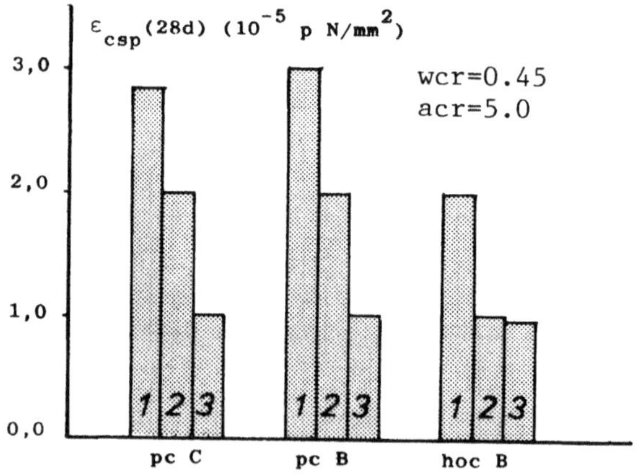

FIG 6. The effect of the three T and RH conditions on the specific creep. (1:T=25°C, RH=50%, 2:summer, 3:winter)

The influence of the average temperature on the magnitude of creep can be described with the Arrhenius equation {7}:

$$\ln \varepsilon_{csp} = (-Q_o/R).\ (1/T) + \ln A$$

A = a constant
Q_o = characteristic activation energy
R = gas constant
T = absolute temperature

The results are shown in FIG A-4.

The calculated values of the characteristic activation energy Q_o are between 20 and 31 (kJ/mol) for pc C and B about 25 (kJ/mol) for hoc B.

It can be concluded that even under normal temperature conditions, it is necessary to introduce the actual temperature in the creep predictions, especially at high creep velocities.

7. Comparing the experimental results in terms of creep coefficients* with calculated creep coefficients according to the Dutch Standards (VB 74), to the ACI committee 209 and to the CEB-FIP 1978, it can be concluded that VB 74 and ACI committee 209 result in upper-limits of the creep even if the actually occuring concrete composition variations are included.
The prediction formula of CEB-FIP results in predictions close to the experimental values. Using these formulas however it must be realized that in practice the influential factors will show variations (see TABLE 3).

T and RH →	constant	summer	winter
VB-74	1,71 (+57%)	1,26 (+48%)	1,14 (+115%)
ACI-comm.209	1,86 (+71%)	1,52 (+79%)	1,45 (+174%)
CEB-FIP	0,92 (-16%)	0,71 (-16%)	0,67 (+ 26%)
experiment	1,09 (0)	0,85 (0)	0,53 (0)

TABLE 3 Calculated and experimental creep coefficients (28d). Between brackets the relative deviations from the experimental values are given.

Note : More details about this research program and the analysis of the results have been presented in {8}.

* creep coefficient (t) = specific creep (t) x modulus of elasticity (t_o)

List of abbreviations

acr	=	aggregate-cement ratio (by weight)
wcr	=	water-cement ratio (by weight)
pc B	=	rapid hardening Portland cement
pc C	=	extra rapid hardening Portland cement
hoc B	=	rapid hardening Portland blast-furnace cement
var	=	variance
^	=	indicates an estimation
σ_o^2	=	variance of the random error

References

1. Benjamin, J., Cornell, C., Gabrielsen B., Stochastic model of creep deflections of reinforced concrete beams. *Proc. of the Intern. symp. on flexural mechanics of reinforced concrete,* Miami 1964, ACI/SP-12.

2. Çinlar, E., Bazant, Z. Stochastic process for extrapolating concrete creep. *J. of the engineering mechanics devision* Dec. 1977, pp.1069-1088.

3. Bazant, Z., Panula, L. Practical prediction of time-dependent deformations of concrete. *Mat. et constr.* vol 11, no.65, 1978, vol 11, no.66, 1978, vol 12, no. 69, 1979

4. Teychenné, D. Recommendations for the treatment of the variations of concrete strength in codes of practice. *BRE current paper* noCP 6/74.

5. Neville, A. Creep of concrete: plain, reinforced and prestressed. *North-Holland publishing company* Amsterdam, 1970.

6. Verslag van de Stichting Betoncontrole, 1978. *Uitgave Betonvereniging Zoetermeer* (Holland) mei 1979. (in Dutch)

7. Wittmann, F. Bestimmung physikalischer Eigenschaften des Zementsteins. *Deutscher Ausschuss für Stahlbeton,* Heft 232, 1974.

8. Cornelissen, H. Creep of concrete - a stochastic quantity. *PhD thesis,* TH Eindhoven, The Netherlands 1979.
(in Dutch, with extended summary in English)

APPENDIX

wcr	acr	pc C			pc B			hoc B		
		21 d	28 d	49 d	21 d	28 d	49 d	21 d	28 d	49 d
0,40	4,5	66,2	59,6	68,2	63,9	58,1	67,6	61,1	58,7	69,1
0,40	5,0	69,0	67,3	74,0	63,3	63,3	70,2	56,0	54,9	61,6
0,40	5,5	70,2	71,8	80,6	51,9	54,3	59,8	55,8	54,4	55,0
0,45	4,5	51,8	53,1	52,4	57,0	53,9	66,9	51,0	48,3	57,4
0,45	5,0	64,0	65,1	70,0	50,0	48,7	55,3	52,4	51,5	59,4
0,45	5,5	64,4	64,4	65,0	61,1	57,6	63,6	53,6	49,3	60,7
0,50	4,5	52,9	45,8	55,8	44,2	43,4	48,7	49,8	45,6	54,7
0,50	5,0	54,8	51,7	59,1	42,2	41,3	50,0	47,4	45,9	53,8
0,50	5,5	55,7	55,7	60,0	43,7	42,4	51,6	49,4	45,3	56,0

TABLE A-1 Compressive cube strengths (N/mm^2) after a curing period of 21, 28 (in water) and 49 days.

spec.surf.Blaine (m^2/kg)	pc C 540	pc B 380	hoc B 470	
composition	%	%	composition	%
C_2S	11,5	12,1	SO_3	3
C_3S	60,0	59,9	SiO_2	27
C_3A	8,0	8,1	Al_2O_3	12
C_4AF	9,6	9,4	Fe_2O_3	2
CaO	1,6	1,5	CaO	45
$CaSO_4$	6,3	6,0	MgO	8
MgO	1,4	1,4	slag content	70%

TABLE A-2 Some typical properties of the types of cement

		pc C				pc B				hoc B			
wcr	acr	n	A	a_o	r(%)	n	A	a_o	r(%)	n	A	a_o	r(%)
0,40	4,5	0,36	-0,65	0,22	99,5	0,36	-0,59	0,26	99,6	0,43	-0,99	0,10	99,7
0,40	5,0	0,42	-0,90	0,13	99,4	0,36	-0,61	0,25	99,8	0,35	-0,71	0,19	99,9
0,40	5,5	0,34	-0,69	0,20	99,8	0,36	-0,62	0,24	99,8	0,34	-0,74	0,18	99,9
0,45	4,5	0,37	-0,53	0,30	99,8	0,34	-0,46	0,35	99,9	0,37	-0,68	0,21	99,9
0,45	5,0	0,34	-0,52	0,30	99,8	0,31	-0,39	0,41	99,9	0,40	-0,79	0,16	99,9
0,45	5,5	0,33	-0,55	0,28	99,8	0,35	-0,54	0,29	99,9	0,37	-0,76	0,17	99,9
0,50	4,5	0,37	-0,36	0,44	99,9	0,34	-0,22	0,60	99,9	0,37	-0,60	0,25	99,9
0,50	5,0	0,36	-0,41	0,39	99,9	0,38	-0,38	0,42	99,9	0,41	-0,81	0,15	99,9
0,50	5,5	0,36	-0,46	0,35	99,9	0,34	-0,31	0,49	99,9	0,35	-0,65	0,22	99,8

TABLE A-3 Calculated coefficients with respect to

$\varepsilon_{csp}(t)=a_o t^n$, or $\log \varepsilon_{csp}(t)= n \log t+A$ *

r = correlation coefficient. T=25°C;RV=50%

		pc C				pc B				hoc B			
wcr	acr	n	A	a_o	r(%)	n	A	a_o	r(%)	n	A	a_o	r(%)
0,40	4,5	^{z}0,25	-0,61	0,25	99,7					^{w}0,15	-0,59	0,26	94,2
0,40	3,0					^{w}0,13	-0,45	0,35	98,8	^{z}0,25	-0,72	0,19	99,7
0,40	5,5	^{w}0,18	-0,65	0,22	97,2	^{z}0,27	-0,57	0,27	99,8				
0,45	4,5	^{w}0,22	-0,57	0,27	99,4	^{z}0,32	-0,50	0,32	99,9				
0,45	5,0	^{z}0,30	-0,54	0,29	99,6					^{w}0,14	-0,47	0,34	94,7
0,45	5,5					^{w}0,18	-0,41	0,39	99,2	^{z}0,25	-0,73	0,19	98,9
0,50	4,5					^{w}0,23	-0,20	0,63	99,7	^{z}0,25	-0,49	0,32	99,8
0,50	5,0	^{w}0,23	-0,34	0,46	99,1	^{z}0,31	-0,33	0,47	99,9				
0,50	5,5	^{z}0,33	-0,46	0,36	99,8					^{w}0,18	-0,53	0,30	96.9

TABLE A-4 Calculated coefficients with respect to

$\varepsilon_{csp}(t)=a_o t^n$, or $\log \varepsilon_{csp}(t)=n \log t+A$ *

r = correlation coefficient. z=summer; w=winter

*(t in hours, ε_{csp} in 10^{-5} per N/mm^2)

	W	W2
water-cement ratio		
$W = (wcr - 0{,}45)/0{,}05$ wcr = 0,40	-1	1
$W2 = 3\,W^2 - 2$ wcr = 0,45	0	-2
wcr = 0,50	1	1
	T	**T2**
aggregate-cement ratio		
$T = (acr - 5{,}0)/0{,}5$ acr = 4,5	-1	1
$T2 = 3T^2 - 2$ acr = 5,0	0	-2
acr = 5,5	1	1
	C	**C2**
type of cement		
pc C	-1	1
pc B	0	-2
hoc B	1	1

TABLE A-5 Coded levels of the factors related to the wcr, acr and to the type of cement

	* ε_{csp}(7 d)	ε_{csp}(14 d)	ε_{csp}(21 d)	ε_{csp}(28 d)	$(\widehat{var}\ b_i)/\hat{\sigma}_o^2$
i	b_i	b_i	b_i	b_i	
0	1,743	2,235	2,585	2,867	0,0185
1	0,522	0,666	0,768	0,850	0,0278
2	0,054	0,075	0,091	0,104	0,0093
3	-0,153	-0,205	-0,243	-0,275	0,0278
4	-0,282	-0,353	-0,403	-0,442	0,0278
5	-0,214	-0,264	-0,299	-0,326	0,0093
6	-0,115	-0,143	-0,162	-0,177	0,0417
7	-0,215	-0,270	-0,309	-0,339	0,0417
8	-0,111	-0,139	-0,158	-0,173	0,0139
9	-0,021	-0,032	-0,041	-0,048	0,0139
10	-0,038	-0,052	-0,062	-0,070	0,0046
11	-0,077	0,097	0,112	0,123	0,0417
12	0,011	0,008	0,004	0,001	0,0139
$\hat{\sigma}_o^2$	0,0137	0,0187	0,0232	0,0279	

TABLE A-6 Estimated parameters of the linear model to describe the specific creep after 7, 14, 21 and 28 days under load, at 25°C and 50% RH.

* (ε_{csp} in 10^{-5} per N/mm^2)

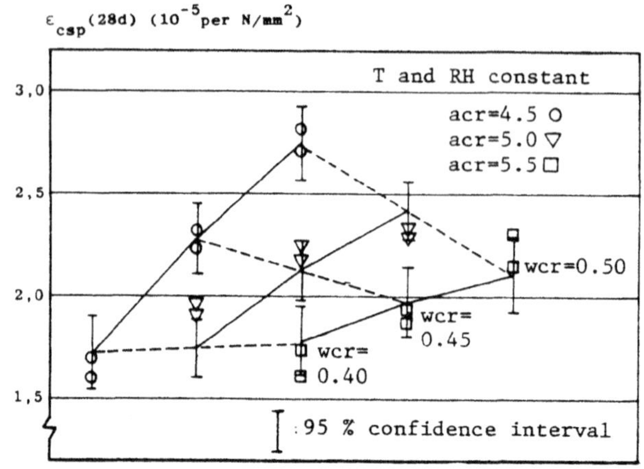

FIG A-1 Relation between the specific creep (loading period= 28d.), the wcr and the acr for hoc B concretes.

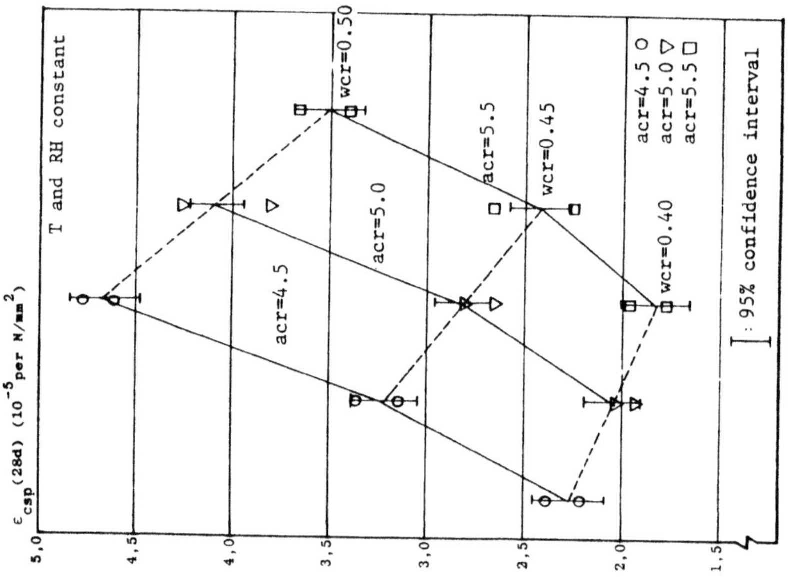

FIG A-2 Relation between the specific creep (loading period= 28d.), the wcr and the acr for pc C concretes.

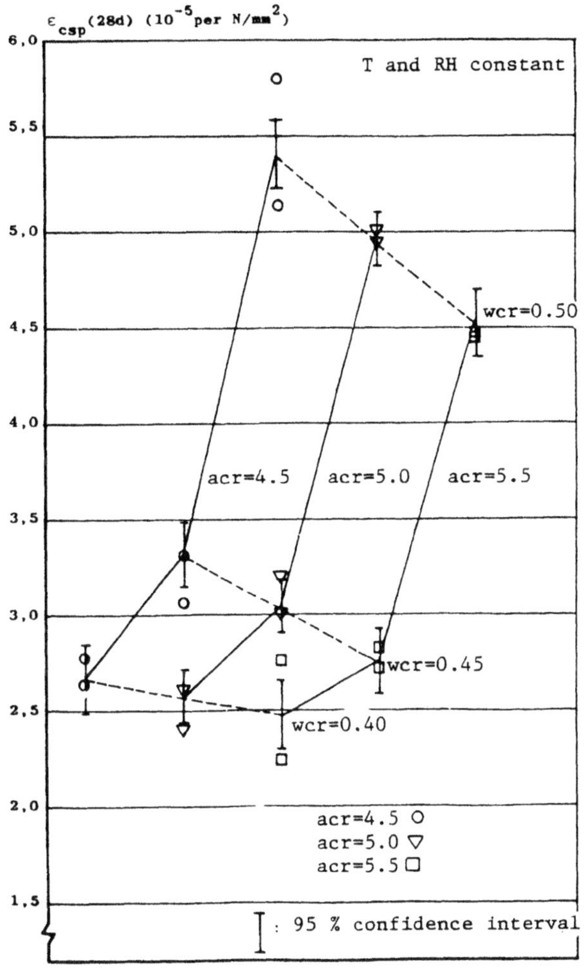

FIG A-3 Relation between the specific creep after 28 days under load, the wcr and the acr for pc B concretes.

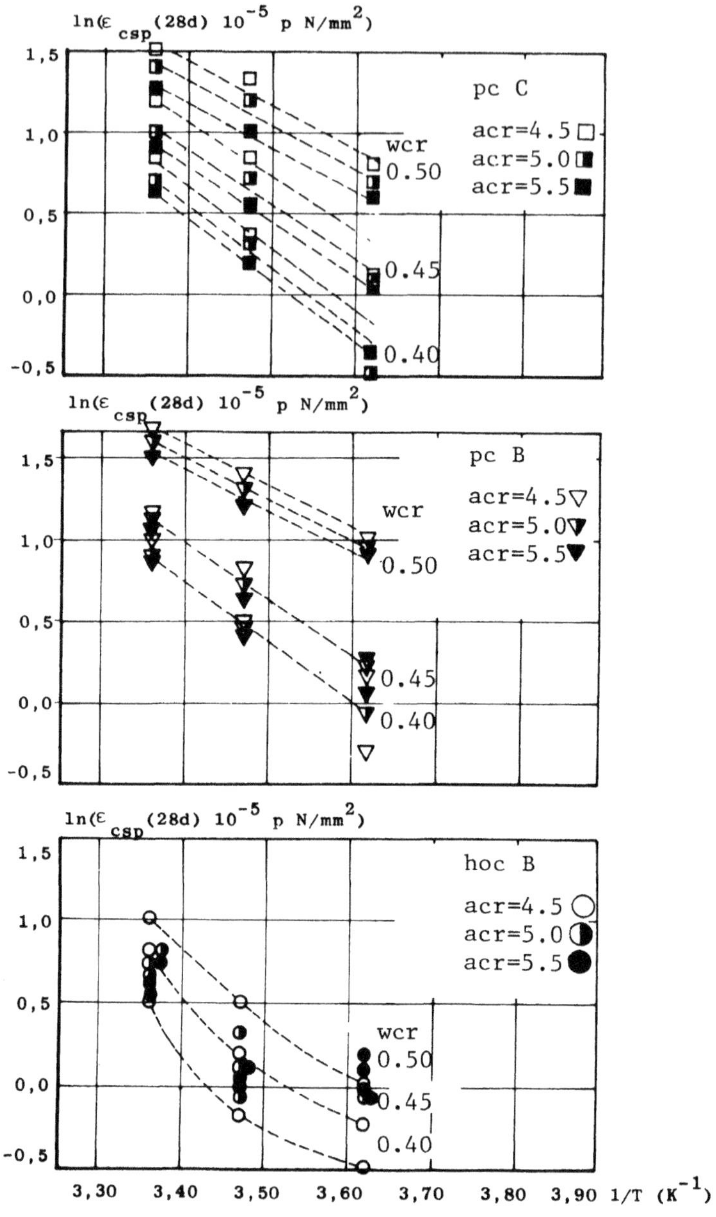

FIG A-4 Relation between the logaritm of the specific creep (28d.) and 1/T, for pc C, pc B and hoc B concretes.

Part III
Experimental Techniques and Results

A CONTRIBUTION TO THE PROBLEM OF REVERSIBLE AND IRREVERSIBLE CREEP OF COMPACT STRUCTURAL CONCRETES

F. KRUML

Institute of Construction and Architecture
of the Slovak Academy of Sciences Bratislava,
Czechoslovakia

ABSTRACT

The variation of reversible and irreversible creep of compact structural concretes with strengths of 13,5 to 45 MPa was studied experimentally as a funktion of age of concrete at the load application.

It is shown that the variation of the reversible creep is influenced by the interaction and the time development of two factors namely of gradual development of microdestructions and the factor of development of hydration products in the time under loading.

The obtained experimental results are compared with the time development of the coefficient of reversible and irreversible creep given in the CEB-FIP 124/125 Recommendations of April 1978.

Der zeitliche Verlauf des reversibilen und irreversibilen Kriechens konstruktiver Betone mit dichtem Gefüge, deren Festigkeit sich von 13,5 bis 45 MPa bewegte, wurde experimental als eine Funktion des Belastungsalters untersucht.

Es zeigte sich, dass der Verlauf der reversibilen Kriechzahl durch Interaktion und zeitlicher Veränderung zweier Faktoren beeinflusst wird und zwar durch den Faktor der nach und nach entstehenden Mikrodestruktionen und durch den Faktor der sich bildenden Hydratationsprodukten in der Zeitspanne der Belastungsdauer des Betons.

Die gewonnenen Laborergebnisse wurden mit dem zeitlichen Verlauf der reversibilen und irreversibilen Kriechzahl, entnommen aus den Empfehlungen CEB-FIP 124/125 vom April 1978, verglichen.

INTRODUCTION

In order to regard the longterm rheological concrete processes in statical calculations, most of the recomendations and prescriptions depart from the total creep value. The CEB-FIP Recomendations {1} of the year 1978 replace the total creep by two of its components namely the reversible and the irreversible parts.

In the world literature, there are only few results dealing with the influence of various factors on the reversible and the irreversible parts of concrete creep. This gap of knowledge should be filled within the next few years and give detailed data for calculation methods allowing to take account of the above mentioned parts of creep.

The paper presents results of experimental studies of time development of reversible and irreversible creep of structural dense concrete depending on concrete strength, loading age and period of loading.

SCOPE AND METHOD OF STUDIES

The experimental studies dealt with concretes within the range of strength of 13,5 up to 45 MPa. The mix consisted of portlandcement of class 400 and river gravel of the Danube. The fresh concrete mix consistence of all the used concrete kinds was chosen as plastic ranging from 4 to 6 VeBe°. The composition of concrete mixes is given in tab.1.

The manufacture of specimens included 48 hours storage at 20°C and 100% of relative humidity and after then storage at 20°C and 60% rel.humidity. The age of loading and unloading of hardened 10/10/40 cm prismatic specimens ranged from 3 to 360 days. The instained load of the specimens amounted to 25% of the prismatic specimen strength. The creep deformation values were obtained in conformity with the generally accepted convention as a difference of long-term deformation of loaded and nonloaded specimens. The strains were studied on two opposite sides of the specimens by means of an applied mechanical straingauge with a measurment base of 300 mm.

Concr.	C	V	Aggregate grain-size					Slump	VeBe
			0-1	1-2	2-4	4-7	7-15	cm	o
B 135	200	180	394	295	197	492	591	1,5	4
B 250	300	170	386	289	193	487	579	0	6
B 450	500	160	348	261	174	435	522	0	4,5

TABLE 1

Concrete composition kg per m^3 of mix.

OBTAINED RESULTS

According to the opinion of the author the origin and the development of reversible and irreversible creep are influenced by two factors. The first is the gradual growth of micro-destruction and the other is the development of hydration products during the period of loading. The internal expansion forces due to unloading of the concrete is resisted by the hydration products developed during the period of loading. In this way a complicated state of stress develops between the aggregate and the hydration products developing prior to and during the period of loading. This state of stress depends on the elastic properties of the aggregate, the ratio of the amounts of hydration products developing prior to and is accompanied by redistribution of stresses due to compression and tension stress in the concrete.

A schematic representation of the cases of time development of reversible and irreversible creep coefficient is shown in Fig.1.

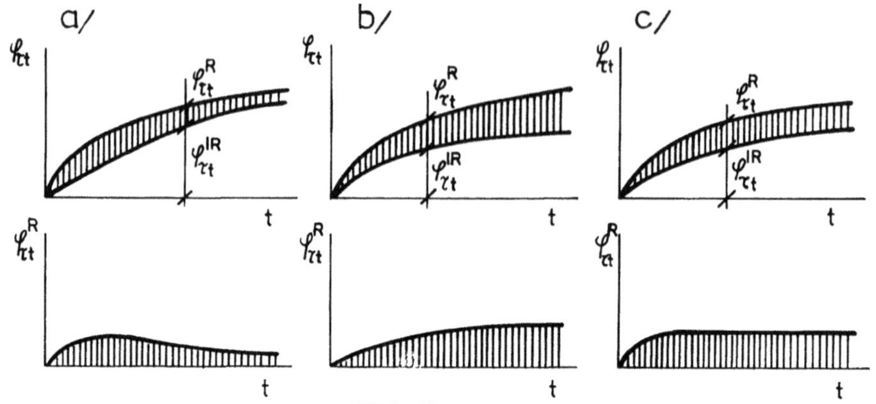

FIG.1

Possible cases of time development of reversible and irreversible creep

In case a/ both the mantioned factors affecting the development of reversible and irreversible creep cooperate.

Case b/ and c/ respectively represent the time variation of coefficient of reversible and irreversible creep under gradual reduction and final extinction of the latter factor.

The function of the time development of the coefficient of reversible $\varphi^{R}_{\tau t}$ and irreversible $\varphi^{IR}_{\tau t}$ creep was expressed by Eqs. (1) and (2)

$$f^{R}_{\tau t} = \frac{\varphi^{R}_{\tau t}}{\varphi^{R}_{\tau \tau+360}} \quad 1./ \qquad f^{IR}_{\tau t} = \frac{\varphi^{IR}_{\tau t}}{\varphi^{IR}_{\tau \tau+360}} \quad 2./$$

The age at load application τ ranged from 3 to 360 days, the duration of the loading $t - \tau$ from 3 to 360 days. An synoptic graphical representation of function $f^{R}_{\tau t}$ is shown by Fig.2 for concretes B 135, B 250 and B 450.

Fig.2 shows also a graph representing the time-function computed from the CEB-FIP draft.

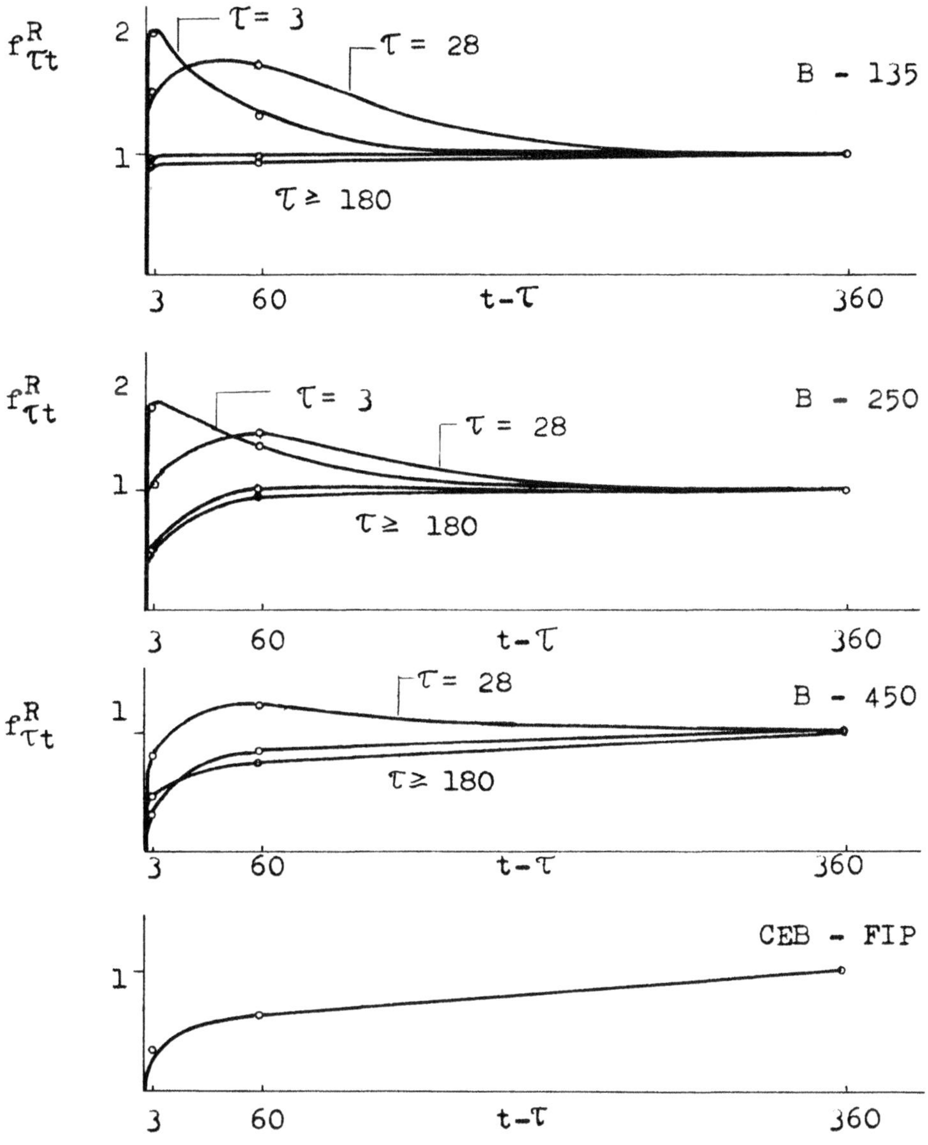

FIG.2

Function of time development of reversible creep coefficient.

$$f^R_{\tau t} = \varphi^R_{\tau t} / \varphi^R_{\tau\tau+360}$$

The obtained results show that in the case of concretes loaded at the age of 3 days the strain development is similar to that shown in Fig.1a. The same but less clear type of development is exhibited by concretes loaded at the age of 28 days from their manufacture. The time development of concretes loaded at higher ages coincide with the curve of reversible creep shown in Fig.1b and 1c.

These results support the author´s opinion concerning the influence of the above mentioned two factors on the development of the reversible and irreversible creep components. The latter factors influence appears mainly in the case of concrete loaded in their early age and disappears with the increating loading age of the concrete.

Fig.2 shows that the time-development of the reversible creep coefficient of concrete loaded in early age $\leq$ 28 days is different from the monotonous rising functions assumed particuraly in the CEB-FIP recomendations in the form of an exponential function attains quickly its maximum value in the case of concrete loaded at their higher age and further on its value remains constant.

For engineering application of the results shows that the function of the time development of the coefficient of reversible creep given in the CEB-FIP recomendation is similar to the results of the author for concretes the degree of cement hydration at the moment of load application is approximately equal to or higher than that of a 28 days old normally hardening concrete.

A survey of the $\varphi^{IR}_{\tau t}$ is given in tab.2 for concretes B 135, B 250 and B 450. The same table gives also the coresponding values of the CEB-FIP recomendations.

The experimental results shows that from the point of view of engineering interpretation the development of $\varphi^{IR}_{\tau t}$ for concretes $135 \leq B \leq 450$, loaded at the age $3 \leq \tau \leq 360$ days do not differ substantially from each other. The same may be found to be true for values in the CEB-FIP recomendations.

Durat. of load. $t-\tau$ days	Experim. values min.-max. τ				CEB-FIP values τ				Mean values Exp.		CEB-FIP
	3	28	180	360	3	28	180	360	$\bar{x}$	v %	$\bar{x}$
0	0,00	0,00	0,00	0,00	0,00	0,00	0,00	0,00	0,00	-	0,00
10	0,26 0,30	0,23 0,26	0,25 0,33	0,24 0,38	0,27	-+	-+	-+	0,28	17	0,27
100	0,66 0,74	0,52 0,84	0,55 0,79	0,36 0,69	0,74	0,65	-+	-+	0,64	20	0,70
200	0,85 0,91	0,75 0,92	0,78 0,84	0,55 0,85	0,85	0,85	0,64	0,63	0,82	12	0,76
360	1,00	1,00	1,00	1,00	1,00	1,00	1,00	1,00	1,00	-	1,00

TABLE 2

Function of time-development of irreversible creep coefficient.
Investigated concrete $135 \leq B \leq 450$.
Note + values could not be exactly evaluated from fig.3,p.250 of reference {1} .

The foregoing statement allows to find the mean value of time variation f^{ir}_{ti} as independent of the parameters studied above.

These data are also given in Tab.2. The obtained mean values have a variation coefficient ranging from 12 to 20%. They justity to state a satisfactory agreement between the function of time development of irreversible creep coefficient found by tests of the author and that of the CEB-FIP recomendations. The analytical studies made by the author of this paper have shown that the values of f^{ir}_{ti} given in Tab.2 may be mathematically expressed by an exponential or a hyperbolical function.

References :

1/ Code model pour les structures en beton CEB-FIP Bull.d´Inf. No.124/125-E/Avril 1978.

CHECKING THE PRINCIPLE OF SUPERPOSITION BY TESTING

MAMILLAN Marc

Chef du Service d'Etudes des Matériaux au Centre Expérimental de Recherches et d'Etudes du Bâtiment et des Travaux Publics
Saint-Rémy-lès-Chevreuse (France)

ABSTRACT : Experiments to check the principle of superposition have been carried out on concrete test pieces subjected to permanent compressive stresses.

A series of test pieces was loaded under a stress σ at 7 days another under the same stress at 28 days and a third series was loaded from 7 to 28 days under a stress σ, the stress then being doubled at 28 days. Deformation after 1700 days loading was measured under three stresses (25, 50 and 75 bars).

The BOLTZMANN principle of superposition is proved valid by interpretation of results taking into amount the different types of testing inaccuracies although there remaind a difference of 10 % between theory and test figures.

RESUME : Une étude expérimentale ayant pour but de vérifier le principe de superposition a été effectuée sur des éprouvettes de béton soumises à des contraintes permanentes de compression.

Une série d'éprouvettes a été chargée sous une contrainte σ à 7 jours, une autre sous la même contrainte à 28 jours, et une troisième série d'éprouvettes chargée de 7 jours à 28 jours sous une contrainte σ, la contrainte étant ensuite doublée. L'étude des déformations après 1700 jours de chargement sous trois contraintes (25,50 et 75 bars) a été effectuée.

L'interprétation des résultats prenant en compte les différentes sortes d'imprécisions expérimentales conduit à la vérification du principe de superposition de BOLTZMANN, bien qu'une différence de 10 % existe entre la théorie et l'expérience.

1 RESTATEMENT OF THE DIFFERENT CONCEPTS OF SUPERPOSITION

1.1 Hypotheses and Concepts

- If we consider S1, S2 and S3 as three perfectly identical bodies (in our research they symbolize the test pieces of identical dimensions drawn from one concrete batch : therefore same origin, same age and mix, placed in an identical exterior medium : temperature and hygrometry constant).

- σ_i, are measurable causes (in this case compressive stresses applied on the test pieces with the same equipment and identical precautions).

- ε_i, are the resultant deformations equally measurable(that is to say the deformations due to the action of these stresses σ_i, therefore the more deferred instantaneous shortenings which are caused by them).

- The load application times are represented by ti; τ defines an instant taken at random, nevertheless after the last loading : $\tau > \sup(t_i)$ which is chosen to carry out measurement of the effect.

  Θ, Θ' also represent time

- $t1 < t2 < \tau < t_\infty$ is chosen, knowing that $t1 > 0$ and that t_∞= infinite date.

1.2 First Case : Simultaneous Actions

Definition of superposition of effects under simultaneous causes.

Let us consider the cases where :

a) from time t1 to t_∞ we apply σ_1 on S1 $\rightarrow$ the effect on S1 at $(t_1 + \Theta) = \tau$ is ε_1.

b) from time t_1 to t_∞ we apply σ_2 on S2 $\rightarrow$ the effect on S_2 at τ is ε_2.

c) from time t_1 to t_∞ we apply $(\sigma_1 + \sigma_2)$ on S3 $\rightarrow$ the effect on S3 at τ is ε_3.

There will be superposition of effects if $\varepsilon_3 = \varepsilon_1 + \varepsilon_2$.

Immediate consequences : proportionality between causes and effect.

If there is superposition of effects, we immediately find :

a) for an action $2\sigma_1$, on S_3 from t_1 to t_∞, that the resulting effect at $\tau = t_1 + \Theta$ is $2\varepsilon_1$ on S_3.

b) the same an action $k\,\sigma_1$ on S_3 from t_1 to t_∞ involves an effect at $\tau = t_1 + \Theta$ equal to $K\,\varepsilon_1$ on S_3.

Therefore the proportionality of the effect to the cause is an immediate consequence of superposition of effects due to simultaneous causes.

1.3 Second Case : Successive Actions

Definition of superposition of effects under successive causes

a) from t_1 to t_∞, under the action of σ_1 on S_1 at $\tau = t_1 + \Theta$ effect ε_1 on S_1.

b) from t_2 to t_∞ under the action of σ_2 on S_2, at $\tau = t_2 + \Theta$ effect ε_2 on S_2.

There will be superposition of effects if : when we apply on S_3 from t_1 to t_2 σ_1 then from t_2 to t $\sigma_1 + \sigma_2$ the effect at τ on S_3 is $\varepsilon_1 + \varepsilon_2$.

2 THE BOLTZMANN PRINCIPLE OF SUPERPOSITION

2.1 Statement of the BOLTZMANN Theory

The first scientist to state this principle was BOLTZMANN in 1874.(1) he was not thinking of concrete in particular but in a more general way of any isotropic material with deferred elasticity.

To quote a translation of his statement from "Zur theorie der elastische Nachwirkung" :"I only assume that the forces which act on the faces of a parallelepiped at a given instant depend not only on the present deformations of the parallelepiped but also on former deformations...

"More precisely, the force necessary to produce a given deformation is smaller when a deformation has occurred previously in the same direction. I will call this fact : the reduction in force due to previous deformations.

If at any instant τ during the elementary time $d\tau$ the body has a deformation $\varepsilon(\tau)$, I assume that the reduction in force which this deformation exerts on the force acting at time t is proportional to $d\tau$, to $\varepsilon(\tau)$ and to a function of time $t - \tau$ separating these two instants.

"To that I add the hypothesis stating that the influence of deformations which have occurred at different times is superposed, that is to say that the lowering of the force having caused a given deformation at a given time will not depend of the states of the body in the interval (.)".

2.2 Interpretation by PERSOZ (2)

A distinction must be made between the two hypotheses formulated in this text :

1) BOLTZMANN states " a priori" the proportionality of forces to deformations as well as the proportionality of a reduction in forces to the previous deformations.

2) The principle of superposition then comes as a secondary hypothesis. Moreover BOLTZMANN announces in a note that perhaps it will not hold for large deformations, which experience acquired since has often confirmed.

Therefore this principle of superposition considered above, that is to say without the context, does not imply proportionality of the effect to the cause. It only assumes that the effects of prior deformations are independent of intermediate states of the body and that these effects are cumulative. It is the superposition of effects due to "successive" causes, but not effects due to "simultaneous"causes.

2.3 Application of the BOLTZMANN Principle to Concrete MAC HENRY's Statement

In 1943 (3) MAC HENRY adapted this principle of superposition for concrete, in particular in an article intitled "A new aspect of creep in concrete and its application to design". It runs as follows :

"The deformations caused in concrete at a time t by a variation in stress applied at a time t_0 are independent of the effects of any other stress applied at a time different from t_0, before or after. The variation in stress can be negative or positive but must not approach ultimate strength".

(.) This principle of superposition is fairly plausible for deformations which are no too large, it perhaps no longer applies with very large deformations (note by BOLTZMANN).

3 TESTING

The tests were carried out on 6 batches of concrete ABCDEF where all precautions were taken at the CEBTP laboratories to make test pieces as identical as possible.

Five test pieces were made from each batch :

- 2 test pieces with no external stress, to be used as reference specimens (indicating the shrinkage value)
- a test piece S_1 loaded at 7 days under a stress σ_0
- a test piece S_2 loaded at 28 days under the same stress σ_0
- a test piece S_3 loaded from 7 days to 28 days under a stress σ_0, the stress then being doubled (from the 28th day S_3 undergoes 2 σ_0).

The only variable is the compressive stress σ_0. Tests are carried out in duplicate :

- Series E and F under σ_0 = 25 bars
  A and D under σ_0 = 50 bars
  B and E under σ_0 = 75 bars

The objective is to study the validity of the BOLZMANN principle for these three stresses after 1700 days of testing.

This consists of checking at 1700 days that the deformation

$$\varepsilon_3 = \varepsilon_1 + \varepsilon_2$$

ε_3 = deformation (instantaneous + deferred) at 1700 days following successive loadings at 7 days under σ_0 and at 28 days under $2\sigma_0$.

ε_1 = deformation (instantaneous + deferred) at 1700 days resulting from loading at 7 days under σ_0.

ε_2 = deformation (instantaneous + deferred) at the same date due to loading at 28 days under σ_0.

4 RESULTS

To set up a comparison between :

- the curve of experimental deformation under the addition of the two stresses, one applied at 7 days, the other at 28 days on the same test piece S_3, and

- the theoretical curve obtained from adding the deformations caused by separate loadings : at 7 days on a test piece S_1 and at 28 days on a test piece S_2.

On the same diagram for each stress we have plotted :

- curve R_m (mean shrinkage) representing the shortening of a reference test piece without any external stress.
- total deformation ε_1 of test piece S_1 loaded at 7 days under σ_0 (shrinkage included).
- total deformation ε_2 of test piece S_2 loaded at 28 days under σ_0 (shrinkage included).
- total deformation ε_3 of test piece S_3 loaded from 7 to 28 days at σ_0 then $2\sigma_0$ from 28 days onwards (shrinkage included)
- déformation ε calculated from $\varepsilon_1 + \varepsilon_2$ but which obviously includes only one shrinkage ($\varepsilon_{cal} = \varepsilon_1 + \varepsilon_2 - R_m$).

Expression of results

The results are presented in diagram form :

1 series C $\sigma_0 = 25$ bars
2 series D $\sigma_0 = 50$ bars
3 series B $\sigma_0 = 75$ bars

On the X-axis : time in days expressed on logarithmic scale the time origin corresponds to the day the test piece has been made.

On the Y-axis : total shortenings (in μ/m).

The origin $\varepsilon = 0$ corresponds to a state devoid of any deformation (but not to zero **strain** since in the absence of any exterior stress there is measurable hydraulic shrinkage).

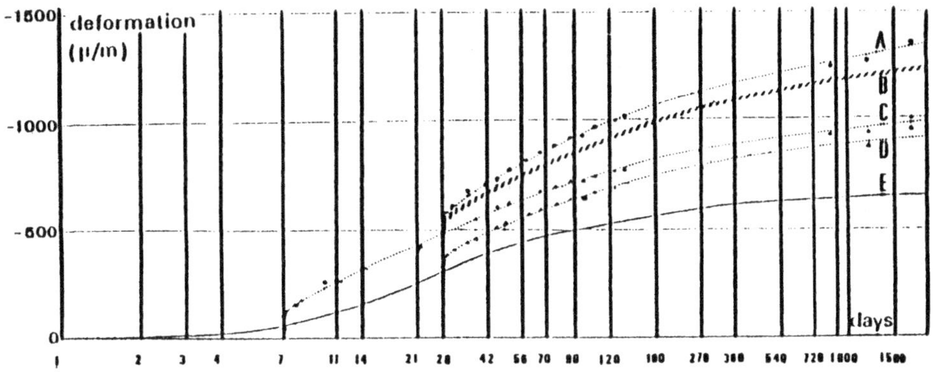

FIG.1

Superposition of the deformations - stress 25 bars(Series C)

A - experimental 25 bars at 7 days + 25 bars at 28 days
B - calculated 25 bars at 7 days + 25 bars at 28 days
C - 25 bars at 7 days
D - 25 bars at 28 days
E - shrinkage

5 ANALYSIS OF RESULTS

Superposition of curve ε_3 and $\varepsilon_1 + \varepsilon_2$

For the six series, therefore for σ_0 = 25, 50 and 75 bars, in all cases the curves calculated according to the BOLTZMANN principle are always lower than the experimental curves.

This seems to prove that this principle of superposition leads to underestimating the real deformation.

If we calculate the difference between the two deferred deformations obtained from calculation and experimental values at 1700 days we obtain :

TABLE 1

Comparison between predicted and experimental data

Series	Stress	deformation $(\varepsilon_\tau - \varepsilon_i - R)$ (μ/m)		difference of deformations Experi_Calcu- mental lated (μ/m	Mean percentage of differences
		Calculated	Experimental		
Series C	25 + 25[b]	623	694	70	9,1 %
Series F		626	742	116	
Series A	50 + 50[b]	1486	1505	19	9,6 %
Series D		1336	1640	304	
Series B	75 + 75[b]	2244	2477	233	10,7 %
Series E		2154	2448	294	

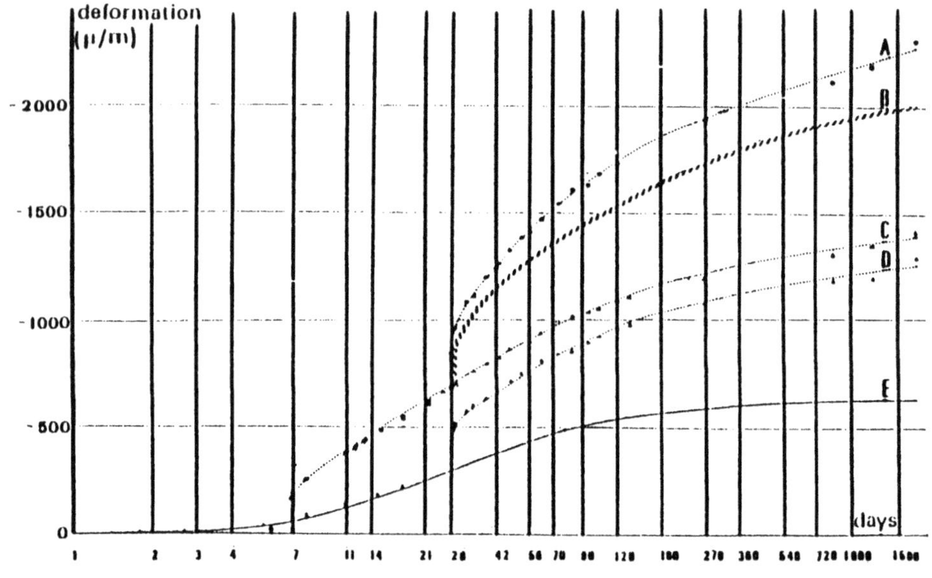

FIG. 2

Superposition of the deformations-stress 50 bars (Series D)

A - experimental 50 bars at 7 days + 50 bars at 28 days
B - calculated 50 bars at 7 days + 50 bars at 28 days
C - 50 bars at 7 days
D - 50 bars at 28 days
E - shrinkage

The deviations are low around 9 ∿ 10 %. But the fact that the predicted values are systematically inferior to the experimental values shows the need for more theoretical explanation.

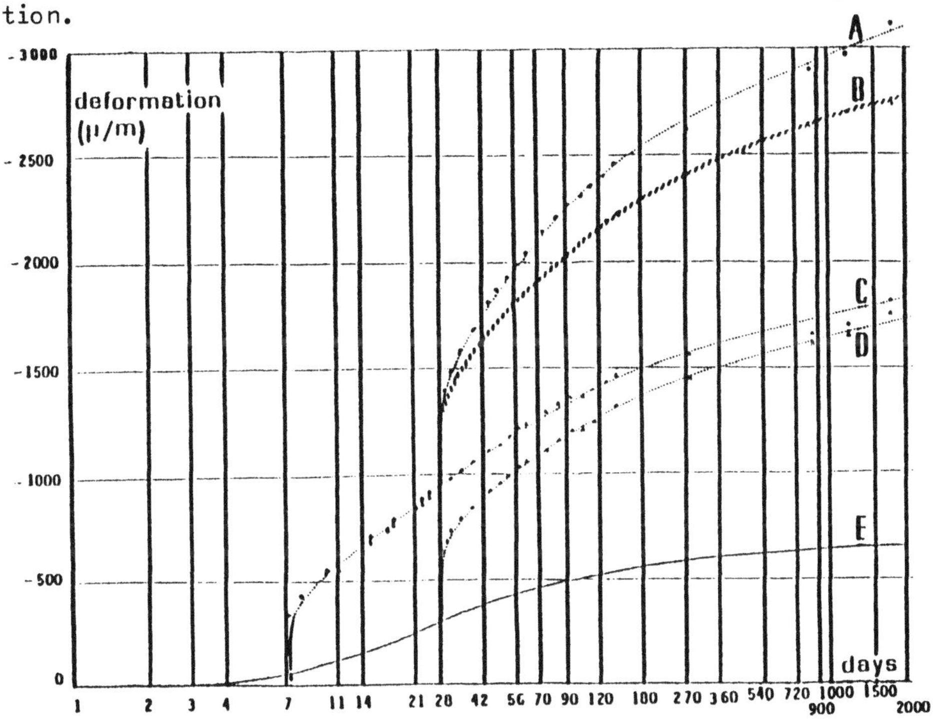

FIG. 3

Superposition of the deformations-stress 75 bars (Series B)

A - experimental 75 bars at 7 days + 75 bars at 28 days
B - calculated 75 bars at 7 days + 75 bars at 28 days
C - 75 bars at 7 days
D - 75 bars at 28 days
E - shrinkage

6 DISCUSSION OF RESULTS

The essential result of this research is a deviation which seems significant (10 %) between the BOLTZMANN theory and experimental values (always higher).

The creep value which is taken into account in the application of the superposition principle is determined from three series of independent tests :

- creep from two different series of test pieces is added together, one loaded at 7 days, the other at 28 days.

- a third series of test pieces is loaded at 28 days under a stress equal to the sum of the other two.

The creep taken into account in each of these series is not obtained by a direct measurement. The total deformation is measured from which shrinkage is substracted (mean of a series of 12 nonloaded test pieces) and the instantaneous deformation is obtained for each series.

Therefore creep figures are determined from three measurements.

We can consider in the most pessimistic hypothesis that all the possibilities of error cumulate.

This succession of calculations with values obtained on different test pieces prevents the physicist aware of the usual dispersion on concrete to consider a 10 % difference in deformation as significative. However for the theorist and the designer it is important to analyse in detail the degree of influence of various parameters.

6.1 Homogeneity of Test Pieces

The quality of test pieces-mechanical strength, instantaneous or deferred elasticity - reveals a certain variation due to the difficulty during testing of reproducing concrete test pieces as identical as possible. As an indication, table 2 gives the results of quality identification obtained on 12 test pieces used to study shrinkage. These test pieces are made from 6 different batches (A,B,C,D,E and F). We chose the criterion of quality homogeneity using non destructive tests (determination is more precise since measurement can be repeated), density, resonance frequency, dynamic modulus of elasticity, ultrasonic testing.

With these tests it is possible to check if all the different batches belong to the same statistical population - if there is any systematic error in the quality of a series of test pieces. Those in series D (loading under 50 bars) have a tendency to be of inferior quality (which is checked in the density, resonance frequency, dynamic modulus of elasticity and ultrasonic testing).

Test pieces in series C (loading under 25 bars) are of slightly improved quality (resonance frequency, modulus of elasticity and ultrasonic testing).

These series do not show a wide variety of quality compared with the mean (-3 % for D, + 6 % for C) and these deviations must be considered quite normal for test pieces made of concrete from different batches. Checking the degree of test piece regularity (since it is normal) does not offer an explanation of the difference between experiments and theory.

TABLE 2

Comparison of the homogeneity of shrinkage specimens

N° specimens	Mass in (gram)	Resonance frequency in Hz	Dynamic modulus of elasticity in bars	Series	Sound velocity in m/s
5	3128	6918	341 000	A	4 515
6	3133	6820	332 000	A	4 590
25	3133	6771	327 000	B	4 570
29	3145,8	6872	339 000	B	4 520
34	3121,4	6960	345 000	C	4 515
37	3135,3	6985	349 000	C	4 625
41	3122,2	6685	318 000	D	4 495
43	3126,7	6589	309 000	D	4 520
52	3144,9	6710	323 000	E	4 500
53	3146,7	6705	322 000	E	4 570
65	3149,8	6685	321 000	F	4 500
66	3150,5	6700	322 000	F	4 515

6.2 Comparing Development of Shrinkage Measurement

We study the development of shrinkage measured on 12 reference test pieces representing the six batches (2 test pieces per batch).

The scatter in measurement is normal. The maximum deviation for a given age is 9.4 %.

7 PROPOSAL FOR A PHYSICAL EXPLANATION OF RESULTS

It is possible to propose an explanation based on physics to justify the fact that the value of deformation observed from testing a specimen subjected to the two loading is higher than the theoretical value calculated from deformationsof two test pieces loaded separately.

Although the concrete deformations remain proportional to the applied stresses in the field of test undertaken, it is possible to formulate the hypothesis of the existence of an irreversible maximum limit deformation.

For example : if we assume that the seat of creep is located in the mortar and at bonds between mortar and aggregates, we can suppose that for each applied stress during a certain time, there occurs in the mortar and at the bond a certain amount of settlement debording or sliding.

It is possible to accept the possibility that there exists for concrete a shortening limit in the reversible field.

When a certain shortening value is reached or exceeded, the displacement of hydrated grains against each other becomes an irreversible arrangement. The results of the present test seem to indicate that this phenomenon occurs with a low loading rate (at (25 + 25)b it seems to be already reached).

Other hypotheses can also be set forward :

- hydraulic shrinkage which is conventionnaly defined as a mean value equal in all cases to be able to subtract from total shortenings the value of deferred deformation, is perhaps not identical under load and without load. We cannot ignore the fact that with different tensile effects, the internal displacement of water influence evaporation, hydration causing hydraulic shrinkage according to the loading rate. Shrinkage after loading is perhaps modified by the application of stress.

A test piece loaded at 7 days will perhaps not undergo the same hydraulic shrinkage as another test piece from the same batch loaded at twenty eight days, nor a test piece loaded at seven days and then at twenty eight days.

There is then a possibility that by considering, shrinkage independently of stress at a given age we commit an error leading to deformation deviations for test pieces under different loads (see Bibliography R. L'HERMITE (4), MITZEL (5), conclusions on the relation between shrinkage and creep according to tests carried out by R.L'HERMITE (4), MITZEL (5), also Z.P. BAZANT (6) and (7)).

8 CONCLUSION

The deviation obtained between deformation under testing and theoretical deformation with application of the superposition principle is 10 %. We think it is possible to conclude that the BOLTZMANN principle is proved correct, the deviation we found through testing is acceptable for calculating reinforced and prestresced concrete designs.

However, it is fitting to note that shortening obtained experimentally is slightly more than the deformation deduced form applying the superposition principle, so this deviation is not on the safety's side.This applies to the three stress rates considered; interpretation and analysis of results has pointed to the conclusion that this difference is due to quite normal dispersion.

We consider that a theory proved correct to 10 % by testing is a very satisfactory result, but we must underline the fact that at the present time and taking into account physical measurement means it is impossible to obtain creep results with better accuracy than 10 %.

In these conditions it is essential that theorists take into consideration the low accuracy of creep measurements. Indeed is this uncertainty always compatible with the very precise calculations in a mathematical expression which adapt a function to experimental results ? Can we satisfactorily judge that a law is more representative than another to forecast deferred deformation without taking in account dispersion of test results ?

At the present time there is an increasing tendency for research scientists to use mathematical, computerized interpretation of creep data.

This trend is quite normal, taking into consideration the development of the enormous possibilities offered by minicomputer, and creep is a favorite subject for making full use of results obtained at different ages.

But we must also consider that the ways of measuring physical properties have not progressed in the last twenty years in such a spectacular way as calculating tools.

Keeping thus test uncertainty in mind, is it always really necessary to set up a mathematical law ? Is a mathematical function always really representative in the static sense of the word ?

In the light of this restriction it is difficult to prove that progress in the field of creep can be considerable solely because of development in calculating tools. On the other hand, interpretation of results could be possible by improving the testing procedures to bring about internal modifications in physical and chemical structure.

Parallel to the use of proverful data processing means, it would be advisedly to promote further work based on knowledge of research on local micro-deformations, internal water transfer, determination of the state of hydration of a phase, etc...

In the years to come using testing to prove assumptions proposed to explain factors influency on deferred deformations of concrete should be based on wider development of research on the physical and chemical means of measuring the structure of concrete.

Bibliography

1. Boltzmann - Zur Theorie der Elastische Nachwirkung Sitz ung Sb. Adad.Wiss. Wien. Ber (1874) 70, 275.

2. B. Persoz - Introduction à l'étude de la Rhéologie Dunod, Paris 1960 (UTI - 23662)

2. B. Persoz - Le Principe de Superposition de Boltzmann U.T.I.(Paris) 15085 R.

3. Mac Henry - "A New Aspect of creep in concrete and its Application to Design" - Proceedings of the 46ème Annual Meeting - Vol.48 - 1943 (pages 1069-1088).

4. L'Hermite-Mamillan - Nouveaux résultats de recherches sur la déformation et la rupture du béton - Annales ITBTP -Paris (Mars-Avril 1965 n° 207-208).

4. L'Hermite - Volume changes of concrete - Proc.Fourth Int. Symp.on the Chemistry of Cements (1960)

4. L'Hermite - Que savons-nous de la déformation plastique et du fluage du béton - Annales ITBTP (Paris) Sept.1957- BBA/46 et Fév.59- BBA/52.

5. Mitzell - Sur l'hypothèse de l'égalité des déformations de fluage en compression et en traction ainsi que le principe de superposition du fluage et du retrait. Annales ITBTP Paris - Juillet 68 n° 241.

6. Z.P.Bazant - Theory of creep and shrinkage in concrete structures. A precis of recent developments (Mechanics Today - Vol.2 - Nemar.Nasser Pergamon Press - Chap.I - Pages 1 à 93).

7. Z.P.Bazant - Experimental study of creep of hardened Portland.

DRYING OF CONCRETE
CONSEQUENCES FOR THE EVALUATION OF CREEP TESTS

Paul Acker
Ingénieur E.C.P.
Laboratoire Central
des Ponts et Chaussées
Paris, France

ABSTRACT

The general kinetics of concrete drying is now known and can be determined for each particular case by gamma-ray absorption measurements. Time-dependent variations in water content and its distribution within a concrete element is such that, in most cases, specimens subjected to creep testing are far from being homogeneous from the viewpoint of water content.

It is possible by means of different hypotheses to deduce the new stress distribution from the difference in water content, i.e. the difference in behaviour between the internal concrete which remains saturated for a long time and the peripheral concrete which dries very quickly.

The significance of this rapid distribution has been confirmed by experimental observations.

As a result, it can be said that:

. in most creep tests, specimens are not subjected to a continuous uniform stress;

. the phenomena involved are not all the same for a loaded specimen and for a control specimen;

. the cracking of structures in service is not necessarily related to tensile stress in the considered zone.

RESUME

La cinétique générale du séchage du béton est aujourd'hui connue et peut être déterminée dans chaque cas particulier par des mesures d'absorption des rayons gamma. L'évolution de la teneur en eau et de sa répartition dans un élément en béton est telle que dans la plupart des cas, les éprouvettes soumises à l'essai de fluage sont loin d'être homogènes du point de vue hygrométrique. De la différence d'état hygrométrique, donc de comportement entre l'intérieur, qui reste longtemps saturé, et de la périphérie, qui sèche très vite, on peut, moyennant différentes hypothèses, déduire la redistribution des contraintes qui en résulte. L'importance de cette redistribution et sa rapidité ont pu être confirmées par quelques constatations expérimentales. Il en résulte que:

. dans la plupart des essais de fluage, les éprouvettes ne sont pas soumises à contrainte constante;

. les phénomènes en jeu ne sont pas de même nature entre l'éprouvette chargée et l'éprouvette témoin;

. la fissuration des ouvrages en service n'est pas nécessairement liée à la présence de traction dans la zone considérée.

0. INTRODUCTION

In 1979, as part of a study on the additivity of creep [1], we had the opportunity to partially unload a concrete specimen. What was involved was a cylinder of 16 cm diameter by 100 cm compressed under 10 MPa for 1 month and then under 5 MPa for 2 months in an air-conditioned room (20°C, 50% RH). When the load dropped from 10 to 5 MPa in a continuous and monotone manner, we witnessed the appearance of transverse cracking on the surface of the specimen, cracking which was surprising on a specimen still carrying 10 metric tons!

The explanation we arrived at after analysis is the following:

0.1. The drying of concrete occurs much faster on the surface than inside the component.

0.2. During drying, the concrete has a much faster creep than "confined" concrete (i.e. impervious to air and water).

0.3. The creep testing fixture imposes a uniform strain on the specimen.

0.4. This involves a redistribution of stresses between the surface and the inside, beginning with a homogeneous initial distribution.

0.5. From a certain amplitude, partial load shedding can then place the surface under tension.

According to the estimations presented here, this redistribution is significant (of the order of ± 50%) which explains the experiment described above. The consequences are of two kinds:

0.6. Most creep tests carried out by different countries are not "pure" tests and the complete interpretation of these tests calls for knowledge of the water content inside the specimens.

0.7. Conversely, the interpretation of cracks in structures in service is contingent upon knowledge of the water content and its distribution: a surface crack does not necessarily mean that the section (or fibre) considered is under tension throughout its thickness.

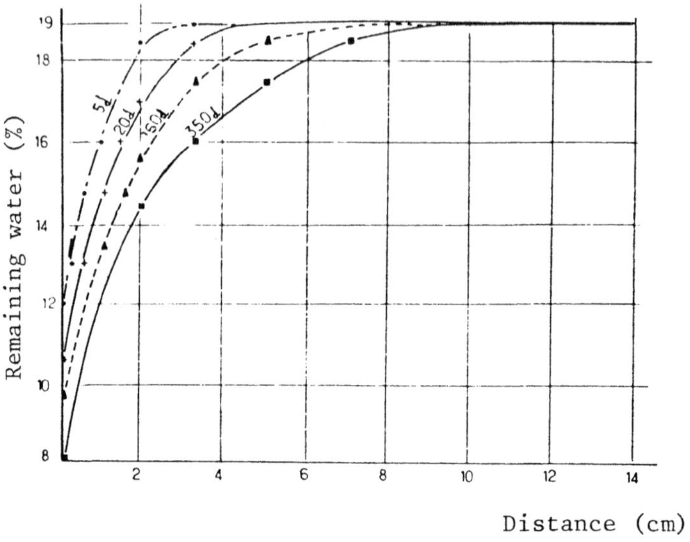

Figure 1: Distribution of free water remaining in the concrete during its drying at different ages.
(Prisms of concrete made impervious on five sides and kept at 20°C and 50% relative humidity)
(After L'Hermite and Mamillan [2])

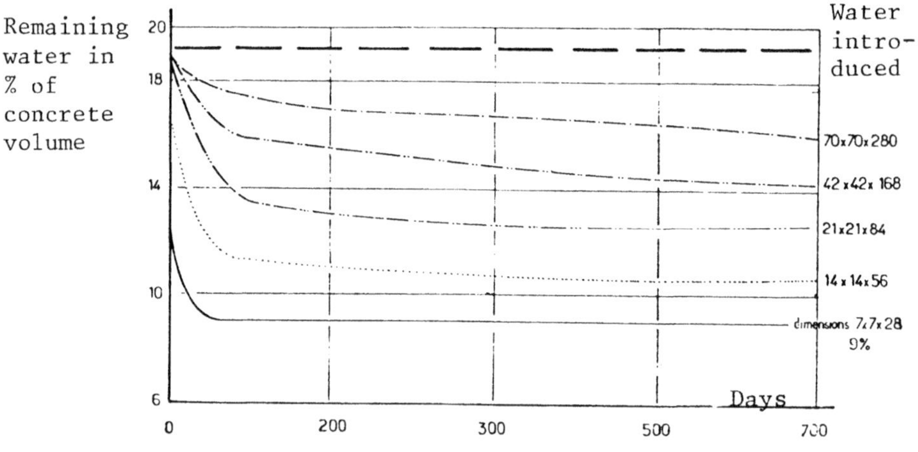

Figure 2: Evolution of the total amount of water remaining in concrete prisms as a function of time for different prism thicknesses.
(After l'Hermite and Mamillan [2])

1. DRYING OF CONCRETE

Occurring very rapidly on the surface, the drying of concrete quickly becomes very slow in depth. By means of dielectric probes [2], L'Hermite and Mamillan were able to show the pattern of this phenomenon (Figures 1 and 2). As of the fifth day, in an air-conditioned room (50% relative humidity), drying covers a depth of 3 cm but it is necessary to wait a year for drying to begin at a distance of 8 cm from the surface. The concrete specimens generally used in creep tests (prisms of 14 x 14 cm, cylinders of 15 or 20 cm diameter) require several years before reaching equilibrium with the ambient air in a nonsaturated air-conditioned room.

Drying can be accelerated considerably by increasing the temperature. In fact, the expansion coefficient of water is clearly higher than that of concrete and, at the same time, its viscosity decreases very rapidly with temperature. Thus, at 105°C, it is possible to dry a 16-cm diameter cylinder of 32 cm in one week (total loss: 5% in weight). By gammametry [11], we were able to determine the distribution of water content inside such a cylinder (Figure 3). The weighing of the specimen made it possible to check that its weight loss corresponded substantially to the average of the water contents measured inside the specimen.

2. BASIC CREEP AND DRYING CREEP

Concrete which dries in air is thus generally very heterogeneous from the viewpoint of its water content. To understand the influence of water content on creep, it is important to examine first the case of the material when it is in equilibrium with its ambient environment: the creep obtained under these conditions is called basic creep.

Except in the case of immersion or saturated atmospheres, it is difficult to obtain specimens in hygrometric equilibrium. It is then necessary to either wait a very long time or to heat the specimen beforehand, or to use specimens of very small size. The few, rare results of this type mentioned by Neville [3] (Hansen, Ruetz) show that the drier the specimen, the lower the creep obtained (Figure 4). By completely drying (7 days at 105°C) hollow specimens in mortar (wall thickness: 2 mm), Brown and Hope [4] obtain zero creep under torsion.

In general, creep tests are carried out on specimens during drying. The opposite result is then found: the drier the ambient atmosphere, the greater the creep.

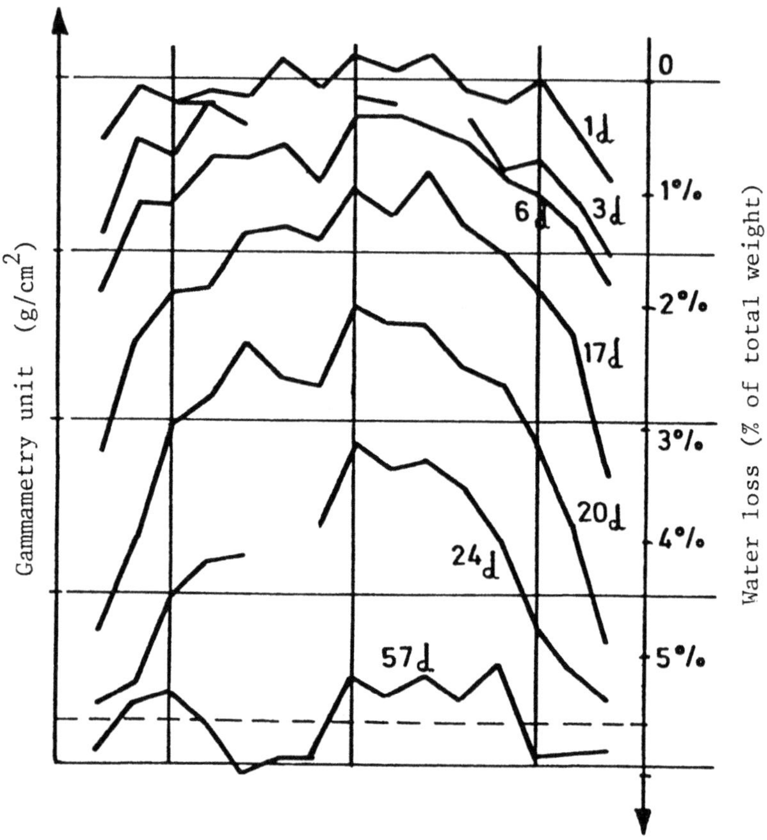

Figure 3: Water loss, expressed in percentage of total weight, following a diameter of a concrete specimen of 16-cm diameter and 15-cm length coated at both ends and kept at 70°C for 17 days and then at 105°C.

(Results obtained at the LCPC by gammametry: measurement every centimetre on a beam of 1-cm diameter).

The last curve, obtained after 57 days, is to be compared with the dotted line which indicates the total loss measured after the same period by direct weighing [11].

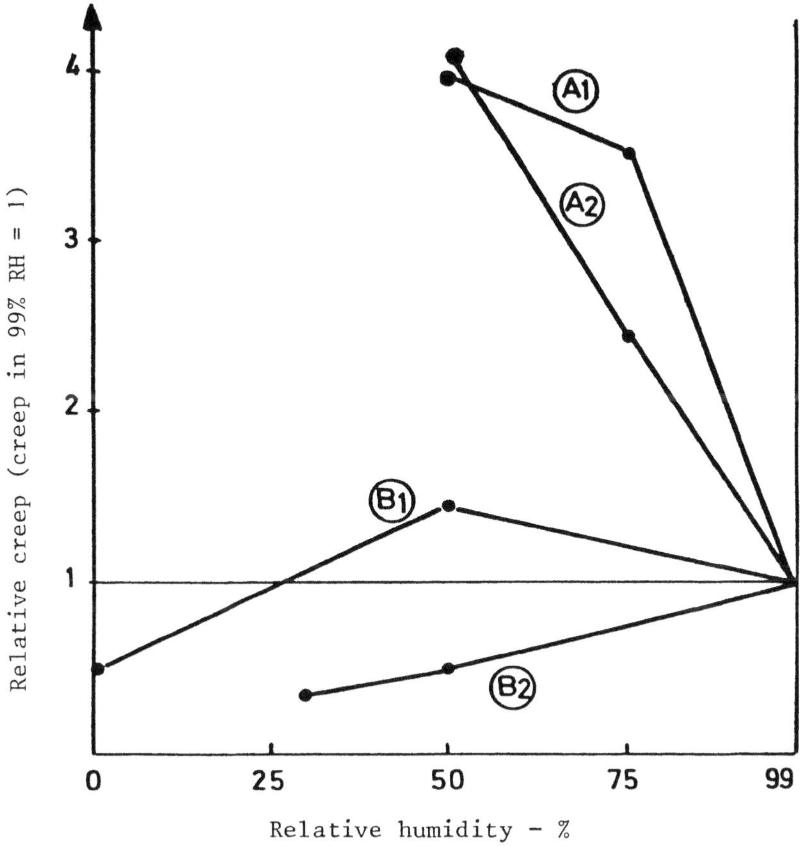

Figure 4: Influence of relative humidity with and without drying

A: Concrete prisms of 7 x 7 x 28 cm kept in water until loading (to 10 MPa after 28 days)
A_1: Relative creep after 7 days of loading.
A_2: Relative creep after 2 years of loading.
(After L'Hermite and Mamillan [6])

B: Hollow cylinders of cement paste 0.7 mm thick loaded after 35 days to 10 MPa for 4 days at 38°C.
The cylinders are placed in the tested environment:
B_1: 1 day before loading
B_2: 34 days before loading.
(After Bazant et al. [5])

The difference between B_1 and B_2 indicates that after 24 hours the cement paste cylinders are still not in hygrometric equilibrium with their environment: the specimens B_1 continue to dry after having been loaded.

This can be explained only by the fact that the part of the creep due to drying, i.e. the variation in water content, referred to as drying creep, prevails over the basic creep.

The shape of the creep versus time curves obtained on specimens of different size confirms the influence of drying: the thinner the specimen, the faster the initial creep. Long-term measurements on some thick components show that creep continues as long as drying is not completed.

The results obtained by Bazant et al. [5] on hollow cement paste cylinders (wall thickness: 0.7 mm!) clearly show the effect of drying (Figure 5). For a specimen kept in a saturated atmosphere until loading in a dry atmosphere (50% RH), creep is 5 to 6 times higher after a few days than when the same specimen has been placed in the dry atmosphere 35 days earlier and, hence, has had time to dry before loading. The initial speed is that of basic creep in a saturated atmosphere but very quickly becomes higher. The creep of a specimen during drying is not intermediate between the specific creeps corresponding to the initial and final conditions: it is much greater than either.

Drying creep occurs when there is a drop in ambient humidity under load or when this drop coincides with the loading, but also when this drop precedes the loading by a period which is too short to allow equilibrium to be reached.

Let us finally note that other tests [6,7] have shown an increase in creep when a rise in humidity occurs during loading: any variation in humidity causes an increase in creep whatever the direction of the disequilibrium.

3. REDISTRIBUTION OF STRESSES

We have just seen that humidity (and its variations) can cause a variation of 1 to 10 in creep: it is by far the most important parameter. And, in general, significant variations in water content occur within the concrete. There are thus probably locally different types of behaviour.

Let us consider a concrete cylinder of 16-cm diameter loaded in a dry atmosphere. Let us assume that it is made up of independent concentric tubes fitted into each other. The inside cylinders, in which drying begins only after a very long period, undergo a basic creep corresponding to a saturated environment.

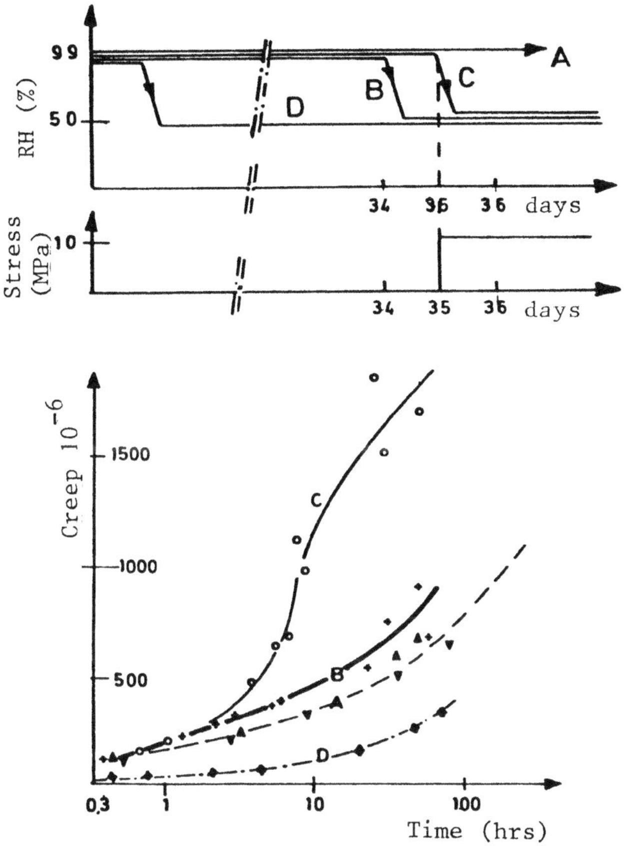

Figure 5: Influence of conservation conditions before and after compression.

Hollow cylinders of cement paste 0.7 mm thick (outer diameter 15 mm, height 92 mm) loaded to 10 MPa after the first 35 days, kept at 38°C under the following humidity conditions:

A: Saturated atmosphere (99% RH).
B: 99% and then, 24 hours before loading, 50% RH.
C: 99% and then 50% applied gradually within the first 8 hours of loading.
D: 50% before and after loading.

Depending on whether drying takes place before (D) or during (C) the loading, creep is divided or multiplied by a factor of two.
The specimen B, whose drying begins 24 hours before loading, is still not in equilibrium with its environment when loading is applied.
(After Bazant et al. [5])

The cylinder on the periphery, whose drying is very fast, has a creep 4 to 5 times faster than the first cylinders. If the loads were applied independently on each tube, the deformations would consequently be 4 to 5 times greater on the periphery than in the centre.

But, in most tests, and in particular in France (Figure 6), loads are applied by metal fixtures whose rigidity in fact imposes a uniform strain on the specimen. During the creep test, there is hence necessarily a transfer of stresses in the specimen, the perphery of the specimen shedding its load onto its core.

In general, in a creep test under unconfined compression, strain is uniform within the section and variable with respect to time, whereas stress varies within the section and it is only its average value which is kept constant with respect to time. From this viewpoint, the relaxation test is more satisfactory: strain can actually be kept uniform in space and in time (but the stress distribution must still be obtained).

4. NUMERICAL EVALUATION

We have attempted to evaluate the extent of this stress distribution by a calculation using the same procedure:

4.1. The specimen (16-cm diameter or 200 cm^2) is discretized into 10 concentric hollow cylinders of equal section (20 cm^2).

4.2. Each hollow cylinder is assigned a creep coefficient corresponding to its water content.

4.3. The calculation is carried out step by step (logarithmic increment: dt = 0.2.t).

4.4. With each time step is calculated successively:

a - the creep increment of each cylinder,
b - their average, which represents the overall strain increment,
c - the stress increments of each cylinder necessary for reestablishing uniform strain.

Only the point a presents a difficulty. It is in fact necessary to calculate the creep under variable stress AND water content. No presently existing model can translate these conditions. We have proposed elsewhere [8] an incremental model of the form:

$$\dot{e} = f(e, \sigma, w, b)$$

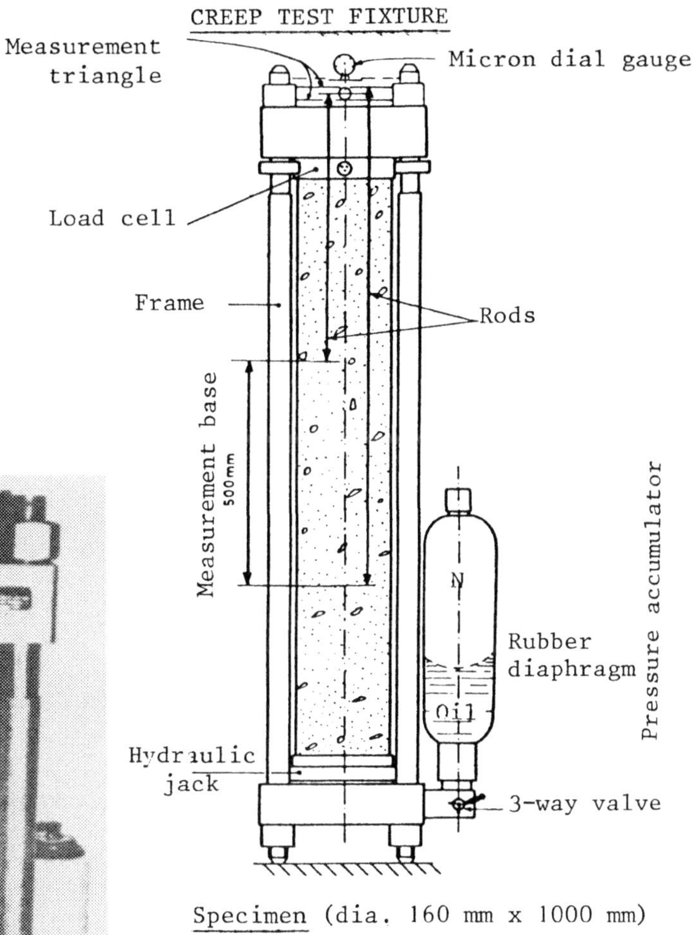

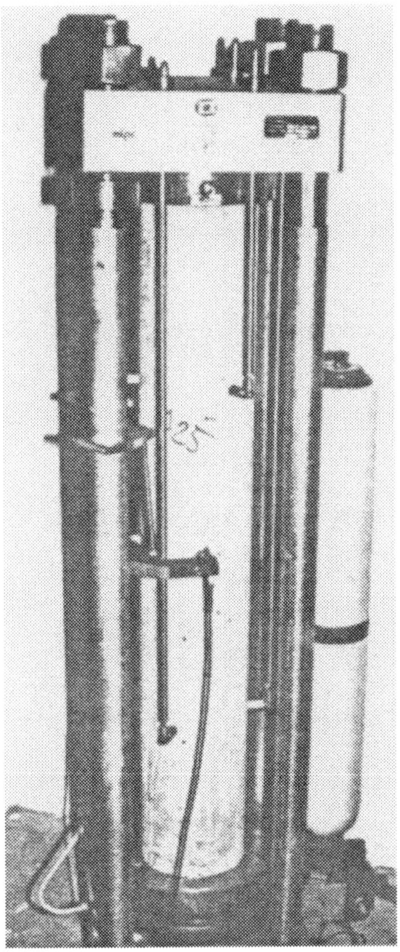

Figure 6: LCPC creep test fixture.

An oleopneumatic accumulator makes it possible to keep the specimen under a constant total load (precision ± 0.5%).

Strains are measured (to within a micron) by a system of six rods anchored to a depth of one-fourth and three-fourths of the length of the specimen.

where:

e is the nonelastic strain (total strain less conventional elastic strain σ/E),

σ is the stress at the considered time,

w is the water content at the considered time,

b is a parameter characterizing the texture and structure of the concrete or, failing this, its age.

An experimental programme is under way at the LCPC for the development of such a model. However, for the calculation which concerns us here, the following observation may be made:

. When w and b vary little with time, the preceding incremental model becomes tantamount to a function of two variables:

$$\dot{e} = f(e,\sigma)$$

which need not be known explicitly when there is a creep law of the form:

$$e = \phi(\sigma, t)$$

. In fact, if $\dot{e}$ depends only on σ and on e, this means that there is a pair of values (σ, t*) in which t* is an unknown such that:

$$e = \phi(\sigma, t^*)$$

. If e and σ are known, it is sufficient to solve, in t*, this latter equation and to write:

$$\dot{e} = \dot{\phi}(\sigma, t^*)$$

(This method, called the "time equivalent" method, has become regulatory in France for the calculation of the "relaxation" of steels under a deformation varying with time [9]).

We used this method, on the basis of the creep law adopted in the new French regulations [9].

The flowchart is given in an appendix.

The results of the calculation are given in Figure 7. We note the following:

4.5. The stress redistributions are very extensive: they reach 60% (less) on the periphery and 40% (more) at the centre.

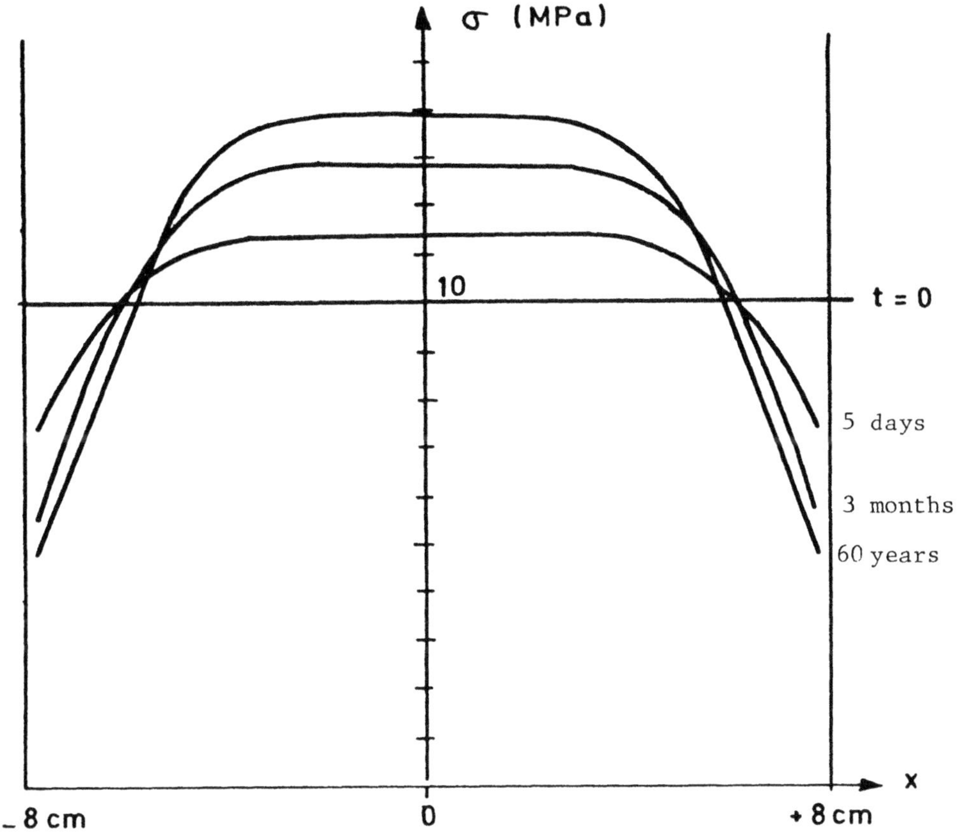

Figure 7: Distribution of stresses along the diameter of a cylindrical specimen subjected to the creep test, calculated at different periods. Initially, the stress is assumed to be uniform and equal to 10 MPa.

4.6. This redistribution begins very early: half the final redistributions occur after 5 days, and three-fourths after 3 months.

4.7. The speed of the phenomenon justifies the subsequent simplifying hypothesis of constant values of w_i with respect to time: in fact, after 3 months, the differences in water content and hence in redistributions tend to be attenuated in relation to our calculations.

4.8. The redistribution curve appears more marked toward the core than indicated by the curve of the water contents: This stems from the round form of the specimen in question, the redistributions taking place in relation to surfaces and hence to squares of the radii.

5. EXPERIMENTAL CONFIRMATION

In order to verify the distribution of stresses within the specimen, we developed an apparatus made up of two mild steel plates and a bed of hard-steel calibrated balls (Figure 9).

The longitudinal deformation of the apparatus (depth of imprint) is of the second order in relation to the diameter of the imprint which constitues a force measurement.

A first specimen was loaded after the first 3 days to 10 MPa in a dry atmosphere (20°C, 50% RH). After 12 days of loading, we were able to obtain an experimental stress-distribution curve.

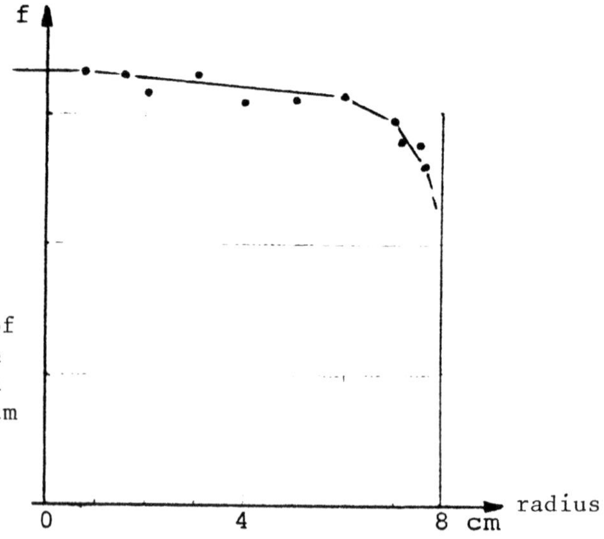

Figure 8: Distribution of stresses measured by the ball imprint method in a concrete cylinder of 8-cm radius after 12 days of loading

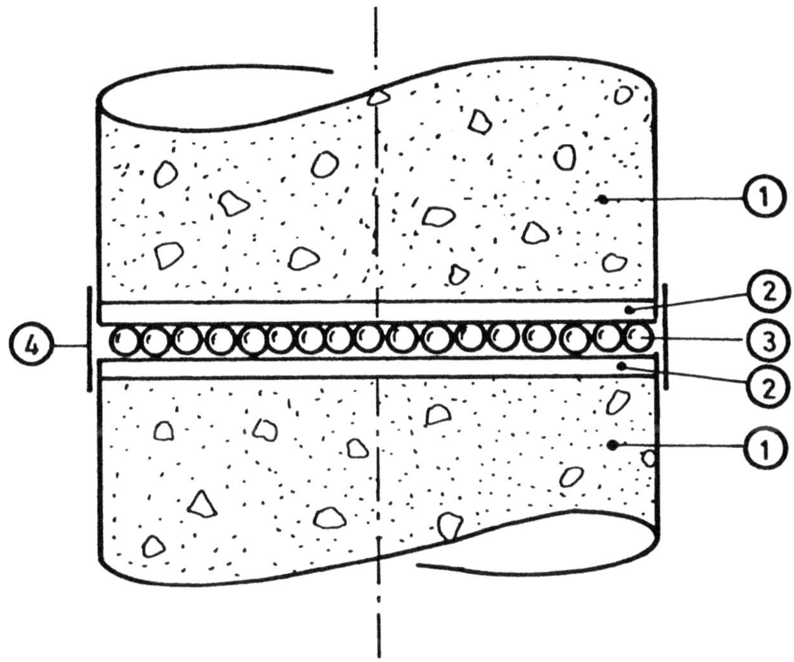

Figure 9: Device making it possible to evaluate the distribution of stresses within compressed specimens. It is made up of:

- two mild-steel plates (2) of 3-mm thickness,
- a bed of hard-steel calibrated balls (3) of 10-mm diameter, inserted between two parts of the specimen, sawn and lapped (1) and held by a clamping collar (4).

6. INTERPRETATION OF SHRINKAGE AND CREEP DATA

The calculations we have presented here are far from being perfect:

6.1. They are based upon a creep law which was developed on heterogeneous specimens, whereas the opposite is what is required: using creep laws on homogeneous specimens, reconstitute the creep (hybrid) of a specimen which dries.

6.2. They assume a uniform modulus, whereas what we have just described is also applicable to the instantaneous behaviour of the concrete: thus, a specimen kept in air probably does not have a homogeneous behaviour.

6.3. They do not take into account shrinkage, which further accentuates the phenomenon. In fact, we have not taken shrinkage into account because we do not have reliable data for a local allowance for shrinkage.

The shrinkage test which generally accompanies (control) creep tests is not representative for at least two reasons:

- The boundary conditions (non-flatness of sections) are different.
- The presence of surface tensile stresses and hence of surface cracking leads to an altogether different distribution of stresses in the section.

The procedure we have undertaken at the LCPC is the following:

- Performance of shrinkage and creep tests on homogeneous specimens (either confined by resin coating, dried or partially dried by heating at 40°C; to reduce the influence of this heating on ageing and natural hydration, a compromise was found in the Y-shape of the specimens).
- Performance of shrinkage and creep tests on Y-specimens during drying.
- Performance of creep tests under hydraulic flux (the periphery which dries is moreover traversed by the water leaving the deeper zones).
- Interpretation of conventional tests by the integration of the results thus obtained.
- And, further, physical study of the role of water in the behaviour of concrete under stress.

This procedure assumes that we are abandoning the linear formalism, moreover called into question elsewhere [1].

The incremental model appears to us to be better suited to the problems involved. In addition, it is better adapted to automatic computation [10] from the viewpoint of the designer.

Finally, it is possible that a similar analysis will allow an explanation of the special delayed behaviour of heat-treated concrete, on the one hand, and lightweight aggregate concrete, on the other. In the first case, heat treatment accelerates greatly the removal of water, while in the second case the light aggregate constitutes a large reserve of free water.

7. CONCLUSION

The analysis shows that concrete which dries in air exhibits a transverse hygrometric anisotropy which leads, especially under loads of long duration, to significant differences in behaviour between the periphery and the core of the same section. In the particular case of unconfined compression, it was possible to show (and to verify experimentally) that a stress distribution of $\pm$ 50% amplitude could occur in a specimen of current size.

The tentative conclusions that may be drawn are the following:

7.1. The interpretation of creep test data must take into account what is known about drying, in particular as concerns:

- the influence of thickness and ambient humidity,
- the need to know the conservation conditions before loading.

7.2. The shrinkage test (control) which generally accompanies creep tests is not representative of these tests. The additivity of shrinkage and creep can result only from a formal definition of creep.

7.3. The interpretation of recovery tests appears impossible to date, at least within the framework of linear formalism.

7.4. The observation of cracking on a structure in service, even if oriented, does not necessarily mean that the section (or the fibre) considered is tensioned.

7.5. Procedures for determining the characteristics of creep in concrete, as defined in national standards and international recommendations, should be amended, in particular as concerns conservation conditions before and during testing.

APPENDIX: REDISTRIBUTION CALCULATION FLOWCHART

σ : initial stress — data

E : deformation modulus

$g = \sigma/E$ — initial deformation

$t = 0{,}001$ — time in hours

→ from $i = 1$ to 10

$\sigma_i = \sigma$ — initialization

$f_i = \sigma_i \cdot w_i \cdot \dfrac{\sqrt{t}}{\sqrt{t} + 50}$ — creep law

→ $dt = 0{,}2 . t$

$f = 0$

→ from $i = 1$ to 10

t_e = solution of following equation: — "time equivalent"

$f_i = \sigma_i . w_i \cdot \dfrac{\sqrt{t_e}}{\sqrt{t_e} + 50}$

$df_i = dt . \sigma_i . w_i \cdot \dfrac{25}{\sqrt{t_e}(\sqrt{t_e} + 50)^2}$ — creep increment

$f_i = f_i + df_i$ — storage

$f = f + df_i/10$ — average deformation

$g = g + f$ — total deformation

→ from $i = 1$ to 10 — redistribution

$\sigma_i = \sigma_i + (f - f_i)/E$

$t = t + dt$

END test

BIBLIOGRAPHY

[1] GAUCHER C.
Tentative de caractérisation du comportement rhéologique non linéaire du béton hydraulique dans la représentation par intégrales multiples.
THESE de Docteur-Ingénieur. ECOLE. NAT. DES PONTS ET CHAUSSEES, 1980 (164 pages).

[2] L'HERMITE R.G., MAMILLAN M.
Répartition de la teneur en eau dans le béton durci
ANNALES de l'I.T.B.T.P. N° 309-310 - 1973 (p. 30-34)

[3] NEVILLE A.M.
Creep of Concrete: Plain, Reinforced, and Prestressed
NORTH-HOLLAND PUBLISHING COMPANY - AMSTERDAM - 1970 (622 pages)

[4] BROWN N.H., HOPE B.B.
The Creep of Hydrated Cement Paste
CEMEÑT AND CONCRETE RESEARCH - Vol. 6 - 1976 (p. 475-486)

[5] BAZANT Z.P., ASGHARI A.A., SCHMIDT J.
Experimental Study of Creep of Hardened Portland Cement Paste at Variable Water Content
MATERIAUX ET CONSTRUCTIONS - Vol. 9 N.52 - 1976 (p.279-290)

[6] L'HERMITE R.G., MAMILLAN M.
Further Results of Shrinkage and Creep Tests
INTER. CONF. ON THE STRUCTURE OF CONCRETE - LONDON 1968 (p. 423)433)

[7] GAMBLE B.R., PARROTT L.J.
Creep of Concrete in Compression During Drying and Wetting
MAGAZINE OF CONCRETE RESEARCH - Vol. 30 N.104 - 1978 (p. 129-138)

[8] ACKER P., BARRAL A.
Déformations différées du béton durci: étude critique des modèles analytiques de comportement
C.R. DE RECHERCHE L.C.P.C. - 1980 - (33 pages)

[9] Conception et calcul des ouvrages en béton précontraint (B.P.E.L.à (à paraître)

[10] ACKER P.
Le problème des pertes de précontrainte par fluage et relaxation combinés
BULLETIN DE LIAISON DES L.P.C. - N.96 - 1978 (p. 129-137)

[11] GABILLY M., LEVILLAIN J.P.
Analyseur gammamétrique de carottes de sol en conteneur
BULLETIN DE LIAISON DES L.C.P. - N.97 - 1978 (p. 113-120)

STRESS RELAXATION OF CONCRETE STRUCTURES AND RELAXATION TESTS OF VERY OLD CONCRETE

Prof. Dr.-Ing. Dr.h.c. Heinrich Trost

Institut für Massivbau, RWTH Aachen
Aachen/Germany

1 STRESS-STRAIN-RELATION OF CONCRETE

Using the principle of superposition, we may write for the total strain ε at any time t:

$$\varepsilon(t) = \frac{\sigma(\tau_o)}{E(\tau_o)}\left[1+\varphi(t,\tau_o)\right] + \int_{\tau=\tau_o}^{t} \frac{\partial\sigma(\tau)}{\partial\tau}\cdot\frac{1}{E(\tau)}\left[1+\varphi(t,\tau)\right]\cdot d\tau \qquad (1a)$$

and with a constant modulus of elasticity

$$\varepsilon_t = \frac{\sigma_o}{E}(1+\varphi_t) + \frac{\sigma_t-\sigma_o}{E} + \frac{1}{E}\int_{\tau=\tau_o}^{t} \frac{\partial\sigma(\tau)}{\partial\sigma}\cdot\varphi(t,\tau)\cdot d\tau \qquad (1b)$$

With this principle of superposition only, one has to solve creep problems (i.e. calculation of deformations under a known stress history) as well as relaxation problems (i.e. determination of the stress at any time under specified conditions of strains or deformations).
For relaxation problems it is convenient to rearrange the integral equation (1) to obtain the following algebraic one:

$$\varepsilon_t = \frac{\sigma_o}{E}(1+\varphi_t) + \frac{\sigma_t-\sigma_o}{E}(1+\rho\cdot\varphi_t) \qquad (2)$$

The establishment of this relation is possible by the definition of the relaxation coefficient ρ [1;2;3] or aging coefficient χ [4]

as snown below (see also |5;6|:

$$\rho = \frac{\int_{\tau_o}^{t} \frac{\partial\sigma(\tau)}{\partial\tau}\cdot\varphi(t,\tau)\cdot d\tau}{(\sigma_t-\sigma_o)\cdot\varphi_t} = \frac{\sum_{\tau_i} \Delta\sigma(\tau_i)\cdot\varphi(t,\tau_i)}{(\sigma_t-\sigma_o)\cdot\varphi_t} \tag{3}$$

Thus the numerical value of $\rho(0{,}5<\rho\leq 1{,}0)$ is calculated as the ratio of the sum of products of the stress steps $\Delta\sigma(\tau_i)$ and their respective creep function $\varphi(t,\tau_i)$ to the product of change of stress $\sigma_t-\sigma_o$ and the creep coefficient φ_t relevant to the stress $\sigma_o=\sigma(\tau_o)$.

Now introducing the relaxation condition $\varepsilon_t = \varepsilon_o$ = const. or $E\cdot\varepsilon_t-\sigma_o = 0$ in eq. (1) we arrive at the following relation for the relaxation with help of the numerical integration:

$$0 = \sigma_o\cdot\varphi_t+\sigma_t-\sigma_o+\sum_{\tau_i} \Delta\sigma(\tau_i)\cdot\varphi(t,\tau) \tag{4}$$

With this very old concrete the creep functions $\varphi(t,\tau_i)$ are obtained by moving the original creep function φ_t on the time axis into the respective step intervals τ_i. In this manner it is possible to evaluate the relaxation equation (4) as well as calculate the relaxation coefficient ρ in eq. (3) with a quasi-exact numerical precision.

The general equation (2) is simplified with the introduction of the relaxation condition $E\cdot\varepsilon_t-\sigma_o = 0$ into the form:

$$0 = \sigma_o\cdot\varphi_t+(\sigma_t-\sigma_o)(1+\rho\cdot\varphi_t)$$

from which we derive immediately the required relaxation of stress as:

$$\sigma_t/\sigma_o = 1-\frac{\varphi_t}{1+\rho\cdot\varphi_t} \tag{5}$$

For this very old concrete the exact value of the relation coefficient ρ has been determined as $0{,}95<\rho<1{,}0$. It will be a good approximation to assume $\rho = 0{,}8$ for all concrete structures of normal concrete loading ages and with usual creep coefficients $1{,}5<\varphi_{t,\infty}\leq 4$.

In the case of a time dependent imposed strain which is affinitive to the shape of the creep curve, $\varepsilon_t = \varepsilon_\infty/\varphi_\infty\cdot\varphi_t$ we obtain the quotient of the actual stress to the elastically calculable stress from eq. (2) since $\sigma_o = 0$:

$$\frac{\sigma_t}{E \cdot \varepsilon_t} = \frac{1}{1+\rho \cdot \varphi_t} \tag{6}$$

In this case the numerical value of ρ is identical with the calculated value of ρ from eq. (3) for the suddenly imposed strain.

2 RESULTS OF CREEP AND RELAXATION TESTS

Creep and relaxation tests were performed on about 40 year old concrete specimens. The aim of these experiments was to check the following:

1. validity of the principle of superposition
2. applicability of an algebraic σ-ε-t relation to solve relaxation problems.

Table I shows the measurement data.
Figure 1 (for concrete 1) and figure 2 (for concrete 2) indicate:

a) the development of the measured creep φ_t and the measured stress relaxation σ_t/σ_o

b) the theoretical shape of the relaxation curve σ_t/σ_o according to the theory of superposition using the measured values of φ_t in evaluating equation (4) or evaluating equation (5) with the value of ρ from eq. (3).

Table I: Stress and strain data of relaxation test specimens

Specimen	Initial stress σ_o	Initial stress-strength ratio $\alpha=\sigma_o/\beta_{CM}$	Specific elastic strain $\bar{\varepsilon}_o$ at loading	unloading	Relaxed stress at time t σ_t/σ_o	Period t under imposed strain
	N/mm²	-	10^{-6} (mm²/N)	10^{-6} (mm²/N)	-	days
1	2	3	4	5	6	7
1.9	14,o	o,3o	29,6	28,2	o,61	317
1.1o	23,3	o,45	28,6	- [1]	o,64 [1]	2o6
2.2	14,1	o,4o	36,7	- [1]	o,66 [1]	3oo
2.3	15,1	o,4o	36,2	36,5	o,68	2o5

[1] Specimen 1.1o and 2.2 still under load

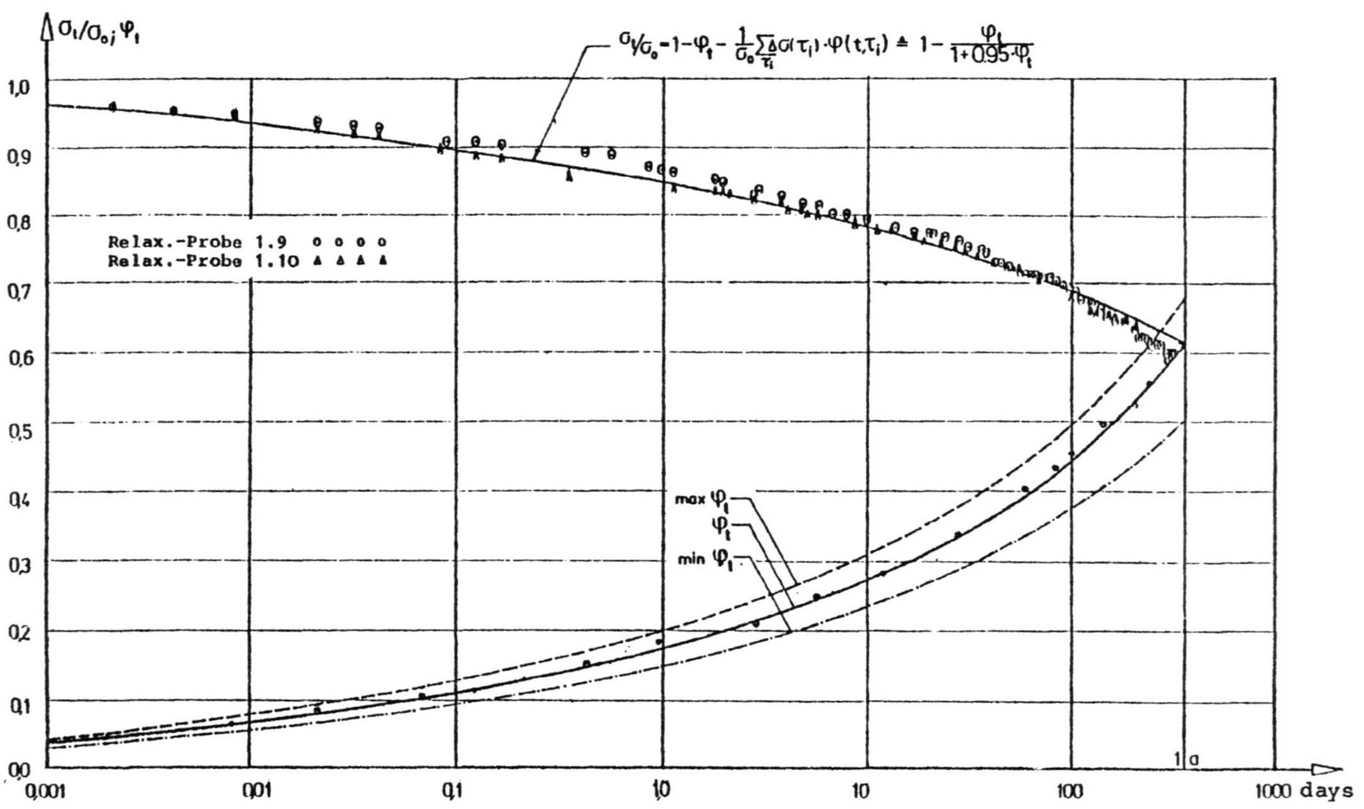

Fig. 1: Results of the creep and relaxation experiments on concrete 1
a) development of mean creep coefficient of 6 specimens with observed bounds
b) observed stress relaxation of 2 specimens in relation to the theoretical one as per principle of superposition using the mean development of creep coefficient

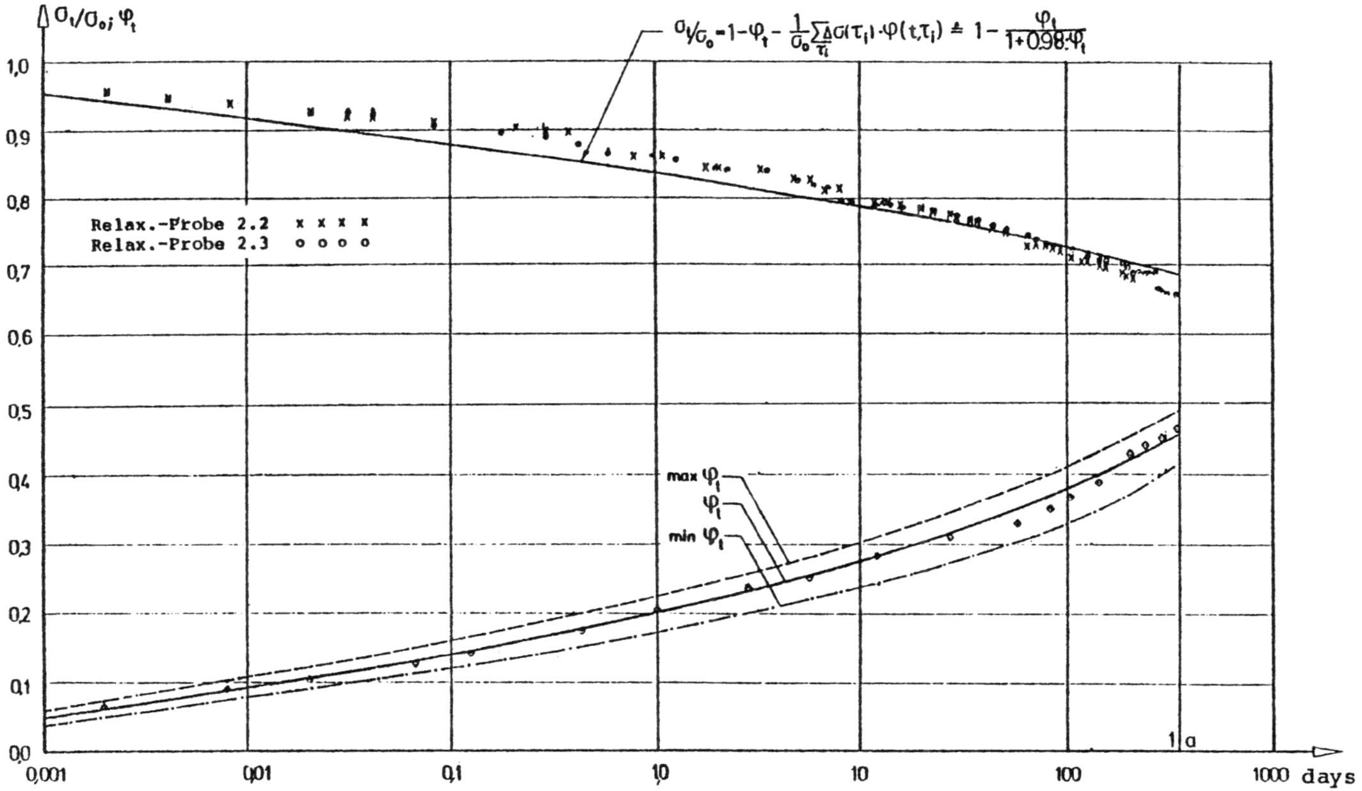

Fig. 2: Results of the creep and relaxation experiments on concrete 2
a) development of mean creep coefficient of 3 specimen with observed bounds
b) observed stress relaxation of 2 specimens in relation to the theoretical one as per principle of superposition using the mean development of creep coefficient

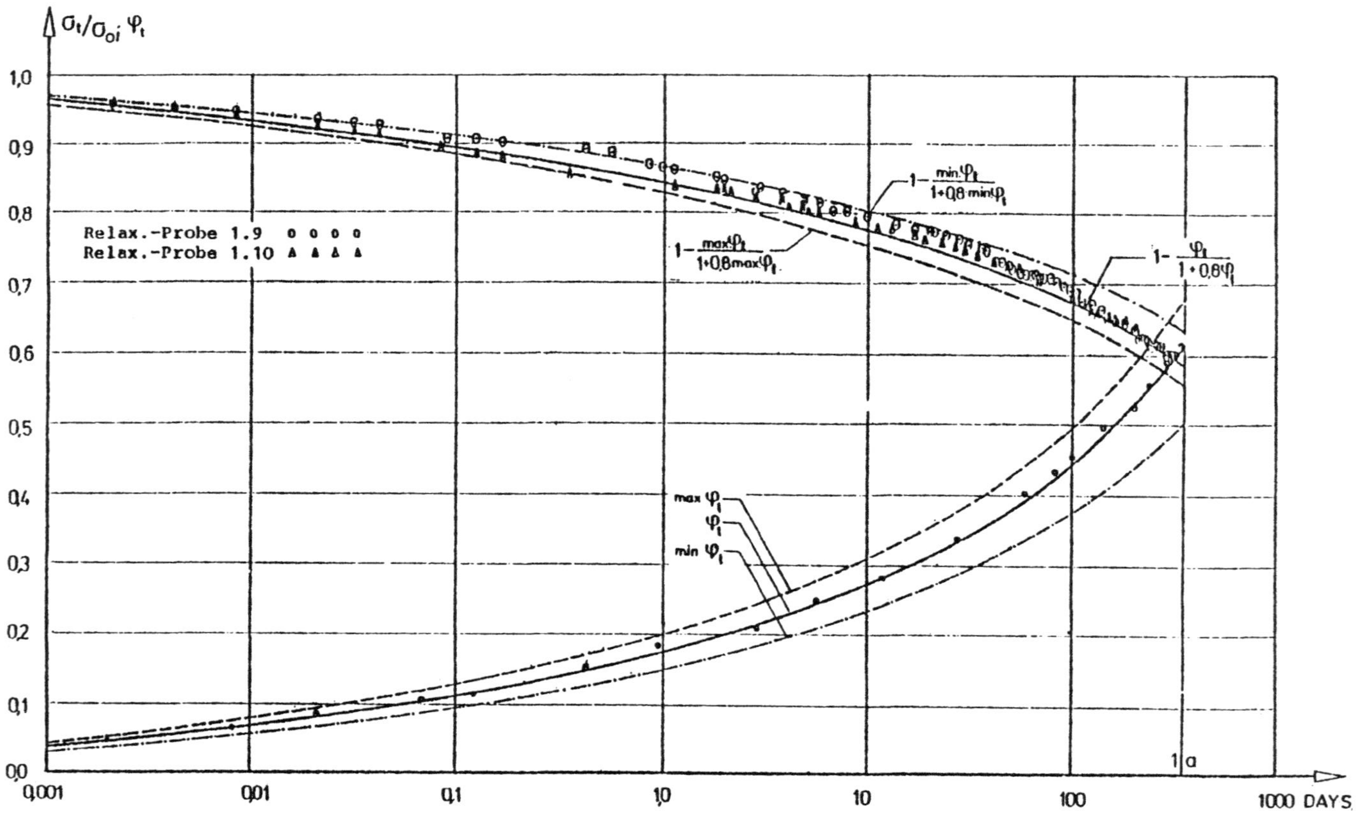

Fig. 3: Developement of the mean creep coefficient of concrete 1 with observed bounds and the measured stress relaxation in relation to the theoretical stress relaxation $\sigma_t/\sigma_o = 1-\varphi_t/(1+\rho\cdot\varphi_t)$ with the approximation $\rho = o.8$

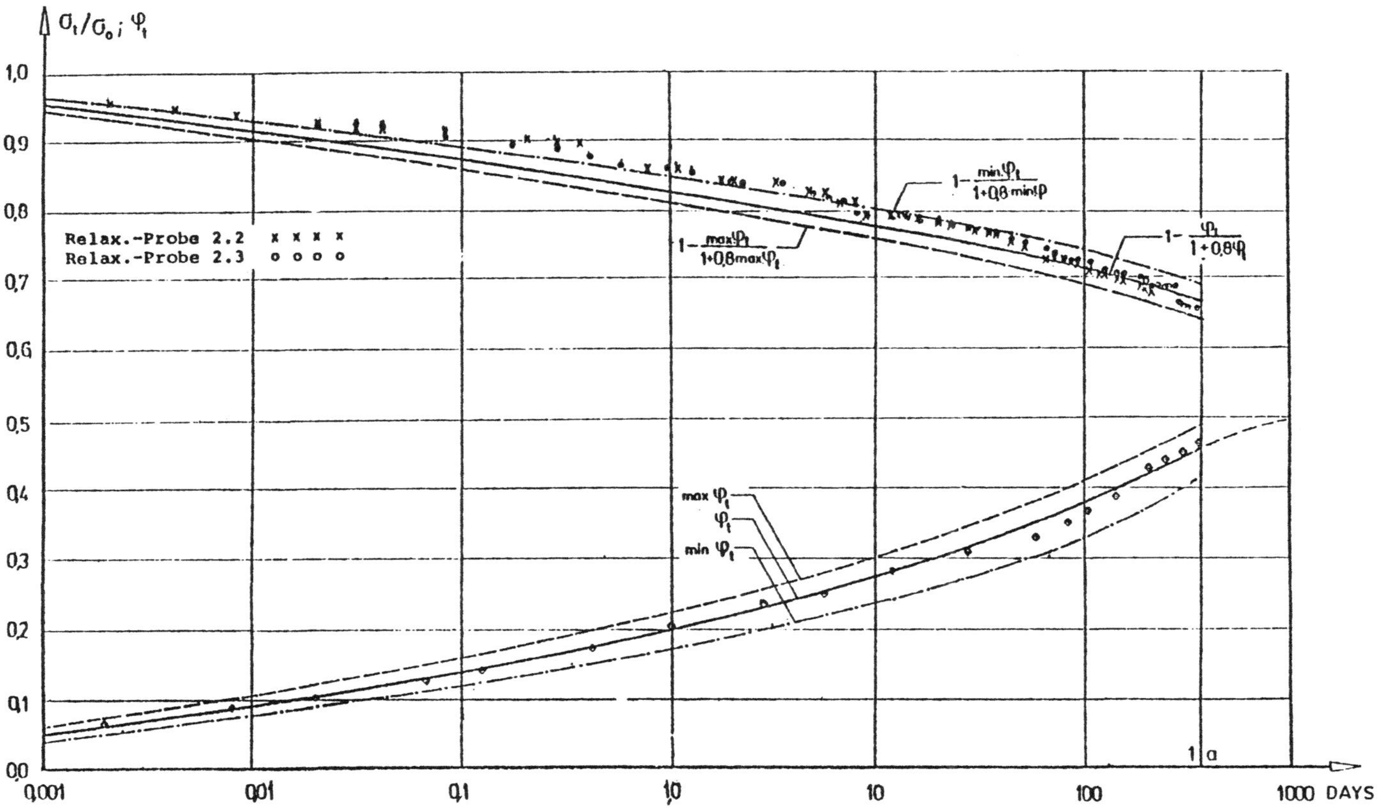

Fig. 4: Development of the mean creep coefficient of concrete 2 with observed bounds and the measured stress relaxation in relation to the theoretical stress relaxation $\sigma_t/\sigma_o = 1-\varphi_t/(1+\rho\cdot\varphi_t)$ with the approximation $\rho = o.8$

Figure 3 (for concrete 1) and figure 4 (for concrete 2) contain:

a) the development of the measured data for φ_t and σ_t/σ_o

b) the development of the stress relaxation with the simplification $\rho = 0,8$ in eq. (5) $\sigma_t/\sigma_o = 1-\varphi_t/(1+0,8\varphi_t)$.

Finally we can summarize the results of the creep and relaxation tests as follows:

1) On the basis of the theory of superposition it is possible to calculate the relaxation of stresses in a practically exact manner.

2) The solution of relaxation problems in concrete structures (e.g. the effect of a sudden or a time-dependent imposed strain) can be treated by a simple algebraic σ-ε-t relation as given in eq. (2).

3) For the application of the principle of superposition there is no need to split up creep into its flow or irreversible part φ_f and its delayed elastic or reversible part φ_v.

Further informations can be taken out:
- Heft 295: DEUTSCHER AUSSCHUSS FÜR STAHLBETON
  Verlag W. Ernst & Sohn, Berlin 1978
- (Vol. 295 of German Committee for Reinforced Concrete).

References

1. Trost, H., "Auswirkungen des Superpositionsprinzips auf Kriech- und Relaxationsprobleme bei Beton und Spannbeton." *Beton- und Stahlbetonbau* (1967) 230-238, 261-269.

2. Trost, H., "Folgerungen aus Theorien und Versuchen für die baupraktische Untersuchung von Kriech- und Relaxationsproblemen in Spannbetontragwerken." *Preliminary Publication, Symposium on Design of Concrete Structures for Creep, Shrinkage and Temperature Changes,* Madrid 1970 (International Association for Bridge and Structural Engineering, Zürich, 1970) pp. 191-198.

3. Neville, A.M. and W. Dilger, *Creep of Concrete; Plain, Reinforced, and Prestressed,* (North Holland Publishing Co., Amsterdam, 1970) chapters 17 to 20.

4. Bazant, Z.P., "Prediction of Concrete Creep Effects Using Age-Adjusted Effective Modulus Method." *Journal of the American Concrete Institute,* Detroit, April 1972, pp 212-217.

5. Favre, R., M. Koprna and A. Rdajicic, *Effets différées, Fissuration et Deformations des Structures en Béton,* (Editions Georgi, Lausanne, 1980).

6. Chiorino, A.M. et al., *Structural effects of time-dependent behaviour of concrete,* (CEB-Bulletin d'Information Nr.136, Paris 1980).

RHEOLOGICAL PROPERTIES AT HIGH TEMPERATURE OF A CONCRETE WITH A CRUSHED LIMESTONE AGGREGATE AND BLAST FURNACE CEMENT

Professor Dr. René Walther and Thomas Pareth

Institute of Structural Engineering - Reinforced and Prestressed Concrete - Swiss Federal Institute of Technology, Lausanne

ABSTRACT In connection with the development of the pressure vessel of a nuclear reactor, extensive tests have been carried out to determine the rheological properties of a concrete subjected to a temperature of 200°C. The paper includes a description of the tests along with the presentation and discussion of the results.

RESUME Dans le cadre du développement d'un caisson de réacteur nucléaire, des essais ont été effectués pour déterminer les propriétés rhéologiques du béton soumis à une température de 200°C. L'article comprend une description des essais, ainsi qu'une présentation et une discussion des résultats.

Introduction

The experimental results presented here derive from a German-Swiss research programme concerning the development of a pressure vessel for a nuclear reactor using high temperature helium as its primary gas (HHT). The concrete in question serves both as structural concrete at normal temperatures and as an intermediary between the liner at approximately 200°C and the cooling system protecting the structural concrete at 65°C. Since comparatively little is known about the rheological behaviour of concrete at such high temperatures, an extensive test programme has been carried out at the Swiss Federal Institute of Technology in Lausanne (EPFL) under the direction of Bonnard & Gardel, Consulting Engineers, Lausanne. The rheological behaviour of this hot concrete layer largely governs the design of the liner and could even determine the feasibility of the reactor. On the one hand, creep and relaxation of the concrete reduces the maximum stress at singular points, but on the other hand, the loss in resistance due to the elevated temperatures, as well as the prestressing losses due to creep, must be confined to acceptable limits.

The concrete concerned is the same as that used in the construction of the pressure vessel of the Schmehausen reactor (type THTR). It is a concrete with a crushed limestone aggregate and blast furnace cement.

Programme and Installations of the Test

Up to the present time, investigations of concrete at high temperatures have been conducted using relatively small specimens, as they had to be heated in ovens. Since the precision of rheological property tests on small specimens is somewhat questionable, special loading frames have had to be developed to accommodate the large, hollow cylindrical specimens used here ($\emptyset_e$ = 27 cm, $\emptyset_i$ = 7 cm, h = 75 cm). In order to collerate the results with standard tests, smaller cylindrical ($\emptyset$ = 15 cm, h = 30 cm) and prismatic specimens (12 x 12 x 36 cm) are employed throughout the test series. The mechanical characteristics (Young's Modulus, compressive strength, direct tensile strength and tensile-flexion strength) along with the shrinkage and creep are determined for the three specimen types at normal temperature, using standard methods and installations.

Each series of creep and shrinkage tests at high temperature is carried out by means of six hollow cylindrical specimens placed vertically one upon the other in the loading frame. Before being set up in the loading frame, these specimens are conserved in water for 90 days. Shrinkage is also measured during the drying period of 7 days at 50°C. Then, before the specimens are put under load for the creep tests, the temperature is increased in stages, with a gradient of less than 5°C/hr, until the level of 200°C is attained.

	COMPARISON TESTS AT NORMAL TEMPERATURE															
	MECHANICAL CHARACTERISTICS									SHRINKAGE			CREEP			
SERIES Nº	1		2	3		4	5		6	7		8	9		10	
SPECIMEN TYPE	A	C	B	A	C	B	A	C	B	A	C	B	A	A	B	C
NUMBER OF SP.	9	15	4	7	14		15	14	2	7	7	6	6	1	6	8
CURING	IN WATER AT 18°C BEFORE BEGINNING OF THE TEST max. 90 days															
AGE (days)	28			56			90			90			90			
TEST TEMPERATURE (°C)	20°									20°						
STRESS σ (β_{CN})										0		0	0.2		0.2	
DUR. OF TEST	(days)									360		360	360		147	

	TEST AT HIGH TEMPERATURE																	
	SHRINKAGE				CREEP													
SERIES Nº	11		12		13			14			15			16		17	18		
SPECIMEN TYPE	B	C	B	C	B	A	C	B	A	C	B	A	C	B	C	B	B	A	C
NUMBER OF SP.	6	14	6	13	6	7	7	6	7	10	6	7	7	9	3	6	6	6	8
CURING	IN WATER AT 18°C BEFORE BEGINNING OF THE TEST max. 90 days																		
AGE (days)	90				90			90			90			530			90		
TEST TEMPERATURE (°C)	150 200 25x		200		150 200 25x	20°			20°		200 150 200 25x	20°		200		200	100 200	20°	
STRESS σ (β_{CN})	0		0		0.2			0.2			0.4			0.4		0.2	0.2 0.4		
DUR. OF TEST	50		65		68			360			310 50			140			34 30 51		

TYPE OF SPECIMEN

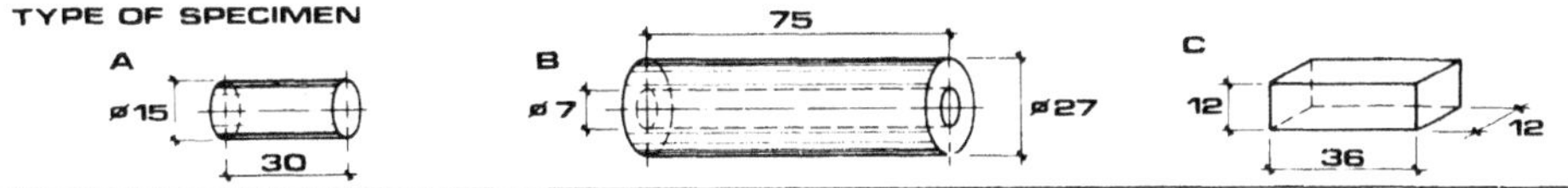

Fig. 1. COMPLETE TEST PROGRAMME

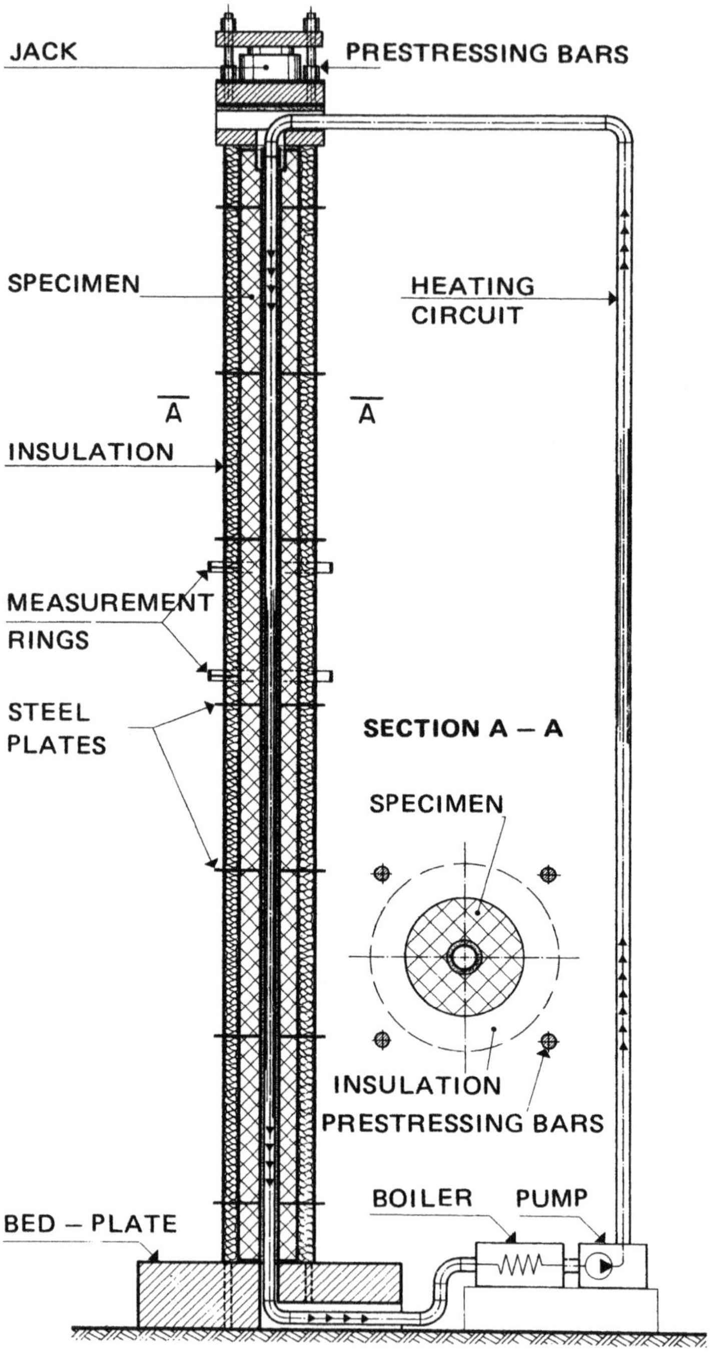

Fig. 2. LOADING FRAME FOR THE WARM CONCRETE

The essential parameters varied of the different series are the temperature level (constant 200° or two-day cycles 200 - 150 - 200°C) and, for the creep tests, the stress level (9,2 N/mm^2 = 0,2 f_{cN} and 18,4 N/mm^2 = 0,4 $\cdot f_{cN}$). At the end of each high temperature test the specimens are subjected to mechanical tests at normal temperatures. At this time, the same mechanical properties are obtained from other, standard specimens coming from the same concrete batch, but not having been subjected to high temperature. FIG. 1 represents the whole of the executed test programme.

The concrete has a crushed limestone aggregate (Dmax = 22,4 mm) and blast furnace cement (370 kg/m^3 HOZ 45L, according to DIN). The water-cement ratio was 0,47 and the fresh concrete had the following averaged properties: Vebe $\bar{t}$ = 7 sec; slump $\bar{s}$ = 1,4 cm; $\bar{\gamma}$ = 2,41 kg/cm^3; air content 1,6 % by volume.

The high temperature tests are carried out on three similar loading frames, one of which is shown in FIG. 2. Six test pieces are placed one upon the other, with a steel plate inserted between them. The axial force for the creep tests is produced by a jack at the top of the frame and four tensioned bars anchored in the bed-plate. The jack is thermally insulated and connected to a large reservoir of pressurized oil, in order to diminish the variations in pressure arising from the deformations of the concrete due to creep. The heat is produced by an electric boiler and transferred to the concrete by means of oil circulating at 15 cm/sec in a pipe passing through the hollow in the cylinders. In order to improve the heat transmission without affecting the deformations of the concrete, a layer of aluminium powder is placed between the heating pipe and the inner face of the concrete specimen. An external thermal insulation consisting of 10 cm of glass-fibre wool limits the temperature difference between the inner and outer surfaces of the test pieces. The temperature is controlled on one of the six specimens using ten thermocouples.

The relatively large size of the specimens allows the precise measurement of strain. The elongations are measured by dial gauges (1/1000 mm), both between the steel plates separating two specimens (measuring length 75 cm) and between two steel rings screwed into each specimen (measuring length 50 cm). (See FIG.3).

Fig.3. MEASURING APPARATUS

Experimental Results

FIG. 4 and 5 represent the comparison between the compressive strengths and Young's moduli obtained from specimens not having undergone heat exposure (types A/B/C) and specimens having undergone heat exposure (type B). The modulus of elasticity is calculated from the strain at 10 N/mm^2 (secant modulus). All the specimens were tested at normal temperature, the new characteristics of a heat-exposed concrete being equal to those of the same concrete once it has cooled.

In relation to the tensile tests, it is interesting to note the difference in the tension rupture surfaces of the heat-exposed and non heat-exposed specimens (FIG. 6). Where there has been no heat exposure, the rupture occurs across the aggregate and concrete alike, whereas the rupture of the heat-exposed specimens occurs at the aggregate-cement interface.

The shrinkage tests at normal temperature have yielded an $\varepsilon_{cs\infty} = 0{,}4\ ^{o}/oo$ (average value of 17 specimens, taking into account the swelling of 0,05 $^{o}/oo$ due to the storage in water).

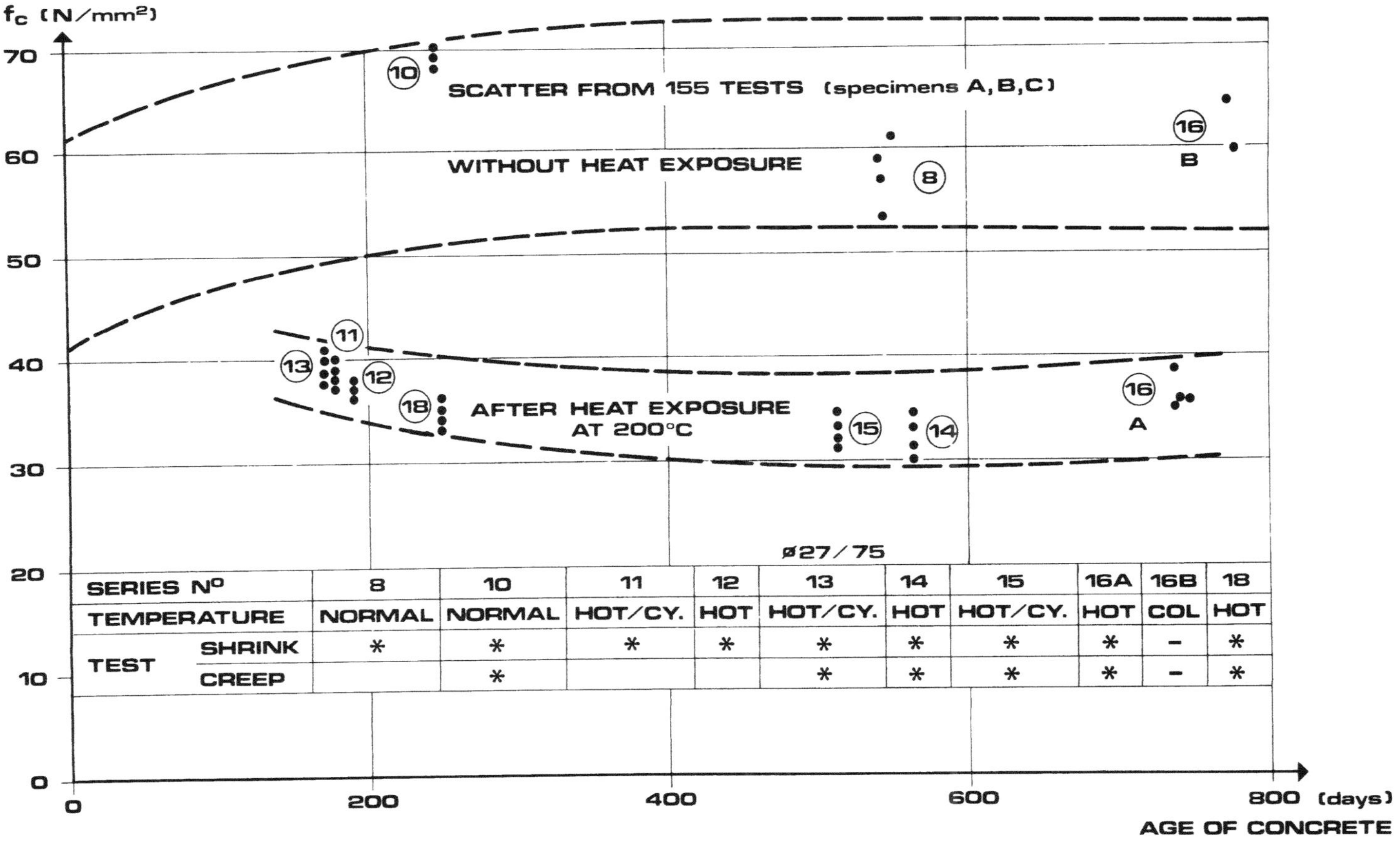

SERIES Nº		8	10	11	12	13	14	15	16A	16B	18
TEMPERATURE		NORMAL	NORMAL	HOT/CY.	HOT	HOT/CY.	HOT	HOT/CY.	HOT	COL	HOT
TEST	SHRINK	*	*	*	*	*	*	*	*	-	*
	CREEP		*			*	*	*	*	-	*

Fig. 4. COMPRESSIVE STRENGTH

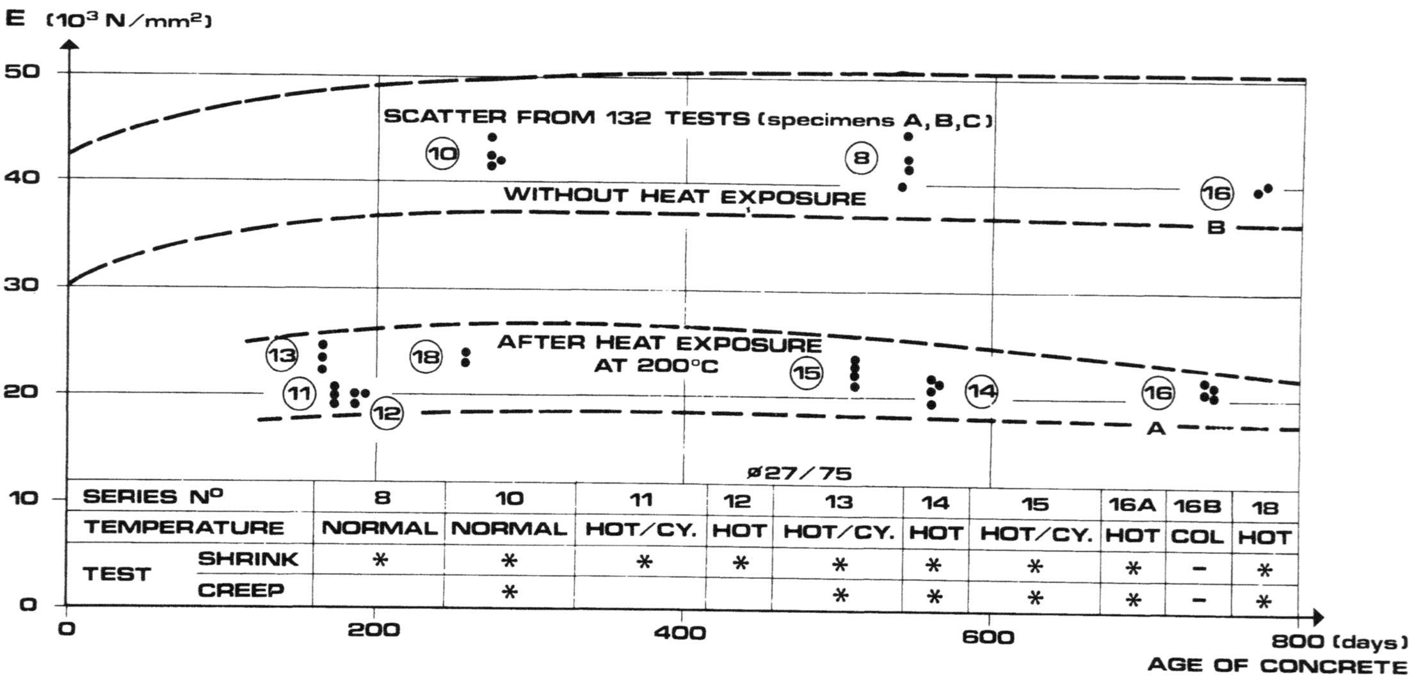

SERIES N°		8	10	11	12	13	14	15	16A	16B	18
TEMPERATURE		NORMAL	NORMAL	HOT/CY.	HOT	HOT/CY.	HOT	HOT/CY.	HOT	COL	HOT
TEST	SHRINK	*	*	*	*	*	*	*	*	-	*
	CREEP		*			*	*	*	*	-	*

Fig. 5. YOUNG'S MODULI

WITHOUT HEAT EXPOSURE

AFTER HEAT EXPOSURE

Fig. 6. COMPARISON OF THE TENSILE RUPTURE SURFACES

The shrinkage tests at elevated temperature (FIG. 7) permit the determination of the influence of shrinkage for the creep tests, in which the measurements cover the two phenomena at the same time. It was not, however, possible to determine the evolution of the shrinkage during the heating process.

FIG. 7 also shows the influence of the thermal cycles on the development of creep under an axial load of 9,2 N/mm^2 (Series 13 and 14), along with the influence of the load level (two series carried out at 200°C with axial loads of 9,2 N/mm^2 and 18,4 N/mm^2 respectively.

The creep at normal temperature is represented in FIG. 8, along with the corresponding shrinkage tests.

The results obtained from the test in which the specimen at 100°C was subjected, firstly to an axial force of 9,2 N/mm^2 and later 18,4 N/mm^2 before the temperature was increased to 200°C are graphically represented in FIG. 9.

Discussion of the Results

On one hand, modifications in the structure of such a concrete would appear to take place in the cement paste. These are due in part to the instability of the calcium-aluminosulphates-hydrates phase beyond a temperature of 150°C (c.f. [1)]). On the other hand, micro-cracks influence the properties of the concrete as well.

The compressive resistance is not, once the specimen has completely dried, affected by either the temperature cycling or the magnitude of the axial force in the creep tests. It is principally influenced by the temperature level and the duration of heat exposure. According to our preliminary tests, the fact that the specimens were mechanically tested at normal rather than high temperature, has not given rise to any significant differences. The moduli of elasticity (secant) also vary essentially in function with the temperature level, but are additionally dependent on the axial force. Each series of creep tests at elevated temperature yield moduli 13 to 14 % greater than the shrinkage series.

Over the whole of the tests employing a maximum temperature of 200°C, the strengths fell to 50 - 64 % and the moduli of elasticity to 45 - 57 % of the comparative values obtained from the untreated concrete.

During the test periods, both at normal and elevated temperature, the creep of the specimens followed a law of the form $\varepsilon_{cc} = a \cdot t^n$ °/oo ; a and n being constants dependent on the concrete

---

[1)] 10. Forschungskolloquium des Deutschen Ausschusses für Stahlbeton, März 1979, Forschungsberichte; J. Seeberger: *Strukturveränderungen von Beton bei Beanspruchung durch erhöhte Temperaturen.*

type, the age of the concrete and the storage conditions before the test and the levels of the force and the temperature.
t represents the number of days after the loading. When taking into account the same shrinkage curve for all heat exposed specimens, the experiments have yielded the following values for a and n:

Series	Temperature	Force	a	n
10	20°C	9,2 N/mm^2	0,054	0,270
14	200°C	9,2 N/mm^2	0,129	0,262
16	200°C	18,4 N/mm^2	0,133	0,373

The number of specimens and parameters studied is too small to be able to deduce any general laws. The following tendencies can be drawn at the present time from the existing results. "a" increases with the temperature, but is independent of the force. "n" increases with the force, but is independent of the temperature. Creep cannot be considered as being proportional to the force.

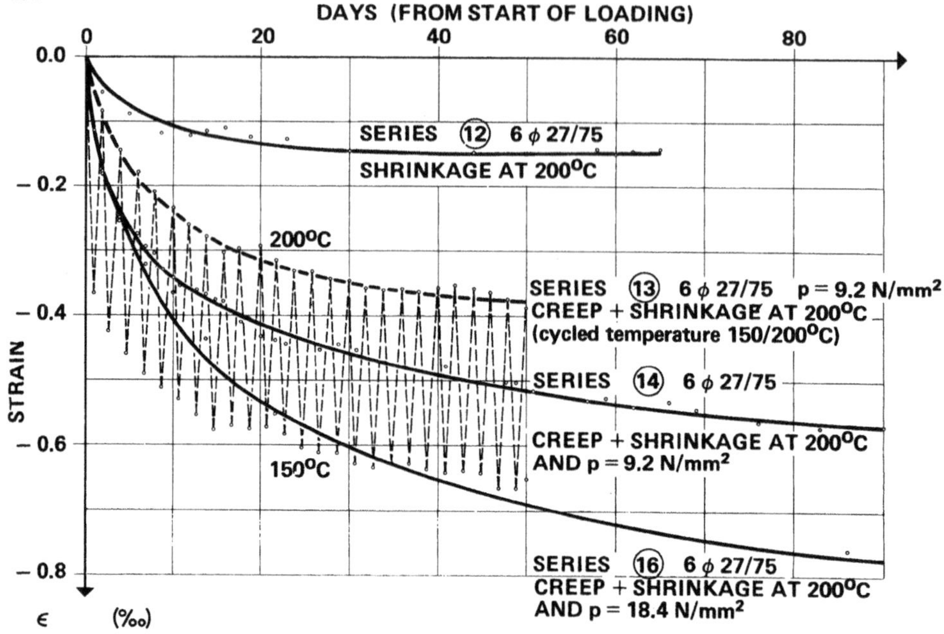

Fig. 7. CREEP AND SHRINKAGE AT ELEVATED TEMPERATURE

(SPECIMENS : HOLLOW CYLINDERS φ 27 / 75 cm)

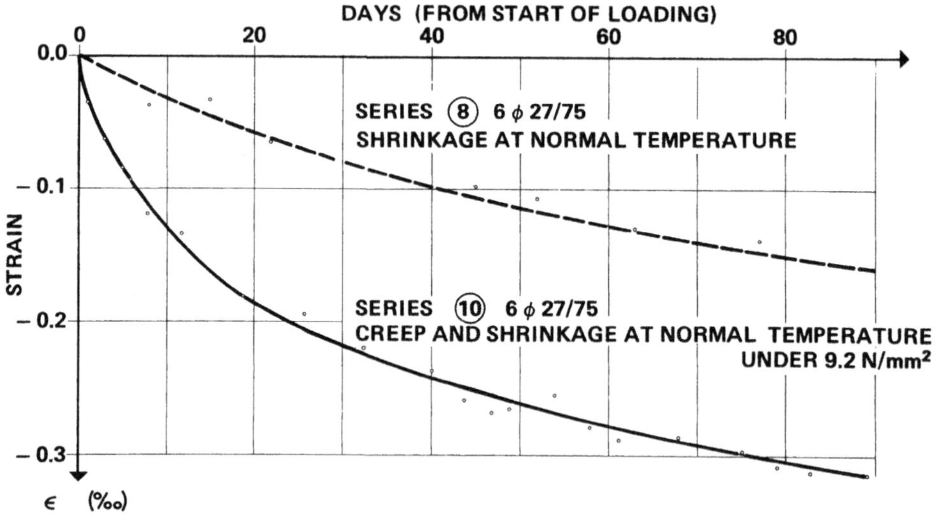

Fig. 8. CREEP AND SHRINKAGE AT NORMAL TEMPERATURE

(SPECIMENS : HOLLOW CYLINDERS φ 27 / 75 cm)

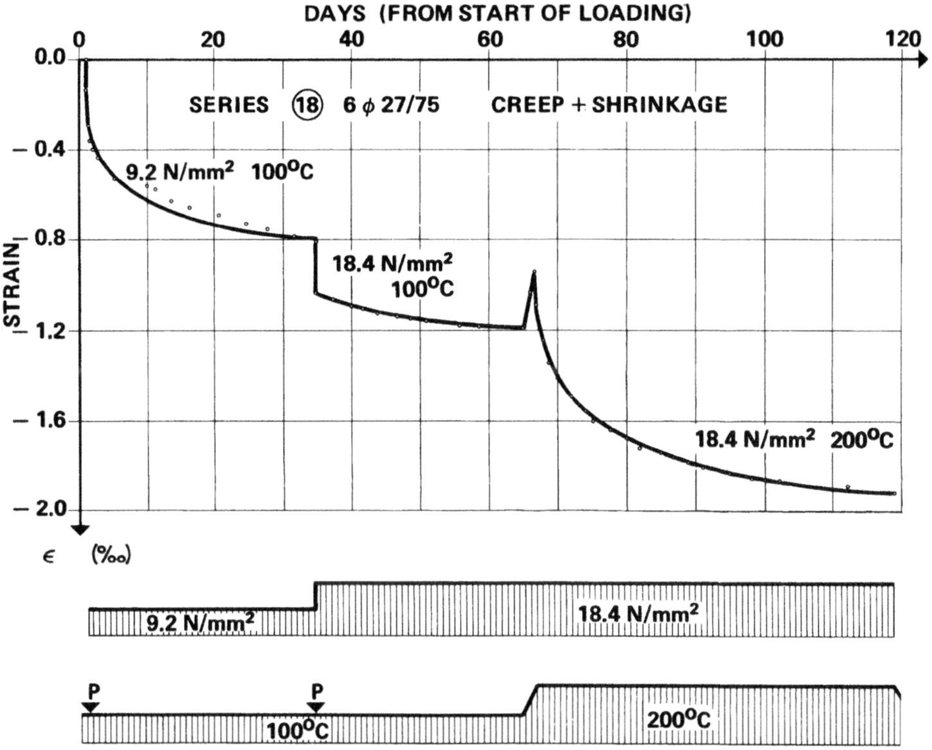

Fig. 9. CREEP AND SHRINKAGE AT DIFFERENT LEVELS OF TEMPERATURE AND LOAD

CREEP EFFECTS UNDER TRANSIENT TEMPERATURE CONDITIONS

U. Schneider, Dr.-Ing. habil.

Institut für Baustoffe, Massivbau und Brandschutz, Technische Universität Braunschweig, Beethovenstr. 52, 3300 Braunschweig, Fed. Rep. of Germany

ABSTRACT

The deformation characteristics of unsealed concrete specimens during heating have been studied in connection with fire research and a hypothetical core failure in a nuclear reactor. A substantial increase of creep rates was observed during the temperature increase in the concrete. It is not possible to compare the non-steady state creep with those creep deformations, which have been observed under isothermal creep conditions. Even if concrete is treated as a material without memory and the appropriate law of superposition is applied the deformations under transient (rapid heating) conditions cannot be described by creep laws being derived under isothermal conditions. The test results suggest a creep law, which is directly related to the concrete deformations being observed under transient temperature conditions.

Das Verformungsverhalten von nicht versiegelten Betonproben unter definierten Aufheizbedingungen wurde im Zusammenhang mit der Feuerforschung untersucht. Während der Aufheizung des Betons ist ein signifikanter Anstieg der Kriechgeschwindigkeiten zu beobachten. Es ist nicht möglich, diese bei veränderlichen Temperaturen gewonnenen Kriechverformungen mit solchen Kriechwerten zu vergleichen, die unter isothermen Bedingungen ermittelt werden. Die Versuche ergaben eine Kriechgleichung, die sich direkt an der unter instationären Temperaturen gewonnenen Kriechverformungen orientiert. In dem Bericht wird das Kriechen von Normalbetonen unterschiedlicher Zusammensetzung während der Aufheizung bis 1000 °C diskutiert.

INTRODUCTION

The creep of concrete at elevated temperatures has been studied many times. The investigations comprise mostly the temperature range from 20 to 150 °C i.e. the service temperatures of concrete structures in nuclear power plants. Those creep studies have been summarized by Geymayer [1] and Eibl et al. [2].

Creep at temperatures exceeding 150 °C has been studied in connexion with fire research. A literature survey in this field has been started by the RILEM COMMITTEE 44-PHT and will be published in two years time [3]. The available data in this field were also briefly summarized in [4]. In the following creep at elevated temperatures and under isothermal conditions will be discussed first. Afterwards the nonsteady state creep will be defined and dealt with.

STEADY STATE AND NONSTEADY STATE CREEP

Steady state creep tests are normally performed with sealed or unsealed specimens. In those tests the concrete specimens are heated up to their individual test temperatures and after stabilizing them for a certain period of time the loads are applied. The spontaneous elastic deformations and the subsequent steady state creep deformations are measured.

Test results of this kind have been reported for normal and lightweight concretes up to 450 °C [4]. The specimens wre oven-dried (105 °C) or standard cured (20 °C/65 % r.h.) before the tests. They were not sealed during the test. FIG. 1 shows some of the test results. The tests indicate a significant increase of creep at temperatures exceeding 300 °C. The creep data were evaluated in a regression analysis. It turned out that the load dependent deformations ε as a first approximation can be described by the well known formula

$$\varepsilon = \sigma \cdot (1 + \varphi(t,T))/E(T) \qquad (1)$$

with T = temperature, t = time, σ = stress and E = modulus of elasticity. As a creep law simple power functions

$$\varphi = a(\sigma,T) \cdot t^{b(T)} \qquad (2)$$

could be employed in the whole temperature range. The coefficient of regression usually exceeded 0,95. The results indicated, that for creep at high temperatures power functions generally are to be prefered. Similar conclusions have been drawn by [5] in connexion with creep at room temperatures. A detailed description of the tests can be found in the literature [4] and [6].

Nonsteady state creep effects are normally observed during heating of loaded concrete specimens. During the heating period considerable deformations may occur. Such deformations are due to the decrease of the modulus of elasticity with increasing temperatures superimposed by primary secondary and possibly tertiary creep effects.

The phenomenon of nonsteady state creep of transient creep was mentioned in the literature first by Hansen [7] in 1966. Later the effect was discussed by Fischer [8], Illston [9], Kordina et al. [10] and Thelandersson [11]. Unfortunately there is no correspondence in the terminology of transient creep of nonsteady state creep effects. For practical reasons we define transient creep (nonsteady state creep) as a sum of deformations, which occur only during heating of loaded concrete specimens (e.g. fire test conditions) and which are not due to thermal expansion or changes of the modulus of elasticity. This definiton is different compared to the term transitional thermal creep, which is used in [9] and [11]. The reasons for this distinction are discussed in the next chapter.

FIG. 2 shows the results of creep experiments with similar concrete specimens being loaded during the heating phase (load level 30 per cent of the ultimate load at 20 °C). The hatched area indicates the area where transient creep deformations occur. The experiments underscore the importance of the transient creep effects. The total deformations derived with the specimens being heated under load after 120 hours are significantly higher than the creep deformations according to the classical isothermal creep test (comp. FIG. 1 and 2). The shapes of the creep curves indicate a different concrete behaviour in both cases too.

THEORETICAL EVALUATION OF TRANSIENT CREEP EFFECTS

The stress dependent deformations of concrete under arbitrary temperature conditions are defined by

$$d\varepsilon = \frac{\partial \varepsilon}{\partial \sigma} \cdot d\sigma + \frac{\partial \varepsilon}{\partial t} \cdot dt + \frac{\partial \varepsilon}{\partial T} \cdot dT \qquad (3)$$

In the case of pure heating according to

$$dT = g'(t) \cdot dt, \quad g'(t) \geq 0 \qquad (4)$$

it has been shown in [4], that equ. (3) and (4) may be combined to

$$d\varepsilon = \left(\frac{\partial \varepsilon}{\partial \sigma}\right)_t \cdot d\sigma + \left(\frac{\partial \varepsilon}{\partial t}\right)_\sigma \cdot dt \qquad (5)$$

Furthermore the relations

$$\left(\frac{\partial\varepsilon}{\partial\sigma}\right)_t = \frac{1}{E(t(T))} \tag{6}$$

and

$$\left(\frac{\partial\varepsilon}{\partial t}\right)_\sigma = \sigma \cdot \left(\frac{\partial J(\bar{\sigma},t)}{\partial t}\right)_\sigma \tag{7}$$

hold.

Equ. (5) can be integrated with regard to equ. (4), (6) and (7). For a constant stress level $\bar{\sigma}$ the following equation has been derived:

$$\varepsilon = \bar{\sigma}/E(T) + \bar{\sigma} \cdot J(\bar{\sigma},T) \tag{8}$$

The total deformation of a heated concrete specimen under constant load may be described by

$$\varepsilon_{tot} = \varepsilon_{th}(T) - \bar{\sigma} \cdot (1 + \varphi(\bar{\sigma},T))/E(T) \tag{9}$$

in which the thermal expansion $\varepsilon_{th}(T)$ is taken as being positive. The φ-function being generally unknown. It can be determined if ε_{tot}, E(T) and $\varepsilon_{th}(T)$ are being measured in separate tests.

EXPERIMENTAL RESULTS

As far as E(T) and $\varepsilon_{th}(T)$ concerns there are many results available. FIG. 3 and 4 summarize some of our own tests with normal and lightweight concretes. The individual concrete mixes were made with Portland cement and different aggregates. It turned out that the thermal expansion being significantly influenced by the type of aggregate. Similar observations have been made with the modulus of elasticity. Certainly the influence of the type of aggregate being less pronounced as in connexion with ε_{th}.

The φ-functions of six different concretes were experimentally verified in a comprehensive test program. FIG. 5 and 6 show a few of the results, which were taken from [6]. Unsealed concrete specimens of 8 cm diameter and 30 cm length were used. The concrete mixes are given in TABLE 1.

From FIG. 5 and 6 it can be seen that there is no clear dependence between φ and $\bar{\sigma}$. In the range of load levels from 10 to 70 per cent approximately

$$\varphi(\bar{\sigma},T) = \varphi(T) \tag{10}$$

holds, which is one of the most surprising results of this investigations. Similar results have been obtained with

- four other types of concretes and
- different rates of heating between 0,5 and 4,0 K/min.

TABLE 1: Composition of the investigated concretes

Concrete Type	BSB 4	KS 3
cement PZ 35 F	350 kg/m^3	340 kg/m^3
calcareous aggr. 0 - 4 mm	-	830 kg/m^3
calcareous aggr. 4 - 8 mm	-	425 kg/m^3
calcareous aggr. 8 - 16 mm	-	570 kg/m^3
total water	175 kg/m^3	185 kg/m^3
siliceous sand 0 - 2 mm	726 kg/m^3	-
basaltic aggr. 8 - 11 mm	1014 kg/m^3	-
basaltic aggr. 11 - 16 mm	185 kg/m^3	-

CONCLUSION

Transient creep effects have been studied in connexion with rapid heating of unsealed concrete specimens as regards fire exposure respectively. A simple creep formula has been developed taking into account elastic and nonelastic transient temperature effects. The formula has been successfully applied to different types of concrete. FIG. 7 and 8 show typical results of the total deformations of loaded concrete specimens being heated at a constant heating rate of 2 K/min. Between the theoretical and experimental results a good agreement was obtained up to 800 °C.

LITERATURE

1. Geymayer, H.G., "The Effect of Temperature on Creep of Concrete. A Lit. Review". Waterways Experiment Station Paper C-70-1, Vicksburg (1970).

2. Eibl, J., et al. "Spannbeton-Reaktordruckbehälter: Studie zur Erfassung spezieller Betoneigenschaften im Reaktordruckbehälterbau", DAfSt., Heft 237, W. Ernst & Sohn, Berlin (1974).

3. Schneider, U., et al. "Effect of Fire on Concrete", RILEM Committee 44-PHT, Draft: Paper I (1980).

4. Schneider, U., "Ein Beitrag zur Frage des Kriechens und der Relaxation von Beton unter hohen Temperaturen", Habilitationsschrift, TU Braunschweig (1979).

5. Wittmann, F., "Über den Zusammenhang von Kriechverformungen und Spannungsrelaxation des Betons", Beton u. Stahlbetonbau, Heft 3, S. 63 - 65 (1971).

6. Schneider, U., et al., "Hochtemperaturverhalten von Festbeton", Unveröffentlichter Arbeitsbericht B3 des SFB 148, TU Braunschweig (1980).

7. Hansen, T.C., et al., "Temperature change effect on the behaviour of cement paste, mortar and concrete under load", J. of ACI, Vol. 63, p. 489 - 504 (1966).

8. Fischer, F., "Über das Verhalten von Zementmörtel und Beton bei höheren Temperaturen", Inst. für Massivbau, Heft 14, TH Darmstadt (1967).

9. Illston, J.M., et al., "Characteristics and prediction of creep of a saturated mortar under variable temperature", Mag. of Concrete Res., Vol. 26, No. 88, p. 169 - 179 (1974).

10. Kordina, K., et al., "Brandverhalten von Bauteilen", Jahresbericht 1973/74 des SFB 148, TU Braunschweig (1974).

11. Thelandersson, S., "Mechanical Behaviour of Concrete under Torsional Loading at Transient High-Temperature Conditions", Inst. of Techn., Bulletin 46, Lund (1974).

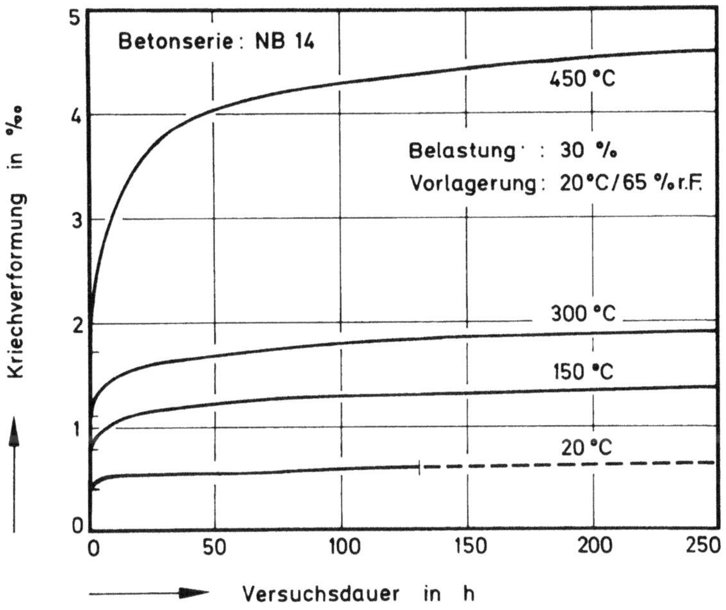

Fig. 1: Steady state creep of concrete at elevated temperatures

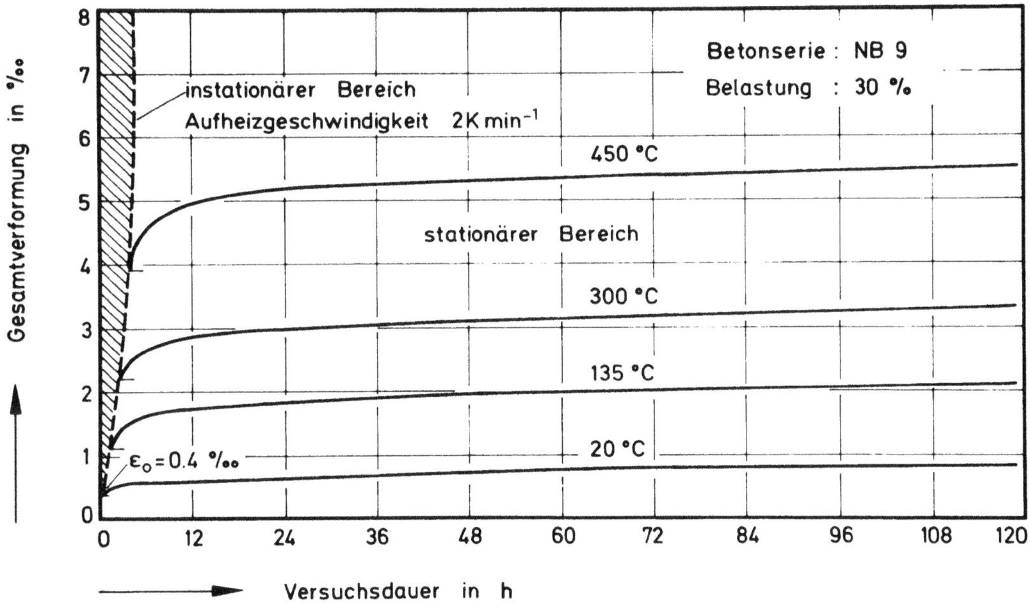

Fig. 2: Nonsteady state and steady state creep of concrete at elevated temperatures

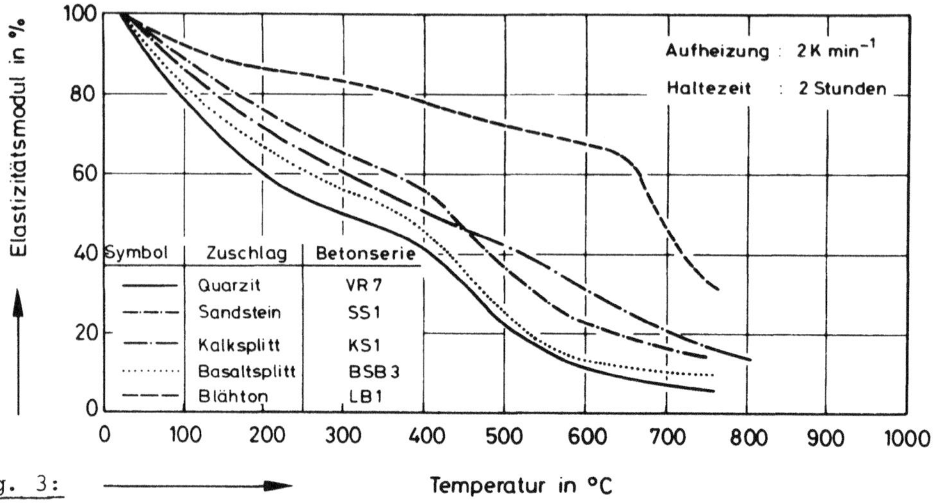

Fig. 3:

Modulus of elasticity of different concretes at elevated temperatures

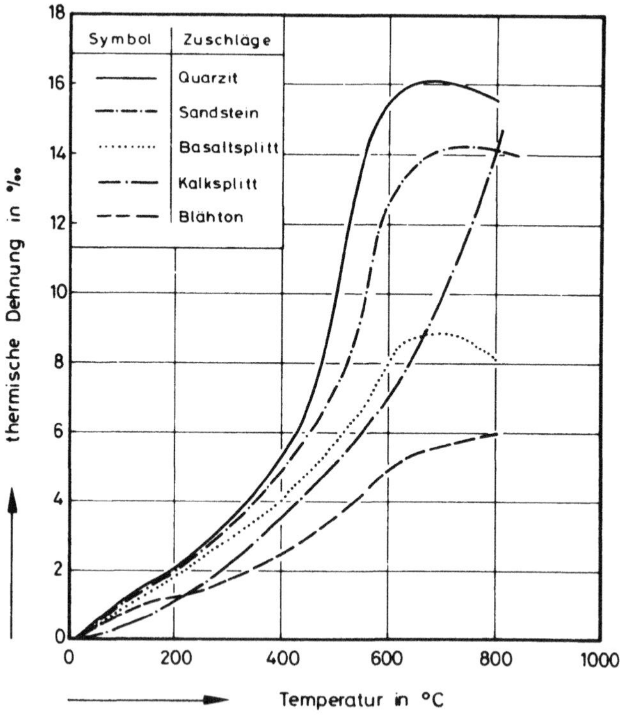

Fig. 4:

Thermal expansion of different concretes at elevated temperatures

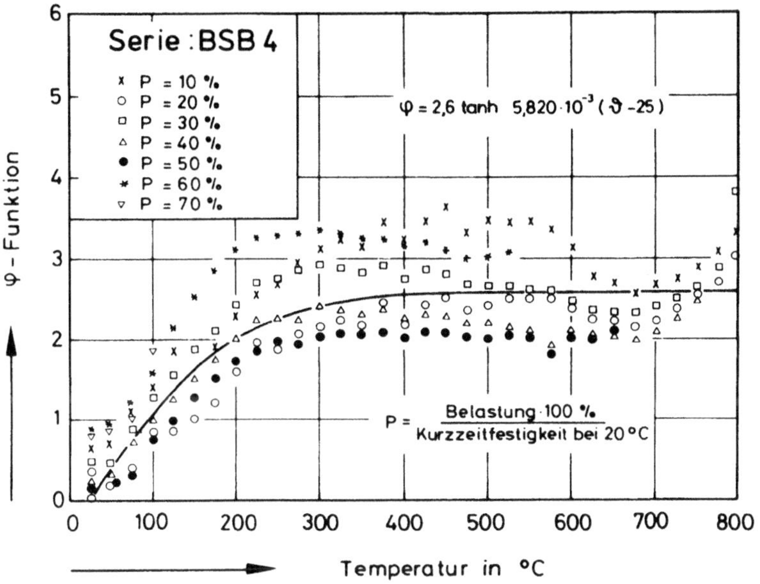

Fig. 5: Nonsteady state creep function φ(T) of a basaltic concrete

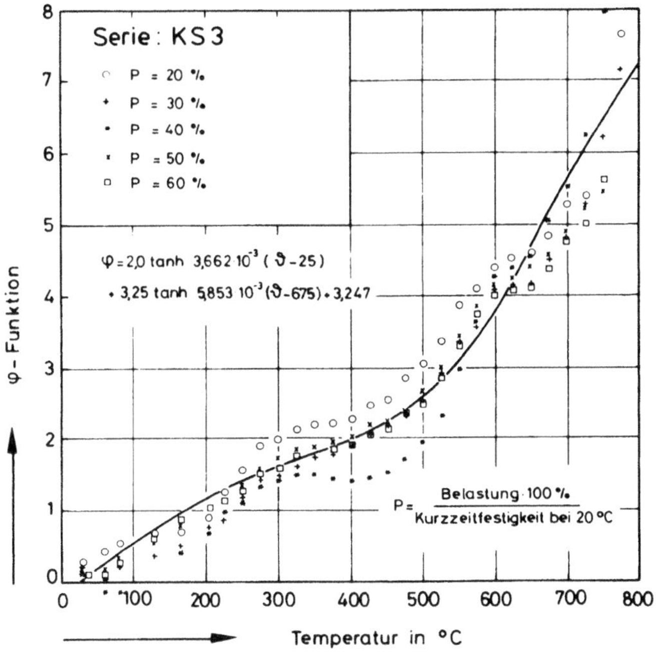

Fig. 6: Nonsteady state creep function φ(T) of a calcareous concrete

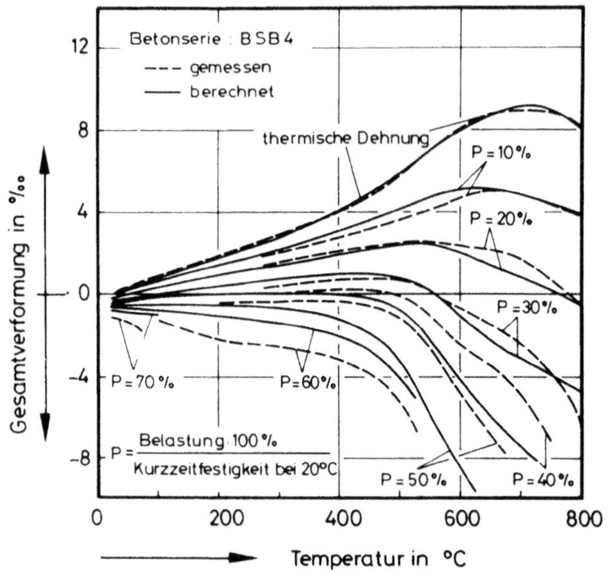

Fig. 7: Experimental and theoretical deformations of a basaltic concrete being heated under load

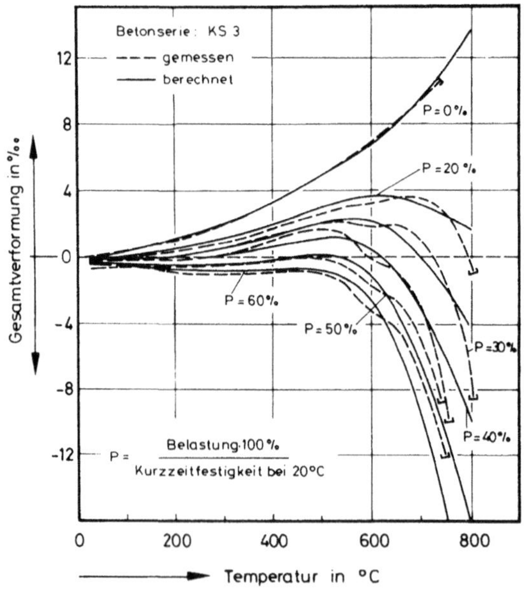

Fig. 8: Experimental and theoretical deformations of a calcareous concrete being heated under load

SHRINKAGE OF CONCRETE IN EXTREME CLIMATE

Klaus Altmann

Technische Universität Berlin
Institut für Baukonstruktionen und Festigkeit

ABSTRACT

Open air measurements made during testing in Jeddah, Saudi Arabia, were taken to demonstrate to what extent the usual stipulated values for limit shrinkage value and the chronological progress of shrinkage as set forth for concrete structures in moderate climates have to be modified for hot and dry regions. In the progress of testing, the influence of various curing and positioning conditions was simultaneously investigated in order to obtain data on intended reduction of the shrinkage deformations. The limit shrinkage values determined thereby for various concrete compositions were significantly greater, and were reached in shorter time, than can be calculated, for example, in formulas given in German standards.

ZUSAMMENFASSUNG

Mehrjährige Freiluftmessungen in Jeddah, Saudi Arabien, sollten zeigen, in welchem Umfang die in gemäßigtem Klima bei Betonkonstruktionen üblichen Vorgaben für das Endschwindmaß und den zeitlichen Ablauf das Schwindens in solchem trockenen, heißen Klima verändert werden müssen. Dabei wurde gleichzeitig der Einfluß verschiedener Nachbehandlungs- und Lagerungsbedingungen untersucht, um Hinweise für eine gezielte Verminderung der Schwindverformungen zu erhalten. Die für verschiedene Betonzusammensetzungen gefundenen Endschwindmaße sind deutlich höher und werden nach kürzerer Zeit erreicht als sie sich z.B. nach Deutschen Rechenvorschriften ermitteln lassen.

1 INTRODUCTION

Extensive construction performed in hot countries has raised the question as to how the particular climate there - characterized as it is by intensive drying-out of concrete parts - affects shrinkage in concrete. Final shrinkage values required in the design of concrete structures, as well as the residual shrinkage values - calculated in dependence on time and stipulated, for example, in German regulations [1] for moderate climates- of course do not apply in such extreme climates. The variables which influence shrinkage - the composition of the concret, humidity, thickness of the prefab units- have, to be sure, been investigated in a great number of studies as regards their qualitative effects (see [2] , for example). Calculation formulas [3] have also been worked out to determine the degree of shrinkage for a particular type of concrete, for particular of the above mentioned parameters, and for a given condition of dryness. The shrinkage deformation to be expected cannot, however, be sufficiently exactly calculated on the basis of the above, since the effect of particularly hot climates with unusually marked solar heating of the concrete has not to date been sufficiently taken into account by these formulas.

In addition to these important limit values for structural purposes, knowledge of shrinkage behavior is also especially significant during execution of construction in the prevention of cracks in the early concrete aging period. Such considerations were raised, for example, in the recent construction of the Jeddah Airport, for which both medium-grade pumped concrete as well as high-quality mixtures were used in great quantities for the construction of precast prestressed concrete units.

In collaboration with the firm HOCHTIEF AG from Essen, West Germany, shrinkage investigations were performed on these concrete types in order to show, in addition to the particular climate influences, the effects of various types of concrete curing [4] . The shrinkage test samples thereby exposed to the weather were subjected to drying conditions closely approximating those affecting normal wall and floor slabs.

2 TEST

The shrinkage tests were conducted on a closed-off part of the new airport site in Jeddah. At this test site, sun and wind were able to have without hindrance the desired effects on the specimes. For purposes of testing, two types of concrete were selected, from which differing shrinkage behaviour could be expected on the basis of their differing compositions (see Table 1 below) and characteristics.

Mix	P 3	F 1
Cement	Greek, Type 1	German, PZ 45 F
	380 kg/m^3	400 kg/m^3
Fine aggregate (0.25 mm)	169 kg/m^3	157 kg/m^3
Water-cement ratio	0.59	0.46
Evaporating water	140 $1/m^3$	100 $1/m^3$

Table 1: Composition of shrinkage test concrete

The first type above was a pumped concrete prepared with a soft consistency (P 3), which could be used for low strength requirements (approx. 25 MN/m^2) for all reinforced wall and floor slab units in architectural engineering work. The other type was a high-grade concrete (F 1) with a required early strength of more than 32 MN/m^2 after one day - a type of concrete which could be used in a prefabricated unit plant, say, for the production of prestressed floor slabs. A shorter hardening time for this type of concrete was able to be achieved by heat treatment at 80° C. The influence expected thereby on shrinkage behaviour was also investigated on concrete samples produced with such treatment.

In contrast to the usual shrinkage tests conducted in laboratories on prismatic samples, slab-shaped samples fitted to the precast units were used in the following sizes: 1.0 x 1.5 m and 12 cm thick for P 3 concrete, and 20 cm thick for F 1 concrete. Aluminium foil applied to the peripheral lateral surfaces ensured that drying out would take place only via the upper and lower surfaces of the samples. In this manner, conditions were created which corresponded to larger floor slabs.

Deformations in the concrete shrinkage samples were measured in each case by means of a dial gauge mounted along the longitudinal axis of the sample. The movable tracer point connected to the dial pointer hung on a thin steel wire which ran through a PVC tube embedded into the longitudinal axis of the specimen, and which was firmly anchored on the opposite samole side. This wire was permanently tensioned by a spring, with the result that any length change in the sample was registered by the dial gauge. Temperature expansions and contractions in the concrete had practically no effect on the measured results, since the mean concrete temperature was always approximately equal to that of the steel wire. Any length changes which nevertheless did occur as a result of temperature were able to be compensated for in the measured results by means of simultaneously measured concrete temperatures.

As is well known, the conditions of drying in concrete have vonsiderable influence on the shrinkage behavior. Variations in such conditons can result, with floor slabs for example, from the effects of sun and shade. This was taken into account in the shrinkage tests for P 3 concrete by means of corresponding mounting configurations. In addition to the A samples mounted in the open sun, the B samples made from the same concrete were therefore shaded by tables in such a manner that the sun could not reach them, although there was no restriction on the movement of air. All samples made of precast unit concrete were subjected only to this shaded configuration.

P 3 concrete was also tested for the effects of curing on shrinkage behavior. For this purpose, the A specimenes placed in the sun were sprayed with a film-forming material (Curing HE) after the fresh concrete had dried. This film had only a weak blocking action against water evaporation and was quickly dissipated by the action of the weather. Other of the P 3 shrinkage specimens were covered with moist jute over the upper side during the first 5 days. These specimens were positioned in the sun, as were the A samples.

3 CLIMATIC CONDITIONS IN JEDDAH

Weather observations carried out in parallel with the shrinkage measurements revealed the maximum daily air temperatures to be 30 °C in winter and between 35 °C and 42 °C in summer. Upon occasion the temperatures reached values over 50 °C. The humidity throughout the year lies at about the saturation point at night. It falls to values between 50 and 20 % realtive humidity after sunrise, as a rule. Throughout several weeks of the summer, however, it does not fall during the day below 70 %.

A typical course of air temperature and air humidity, each during one day in February and August, shows fig. 1. Therefore the air cools off at nearly the same temperature towards night, independent of the season.

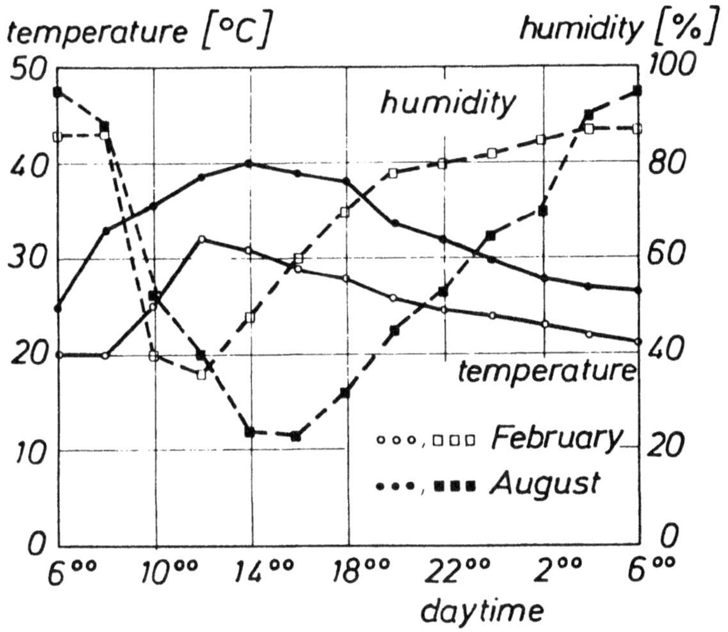

FIG 1: Air temperature and humidity in Jeddah

Throughout the whole year the sun shines for the fairly constant period of 10 to 12 hours per day. During the conduct of testing, it rained only briefly on one day. The wind constantly blows during the day, with a velocity of approx. 6 m/s and especially contributes to the dryring-out effect; as is the case with prefab units, the wind was allowed to blow over the samples without hindrance.

The long period of solar radiation caused the temperatures on the surface of the unshaded concrete specimens to rise to 10 - 15 K above the respective air temperature. In this way, temperatures of approx. 60 °C can occur on the upper soncrete surfaces in summer - a situation which, in conjunction with the strong wind, results in considerably more intensive drying out of the concrete than is ever possible at temperatures 10 K lower in moderate climates. The effect is reinforced by the long periods of sunshine and the lack of rain.

4 RESULTS OF TESTING

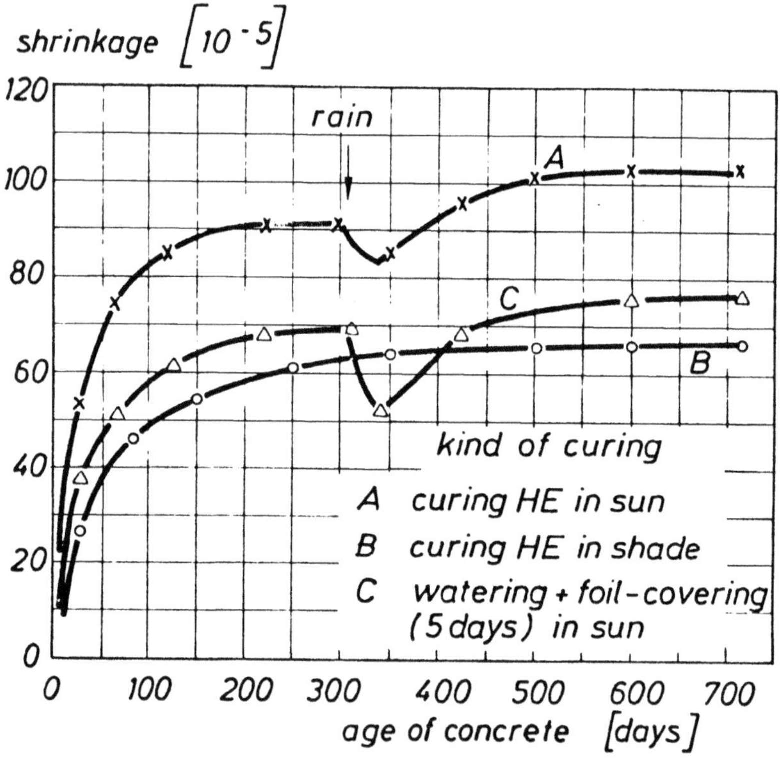

FIG 2: Shrinkage of pumping-concrete in Jeddah

Fig. 2 shows the results of the shrinkage measurements conducted on P 3 concrete on the construction site in Jeddah. In accordance therewith, the A samples - placed in the sun and protected on the upper side by the above-described film - began to shrink considerably faster than the B samples - provided with the same curing procedure, but mounted in the shade. The A specimens reached a shrinkage value almost one-third higher, up to the one occurring rainfall, after about 300 days. At this point in time, shrinkage had almost ceased in the A and C specimens, mounted as they were in the sun; i. e., the concrete was almost completely dried out. In contrast, the B samples mounted in the shade could not yet dry out to such a degree - owing to the lower temperature - as the progressively decreasing shrinkage demonstrated.

The wetting by rain induced a temporary swelling of the A and C specimens, in which case the A samples swelled only half as much as the others. This was apparently due to a lesser absorption of water, in turn the result of the still effective surface film. The penetrated water then dried out, allowing the samples to shrink again beyond the degree already achieved. As explanation for this phenomenon could be considered the fact that the capillary pore spaces had reduced in size by the penetrating growth of cement gel during the first 300 days. The rainwater which had penetrated through then caused the higher shrinkage rates after it evaporated. By virtue of their table covering, the B specimens were also protected from the rainwater, with the result that they showed continuous drying, as the shrinkage curve demonstrates.

All the samples reached the practical limit values after two years of testing, as given by the following respective values for the A, B and C samples: 103×10^{-5} 67×10^{-5}, and 77×10^{-5}, which represent the expected limits. These values - as well as the chronological progression of the phenomenon - clearly demonstrate the great influence exerted by the sun, and show how much more effective a moist cover can be than a sprayed-on curing film.

The shrinkage values measured for the samples of precast unit concrete have been entered in fig. 3. In accordance therewith, the samples not subjected to heat treatment show at the beginning a more rapid increase in shrinkage than do the specimens heated to 80 °C - samples which in this latter case cannot release their moisture so rapidly owing to their more finely configured pore spaces formed already during the process of accelerated hydation.

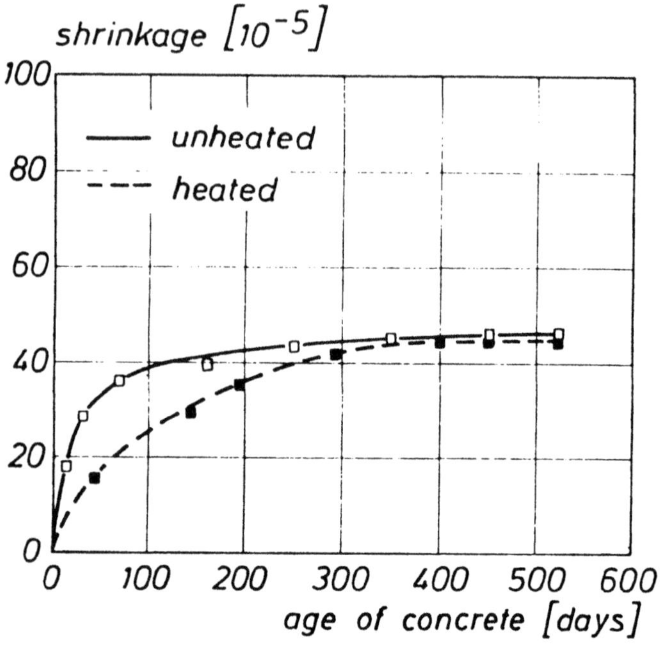

FIG 3: Shrinkage of precast-concrete in Jeddah

These curves are comparable to those for the C specimens of P 3 concrete, as a result of their also being mounted in the shade. The limit value achieved here as early as 500 days after test begin, and independent of the heat treatment, was approx. 47 x 10^{-5}, which was about 10 x 10^{-5} lower than the comparable value for P 3 concrete. This can especially be traced to the greater amount of evaporating water used, i. e., 40 l/m^3 more.

5 SHRINKAGE IN JEDDAH AND IN ACCORDANCE WITH GERMAN STANDARDS

As one aspect of the test objectives, the attempt was made to determine in what manner, say shrinkage deformation as set forth in German standards for concrete construction could be applied to the climatic conditions in Jeddah. According to the latest German regulations [1], limit shrinkage values ar stipulated for prefabricated units to be installed in dry air, and for low-slump concrete, of 37 x 10 and, for high-slump concrete, of 58 x 10 . The chronological behavior of the shrinkage is represented as a dependent of the effective concrete age (t), by the following expression:

$$k_s = \frac{E_s(t)}{E_{s\infty}}$$

With the aid of the pregiven numerical value for E_s (the limit shrinkage value), the determining part of the shrinkage value can in accordance therewith be calculated in advance for a particular point in time t. The effective concrete age can ve determined by the following equation:

$$t = \frac{T_i + 10\ °C}{30\ °C} \cdot \Delta t_i \quad \text{(in days)}$$

If, for example, a mean daily temerature for each day of the year is assumed of T_i = 30 °C, then the effective concrete age is 1.33 times the actual concrete age of Δt_i. In this manner, the more rapid concrete hardening as a result of temperature, is taken into account.

Shrinkage is also determined by the thickness of the concrete unit d_{ef}, which in turn determines the value of the factor k_s which is achieved. This thickness of the concrete building unit can be calculated by the following equation:

$$d_{ef} = \frac{2.A}{U} \quad \text{(in cm)},$$

where

A = the surface area of the concrete cross section
U = the development of the area of concrete cross section exposed to drying

Under consideration of the climatic conditions prevailling in Jeddah, and of the effective thickness of the test specimens in this case, fig. 4 contrasts the chronological progress of concrete shrinkage as given in DIN with the actually observed values measured from the three specimen groups A, B, and C.

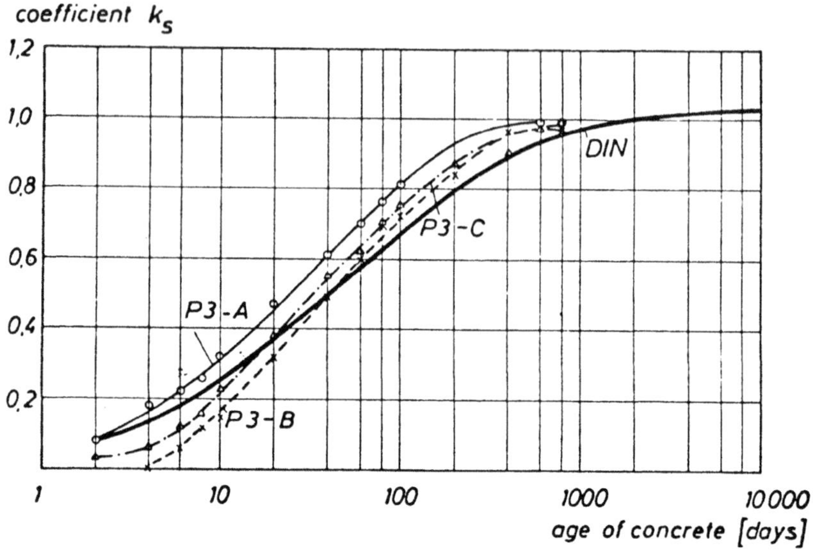

FIG 4: Time-Coeffient of shrinkage from concrete in Jeddah compared to DIN [1]

In accordance therewith, the shrinkage of the A test specimens progressed more rapidly than in DIN, although a protective film had been sprayed on the concrete. Only the moist covering and the shade protection cause the progress to sonsiderably decrease in rate. Aside from this decrease at the beginning, all the test samples progress toward their limit shrinkage values more rapidly than would be expected in accordance with DIN. The intensive drying-out of concrete in Jeddah apparently results therefore in not only greater limit shrinkage values, but also in a modified chronological behavior - a phenomenon which cannot be adequately described here.

REFERENCE

1. DIN 4227, Teil 1, Spannbeton, Dezember (1979)
2. O. Graf, "Die Eigenschaften des Betons" Springer Verlag Berlin Göttingen Heidelberg, 2 Auflage (1960)
3. H. K. Hilsdorf, "Austrocknung und Schwinden von Beton", Berichte aus Forschung und Praxis (1969)
4. U. Stoffregen, "Konstruktion und Ausführung des Hauptempfangsgebäudes für den Flughafen, Saudi Arabien", Vorträge auf dem Betontag 1979, Deutscher Beton-Verein E. V.

DRYING SHRINKAGE OF SYMMETRICALLY REINFORCED CONCRETE AND SOME ESTIMATES OF THE CREEP OF CONCRETE IN TENSION

D.W. Hobbs

Materials Research Department, Cement and Concrete Association, U.K.

ABSTRACT

Results are given of an investigation into the influence of symmetrically placed reinforcement upon the drying shrinkage behaviour of a range of concretes. It is shown that the observed shrinkage is influenced both by the level of reinforcement and the Young's modulus of the concrete, and that the average tensile stress induced in the concrete can be higher than the short term tensile strength of comparable plain concrete specimens. Estimates of the magnitude of the specific creep showed no obvious dependence upon curing period, mix proportions or induced stress level.

Les résultats se présentent d'une investigation sur l'influence de l'armature symétriquement disposée sur le retrait au séchage d'une étendue des bétons. On montre que le retrait observé est influencé par la quantité de l'armature et par la module d'élasticité du béton, et que la contrainte moyenne de tráction inducéé dans le béton peut être plus grande que celle de courte durée des éprouvettes comparables en béton non armé. On estime que la grandeur du fluage spécifique ne montre aucune dépendance évidente de la duration de la cure, des proportions de mélange ou du niveau de contrainte inducée par le retrait.

INTRODUCTION

In structures the drying shrinkage of concrete is restrained by adjacent members, by reinforcement and by base friction. A number of investigations have been made of the effect of restraint upon shrinkage and upon the stresses induced in concrete by the restraint (1-16). In much of the experimental work measurements have been made of the tensile strength and failure strain of 'concrete' prisms containing a symmetrically placed reinforcing bar which is not 'bonded' to the concrete along its central section (5-13). In these tests it was found that the tensile strength was between 1 and 3 MPa and that the failure strain was about 100×10^{-6}. Relatively few data have been published on the shrinkage behaviour of concrete containing continuously bonded reinforcement (1-3, 14-16). Several of these workers (1,3,14,16) found that cracking was not induced at stresses in excess of the short term tensile strength of comparable plain concrete specimens.

Because of the few data published that deal with the effect of continuously bonded reinforcement upon shrinkage it was considered that further work was justified. The experimental work described here is concerned with the influence of symmetrically placed reinforcement upon the shrinkage behaviour of a range of concretes. Attempts are made to estimate the creep in the concrete produced by the induced tensile stresses.

EXPERIMENTAL PROCEDURE

Specimens, approximately 40 x 40 x 250 mm in size, were prepared from a number of different batches as described in Table 1. The reinforced specimens contained a symmetrically placed reinforcing bar which was either plain or threaded. The cement was an OPC and the aggregate was Thames Valley flint gravel and sand, which was used air dried, extra water being added to allow for aggregate absorption.

The specimens were demoulded at 24 hours and cured at 20°C in either a moist curing room or in a small quantity of water. After curing for the periods specified in Table 1 the specimens were each wrapped in a polythene jacket, a 25 mm diameter disc was fixed to one end of the specimen to enable it to stand vertically without support and a fine invar screw adjuster was fixed through the jacket onto the upper end of the specimen. The specimens were then set up in measuring rigs (see Figure 1) and the axial shrinkage measured with inductive displacement transducers with a linear range of 250 µm. The demodulated transducer outputs were recorded immediately after the jacks were removed from each specimen and then the demodulated transducer outputs were scanned sequentially with the outputs being recorded on paper tape whilst the specimens were drying in an atmosphere at 65 ± 1% RH and 20 ± 1°C. The measuring

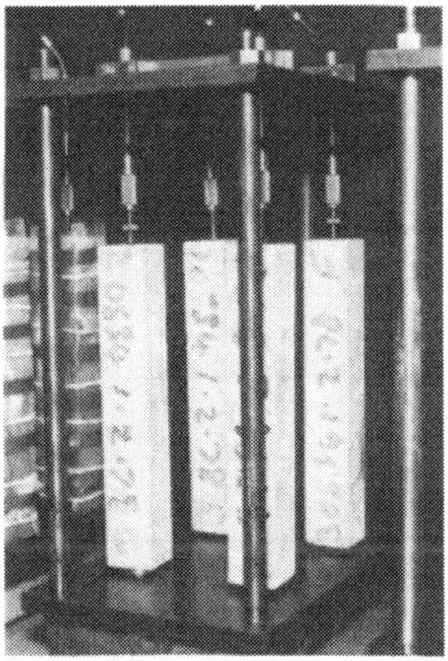

FIG 1 Four shrinkage specimens set up in their measuring rig

apparatus was sensitive to a shrinkage movement of 1 micro-strain.

Measurements were also made of the expansion of both plain and reinforced specimens while they were cured under water for 112 days. The observed expansions were found to be insignificant compared with the subsequent drying shrinkage.

RESULTS

Some of the results obtained are shown plotted in Figures 2 to 4. Each of the points plotted is the mean result obtained from 3 specimens. In Figures 2 and 3, shrinkage is shown plotted against drying period for specimens cured for 1,7,28 and 112 days and in Figure 4 the ratio of the restrained shrinkage to the unrestrained shrinkage after a drying period of 180 days is shown plotted against the level of reinforcement, ρ. These figures show that the shrinkage is not solely dependent upon the level of reinforcement but is also dependent upon curing period and the Young's modulus of the concrete.

The average tensile stress induced in the concrete by the reinforcement is given on the right hand side of Figure 2. The average stresses, supported by the concretes, are in a number of instances, higher than those supported in short-term uniaxial tests. The highest induced tensile stresses were 3.5 MPa for specimens with a water/cement ratio of 0.59 and 4.7 MPa for specimens with a water/cement ratio of 0.35.

TABLE 1 Mix Details, Curing Period and Reinforcement Levels

w/c	a/c	pfa/c	Curing period (days)	Young's modulus E_c (GPa)	Compressive strength f_{cu} (MPa)	Reinforcement levels (%)
0.35	2.55	-	1 7 28 112	16 28 35 37	11.5 48.9 66.2 75.8	0,4,11
0.35	3.3	-	7	31	52.9	0,0.84,1.6,2.2⁺ 3.9, 4.9⁺, 6.8
0.40	4.75	-	7	32	44.9	0,0.84,1.6,2.2⁺ 3.9, 4.9⁺, 6.8
0.59	5.49	-	1 7 28 112	11 17 30 31	3.7 20.2 32.8 37.8	0, 4, 11
0.59	6.99	-	7	20	25.3	0,0.84, 1.6, 2.2⁺ 3.9, 4.9⁺, 6.8
0.59	5.49	0.20	1 7 28 112	- - - -	- - 38.0 -	0, 4

⁺Plain bars

The results obtained on specimens containing 20% of pulverised fuel ash by weight of the cement are shown plotted in Figure 5. A comparison of this figure with Figure 3 plus an examination of Figure 4 shows that the effect of reinforcement on the shrinkage of both concretes is similar except that the shrinkage of the concrete containing pfa is generally higher.

DISCUSSION

Data Normalized with Respect to $\rho E_s/E_c(1-\rho)$

The shrinkage of concrete containing symmetrically placed reinforcement may be represented by (17)

$$S = \frac{s}{1 + \frac{\rho E_s}{(1-\rho)E_c}(1 + \gamma\psi)} \quad . \quad . \quad . \quad . \quad . \quad (1)$$

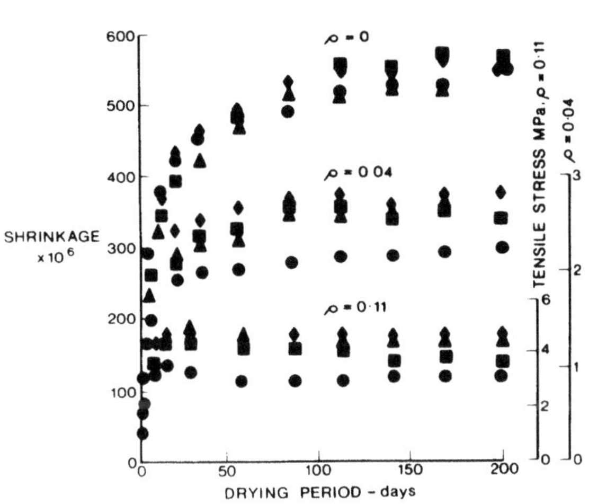

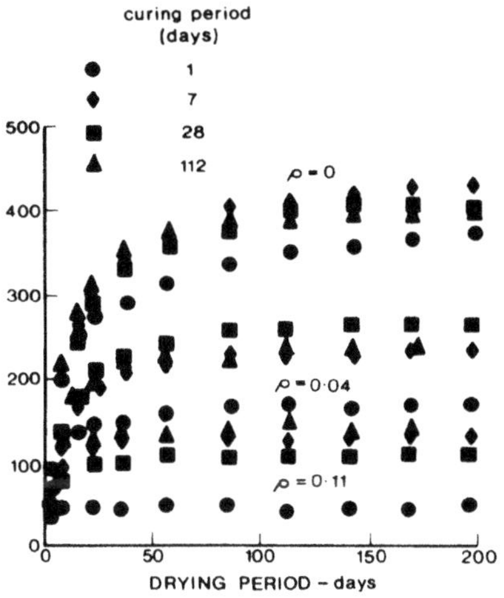

FIG 2 Dependence of shrinkage upon curing and drying period. w/c 0.35, a/c 2.55

FIG 3 Dependence of shrinkage upon curing and drying period. w/c 0.59, a/c 5.49

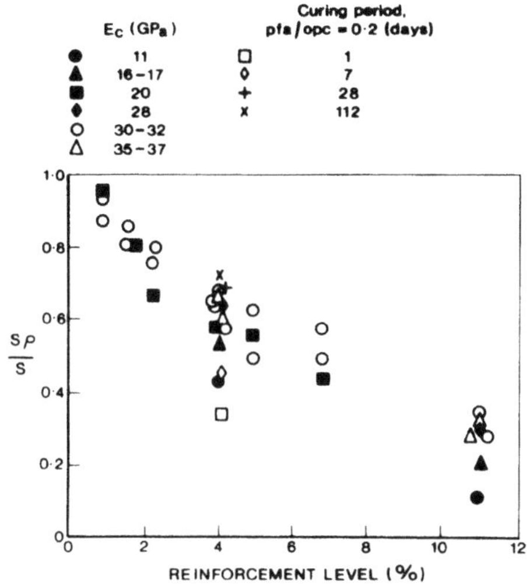

FIG 4 Dependence of shrinkage upon level of reinforcement. Drying period 6 months

where E_c is the Young's modulus of the concrete at the beginning of the drying period, E_s the Young's modulus of the steel, η allows

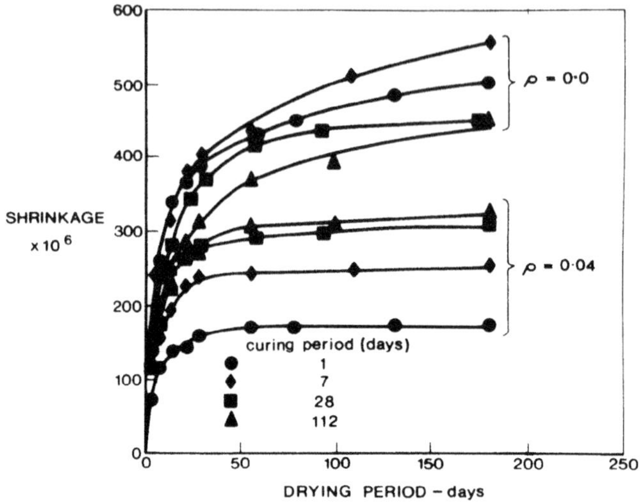

FIG 5 Dependence of shrinkage upon curing and drying period. w/c 0.59, a/c 5.49, pfa/c 0.20

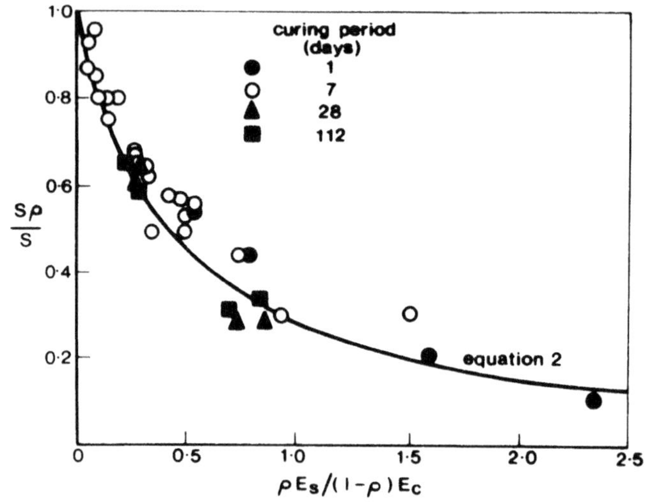

FIG 6 Variation of shrinkage at 6 months after drying began with $\rho E_s/(1-\rho)E_c$

for the change in creep properties of concrete with ageing, and ψ represents a term to 'describe' the creep properties of concrete when shrinkage begins.

In Figure 6 the ratio of the restrained shrinkage, S_ρ , to the unrestrained shrinkage, S, 180 days after the start of drying is shown plotted against $\rho E_s/E_c\ (1-\rho)$. The data may be expressed approximately by

$$S_\rho = \frac{S}{1+\dfrac{(1+k)\rho E_s}{(1-\rho)\,E_c}} \quad . \ . \ . \ . \ . \ . \ . \ . \quad (2)$$

where the 'creep coefficient' k is 1.5. Equation (2) also fits the shrinkage data obtained after other periods of drying. The best-fitting values for k after other drying periods are 0.2 after 7 days' drying, 0.4 after 14 days' drying, 0.8 after 28 days' drying and 1.0 after 56 days' drying. It therefore follows that the influence of the reinforcement upon the shrinkage of concrete may be described quite adequately by a simple effective modulus approach.

Creep in Tension

Although concrete is normally designed to support compressive stress there are a number of situations where concrete is subject to tensile stress. Despite this, few attempts have been made to determine the creep of concrete in tension, possibly because it is so difficult to induce a uniform axial tensile strain in concrete. In most of the published work dealing with tensile creep (see Table 2) the concretes were subject to tensile stresses ranging from 0.6 to 1.3 MPa so if the concrete is drying under load the resulting shrinkage may be several times larger than the creep and this may produce a large error in the calculated creep.

The current work has the advantage that the tensile stresses in the concrete were relatively high but has the added disadvantage that the tensile stresses were variable. The specific creep for the concrete tested may be estimated approximately as follows. Let the tensile stress, due to restraint acting in the concrete specimens be given by

$$\sigma_t = k_1 \ \ln {}^1/_4 \left[1 + (t-T) \right] \quad \text{(compare Fig.7 with Fig. 2 and 3)} \ . \ . \ . \quad (3)$$

where k_1 is a function of the modulus of the concrete and level of reinforcement, t is the age of the concrete and T is the age of the concrete when drying (loading) commences.

TABLE 2 Magnitude of Specific Creep in Micro-strain/MPa for Concretes Subject to Tensile Stresses Ranging from 0.6 to 1.3 MPa

Reference	Age of Loading (days)	Drying Conditions	Specific creep		
			Time under load (days)		
			7	28	180
Illston (18)	7	18°C, 63%RH	20	45	80
	48	18°C, 63%RH	5	10	36
Mamillan (19)	28	21°C, 50%RH	150	220	275
Davis, Davis and Brown (20)	28	27°C, sealed	15	35	45
U.S. Bureau of Reclamation (21)	28	sealed	10	20	-
	90	sealed	5	10	-
Glanville and Thomas (22)	30	Humidity varied	25	60	110

Let the specific creep under a uniform tensile stress be given by

$$\epsilon_{c,sp} = k_2 \quad \ln\left[1 + (t-T)\right] \quad . \quad . \quad . \quad . \quad . \quad . \quad (4)$$

If we now ignore changes in Young's modulus after drying commences then an estimate of the creep may be obtained from the rate of creep method as described in Concrete Technical Paper No 101 (23) namely

$$\epsilon_c = \int_T^{t_T} \sigma_t \frac{\partial \epsilon_{c,sp}}{\partial t} dt \quad . \quad . \quad . \quad . \quad . \quad . \quad . \quad (5)$$

Thus after a drying period, t

$$\epsilon_c = \frac{4}{5} k_1 \; k_2 \quad \ln {}^5\!/\!_4 \left[1 + (t-T)\right] \quad . \quad . \quad . \quad . \quad (6)$$

which from equations 3 and 4 may be written

$$\epsilon_c = \frac{4}{5} \; \sigma_t \; \epsilon_{c,sp} \quad . \quad . \quad . \quad . \quad . \quad . \quad . \quad . \quad (7)$$

Thus the specific creep is given by

$$\epsilon_{c,sp} = \frac{5}{4\sigma_t}\left[S - S_\rho - \frac{\sigma_t}{E_c}\right] . \quad . \quad . \quad (8)$$

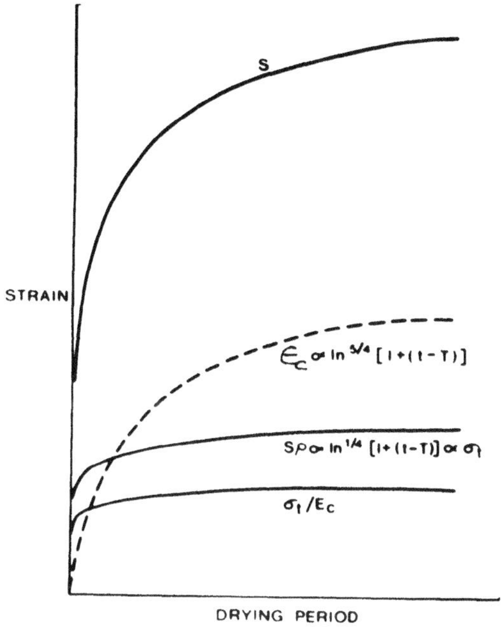

FIG 7 Diagrammatic representation of shrinkage and creep behaviour

where σ_t may be written

$$\sigma_t \;=\; S_\rho \; E_s \rho/(1-\rho) \; . \; . \; . \; . \; . \; . \; . \; . \qquad (9)$$

The magnitude of $\epsilon_{c,sp}$ determined from equations 8 and 9 for the various concretes tested after drying periods of 28 and 180 days are given in Table 3. In the case of the specimens cured for 7 days the $\epsilon_{c,sp}$ values are the mean values for the specimens tested with the three lower levels of reinforcement and the three higher levels of reinforcement.

There is no obvious dependence of the magnitude of $\epsilon_{c,sp}$ upon curing period, mix proportions or induced stress level. The magnitude of $\epsilon_{c,sp}$ for the specimens cured for 7, 28 or 112 days and dried for 180 days ranged from 25 to 100 microstrain/MPa and is therefore similar to that observed by other workers on concretes subject to lower tensile stresses (see Table 2). Thus it may be concluded either that no damage has been induced in the specimens by the sustained stress or alternatively that the level of damage is proportional to the

TABLE 3 Specific Creep (micro-strain/MPa)

Concrete w/c	Concrete a/c	Curing period (days)	Specific creep* After 28 days drying	Specific creep* After 180 days drying
0.35	2.55	1 7 28 112	20(2.2); 45(3.2) 15(2.8); 25(4.7) 25(2.5); 40(4.2) 20(2.5); 30(4.6)	60(2.25); 100(3.0) 40(2.8); 70(4.4) 55(2.6); 100(3.7) 40(2.7); 70(4.5)
0.35	3.3	7	15(1.1); 30(2.9)	35(1.5); 55(3.4)
0.40	4.75	7	20(1.0); 15(2.8)	35(1.3); 25(3.5)
0.59	5.49	1 7 28 112	70(1.3) 30(1.7); 10(3.3) 30(1.9); 70(2.6) 50(1.8); 35(3.5)	150(1.3) 70(1.8); 30(3.6) 50(2.0); 70(3.3) 70(1.7); 50(3.7)
0.59	6.99	7	55(0.8); 40(2.0)	65(0.9); 55(2.2)
Mean values		7,28 and 112	30	55

*The figures in brackets are the average tensile stresses, in MPa, supported by the concretes.

induced stress level even though the induced stress level was in some instances possibly higher than the tensile strength of comparable plain concrete specimens. The average values for the specific creep in tension for specimens cured for 7, 28 and 112 days and dried for 28 and 180 days was 30 and 55 micro-strain/MPa respectively.

CONCLUSIONS

For the particular concretes tested, it was found that

1. The shrinkage of symmetrically reinforced specimens, S_ρ, was given approximately by

$$S_\rho = \frac{S}{1 + \dfrac{(1+k)\,\rho\, E_s}{(1-\rho)\, E_c}}$$

where k is the 'creep coefficient'.

2. The values of the 'creep coefficient' after drying periods of 7, 14, 28, 56 and 180 days were 0.2, 0.4, 0.8, 1.0 and 1.5 respectively.

3. The average tensile stresses induced in the concrete by the reinforcement were, in a number of instances, higher than those supported by concrete in short-term uniaxial tests.

4. The magnitude of the specific creep in tension for specimens cured for 7, 28 and 112 days and dried for 28 and 180 days was estimated to be 30 and 55 micro-strain/MPa respectively.

REFERENCES

1. Goldbeck, A.T. and F.H. Jackson. "Expansion and contraction of concrete and concrete roads". U.S. Department of Agriculture Bulletin No 532 (1917).

2. Matsumoto, T. "Study of the effect of moisture content upon the expansion and contraction of plain and reinforced concrete". Engineering Experiment Station, University of Illinois Bulletin No 126 (1921).

3. Glanville, W.H. "Studies in reinforced concrete. Part II Shrinkage stresses". Department of Scientific and Industrial Research, London. Building Research Technical Paper No 11 (1930).

4. Vetter, C.P. "Stresses in reinforced concrete due to volume changes". Transactions of the American Society of Civil Engineers 98 (1933) 1039-1079.

5. Carlson, R.W. "Attempts to measure the cracking tendency of concrete". Journal of the American Concrete Institute, Proceedings 36 (1940) 533-537.

6. Pickett, G. "Shrinkage stresses in concrete, Part 1" Journal of the American Concrete Institute, Proceedings 42 (1946) 165-204.

7. Pickett, G. "Shrinkage stresses in concrete, Part 2 - Application of theory presented in Part 1 to experimental results". Journal of the American Concrete Institute, Proceedings 42 (1946) 361-398.

8. Beyer, F.R. "Stresses in reinforced concrete due to volume change". Journal of the American Concrete Institute, Proceedings 45 (1949) 713-722.

9. Blakey, F.A and R.K. Lewis "Effect of cyclic wetting and drying of restrained mortar specimens". Highett, Australia, Commonwealth Scientific and Industrial Research Organisation, Division of Building Research. Report No. C2, 4-1 (1958).

10. Blakey, F.A. "Influence of water/cement ratio on mortar in which shrinkage is restrained". Journal of the American Concrete Institute, Proceedings 55(1958) 591-604.

11. Blakey, F.A. and F.D. Beresford. "Cracking of concrete". Constructional Review 32 (1959) 24-28.

12. Blakey, F.A. and R K. Lewis. "Effect of sand characteristics on the cracking of small bars of mortar in which shrinkage is restrained". Civil Engineering and Public Works Review 55 (1960) 398.

13. Blakey, F.A. and R.K. Lewis. "The effect of curing on the cracking of small bars of mortar in which shrinkage is restrained". Civil Engineering and Public Works Review 56 (1961) 473, 475, 477, 639, 641, 643.

14. Campbell-Allen, D. and J.G. Holford. "Stress and cracking in concrete due to shrinkage". Institution of Engineers, Australia, Civil Engineering Transactions (1970) 33-39.

15. Pfeifer, D.W "Reinforced lightweight concrete columns". Journal of the Structural Division. Proceedings of the American Society of Civil Engineers 95 (1969) 57-82.

16. Richart, F.E. and R.L. Brown. "An investigation of reinforced concrete columns". Engineering Experiment Station University of Illinois Bulletin No 267 (1934).

17. Neville, A M. Creep of Concrete : Plain, Reinforced and Prestressed (North-Holland Publishing Company, Amsterdam, 1970).

18. Illston, J.M. "The creep of concrete under uniaxial tension". Magazine of Concrete Research 17 (1965) 77-84.

19. Mamillan, M. "A study of the creep of concrete" RILEM Bulletin No 3 (1959) 15-31.

20. Davis, R.E., H.E. Davis and E.H. Brown. "Plastic flow and volume changes of concrete". Proceedings ASTM 37 (1937) 317-330.

21. U.S. Bureau of Reclamation. "A ten-year study of creep properties of concrete". Concrete Laboratory Report No. SP-38 (1953).

22. Glanville, W H. and F.G. Thomas. "Studies in reinforced concrete - IV Further investigations on creep or flow of concrete under load". London, Building Research Technical Paper No 21 (1939).

23. The Concrete Society. "The creep of structural concrete". Technical Paper No 101 (The Concrete Society, Terminal House, Grosvenor Gardens, London SW1W OAJ, 1974).

DEFORMATIONS OF TRANSVERSE SECTIONS AND DEFLECTION OF FLEXED BEAMS UNDER SUSTAINED LOADING

M. KAVYRCHINE and Z.T. NUNG Service d'Etude des Structures
C.E.B.T.P. - Sai Rémy-lès-Chevreuse

ABSTRACT

A method is proposed for the interpretation of creep on flexed beams. Local deformations measured on longitudinal lines on the beam are taken into account to calculate a deflection curve Cf ; it is supposed that the hypothesis of linearity can be adopted for this calculation, in spite of the big dispersion of the measures. A theoretical value Cth of the deflection is obtained from the consideration of creep laws applied to longitudinal fibers, laws which are derived from tests on samples under uniaxial compression. Comparison between Cf and the measured deflection Cm gives an information on the structural behaviour of the beam, between Cth and Cf and Cm allows to test the validity of applying to flexion creep laws of uniaxial compression.

Four examples are given but more experimental data is needed to be able to have general conclusions.

Many creep tests on flexed beams have been achieved in different laboratories and in our Service d'Etude des Structures du CEBTP. The results show a big dispersion, and it is not easy to find a link between the strains measured in long duration tests in compression on samples and deflections or local deformations measured in tests on flexed beams.

In fact, if we measure the deformations on the faces of the beams, we have a big dispersion around a supposed linear distribution (see FIG.7), we measure both shrinkage and creep effects and it is difficult to trace a direct relation between the deformation of so called "longitudinal fibers" and the strains of a sample under the same history of loading.

On the other hand, deflections are more easy to measure, they are independant of shrinkage if the latter is homogeneously distributed within the beam, they show less variations during time than local measures. We propose then, for the interpretation of flexural tests, to compare three curves :

1. the variations of deflection with time (Cm)
2. the variations of deflection calculated on the basis of measures made on longitudinal fibers on the beam (in our case on the surface of the beam) (Cf)
3. the variations of deflection calculated on the basis of the theoretical values of deformation of longitudinal fibers (based on tests on uniaxially loaded samples) (Cth).

To determine curve (2), we consider the measures made on each line of extensometric bases, and we calculate the most probable linear distribution, (regression straight line). We can then calculate the rotations of the sections of the beam, and hence the deflection Cf.

Curve (1) is directly obtained from the measures of the deflection Cm.

If Cf and Cm are close one to the other, the measure of local deformations give a good information of the structural behaviour of the beam.

If Cth is close to Cf and Cm we have a good correspondance between the creep laws of uniaxially loaded samples and the creep laws in flexion.

In our case we have not always tests on samples having the same transverse section as the beams, and our calculations are based on semi-empirical curves which gave good results, compared with uniaxial tests made in the same testing center, by ourselves or other authors. The beams we are considering have a section of 200 x 300 mm.

We used the following formulas for representation of creep strain
a) after an increase of compressive strain equal to ε_i

$$\varepsilon(t_o t) = \varepsilon_i \cdot 0{,}94\left[0{,}4 + 3 \cdot \frac{100}{t_o + 100}\right] \frac{\sqrt{t - t_o}}{4\sqrt{10} + \sqrt{t - t_o}}$$

where to is the age (in days) at loading
b) after a decrease of compressive strain :

$$\varepsilon(t_o + t) = \varepsilon_i \frac{\sqrt{\log(t_1 - t_o + 1)}}{4}\left[1 - \frac{1}{\sqrt{1 + t - t_1}}\right]$$

when t1 is the age at unloading.

We have had some difficulties in adding steps of loading, so we tried to made logical assumptions on the effect of these steps.

We shall take four examples in which a longitudinal compression was maintained (no tensile stresses, or very low tensile stresses, to avoid cracking) ; the flexural moment was constant between two supports (with cantilevers at both ends) the calculation of Cf is then made on the central part assuming that all rotations of sections are equal. Measures of deflection and of local deformation on the lateral and upper faces of the beams were made ; the measured values of deformation are the mean value of measures made on lines on the same level, and on several parallel lines on the upper and lower face. A measured value is the mean value of twelve measures on lateral faces (eleven + one) and of thirty three on each upper or lower face (see FIG.1). The extensometric bases are made with glued balls.

The given examples differ by the procedures of testing. We consider in each case a limit flexural moment, ML which creates a longitudinal stress at the upper face equal to 7.5 MPa (tension) and at the lower face of 7.5 MPa (compression). The initial prestress gives a compression of 7.5 MPa supposed uniform.

First example (EY1)
The story of the loading is the following for the first beam (called EY1) in function of age t (in days) after casting
t = 30 days prestress
t = 31 days flexural moment ML
t = 282 days flexural moment reduced from ML to zero

When can them calculate the strains sustained in the whole section during the test - if we neglect, in a first step, the loss of prestress, we have, for instance, for the upper face
t = 30 = 0 and then = 7.5 MPa
t = 31 = 7.5 MPa and then = 0
t = 282 = 0 and then = 7.5 MPa

Second example (EY3)
t = 29,75 prestress
t = 30 applied moment ML
t = 282 moment reduced from ML to zero

Third example (EY1 HG 4)
The first beam has been submitted, at age t = 560 days to a loading at ML and unloaded from ML to zero at age t = 673 days

Fourth example (EY9)
The beam was submitted to the following programme of loading :

t =	29,75	prestress
t =	30	Flexural moment 0,7 ML
t =	114	ML
t =	198	1,30 ML
t =	282	zero

The flexural moment was maintened constant between each variation of moment. Tensile strains were induced at the upper face of the concrete, but no cracking was apparent.

We calculated the variation of stress during time at the upper and the lower fibers of the sections, and also in the center of it, using the above quoted formulas. We suppose then that, if the beams would be formed of separate fibers, the length of which can variate independantly, we should have a linear distribution between the center and each face. But the laws of variation of strain after an increase of compressive strain, and after a decrease of this strain, are not the same, and distributions above and under the center are not on the same line. We suppose then that, due to links between the fibers, the transverse section remains plane, which means that an internal redistribution of longitudinal stresses and strains exists. This can be obtained by adding a complementary stress distribution, the sum of which is equal to zero, so to have a linear distribution of stresses. We can then calculate the rotation of each section and then the deflection in the center of the beam which we call Cth (curve 3)..

Our calculation necessitates some complementary studies especially on the following points :
- the prestress is not exactly axial (and the compression not uniform)
- the loss of prestress has been taken into account in a global manner
- the effect of transverse deformation on the behaviour of longitudinal fibers has to be checked.

The curves (1Cm) (2 Cf) (3 Cm) are given for our different examples on figures 2,3,4,5.

The creep laws that we adopted give small differences with linearity, as shown on table 6. It is practically not necessary, with our examples, to consider laws of different forms for creep under increasing or decreasing loads.

We cannot arrive for the moment at a conclusion on the validity in each case of the formulas we have taken for creep representation : some big differences appear, and we need more statistical data that we shall try to obtain from other tests.

In our example of a beam submitted to several successive loadings (EY9) the final deflection was close to the theoretical one, when one loading only gives result lower than the theory. A succession of loadings give, in our case a final value higher than the simple addition of the effect of each loading.

Concerning the structural behaviour of the beam, we note that the curves Cf and Cm are generally close ; the local deformations present always a big dispersion around linearity (See FIG. 7) but the comparison of curves (1 Cm) (2 Cf) shows that the hypothesis of linearity can be adopted for the calculation of the deflection from the local longitudinal deformations.

List of figures

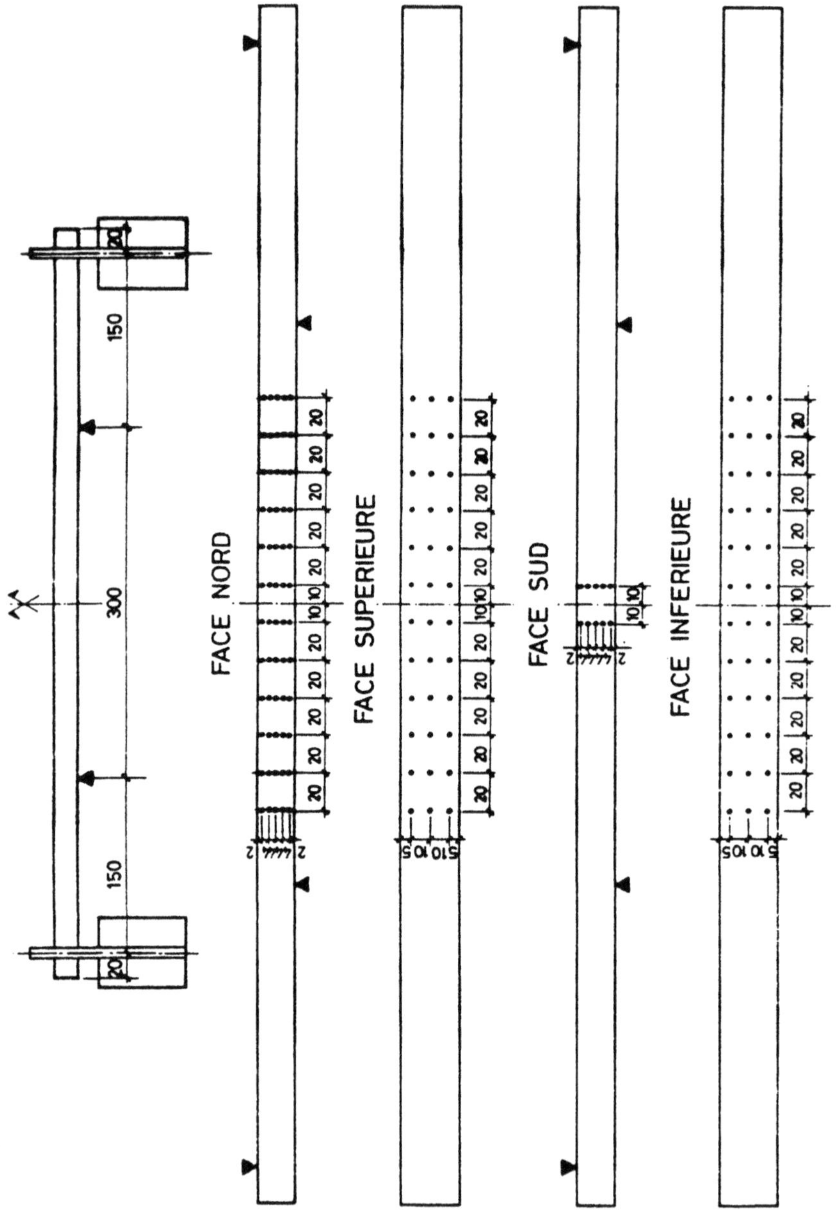
FACE NORD
FACE SUPERIEURE
FACE SUD
FACE INFERIEURE
150
300
150

Figure 1

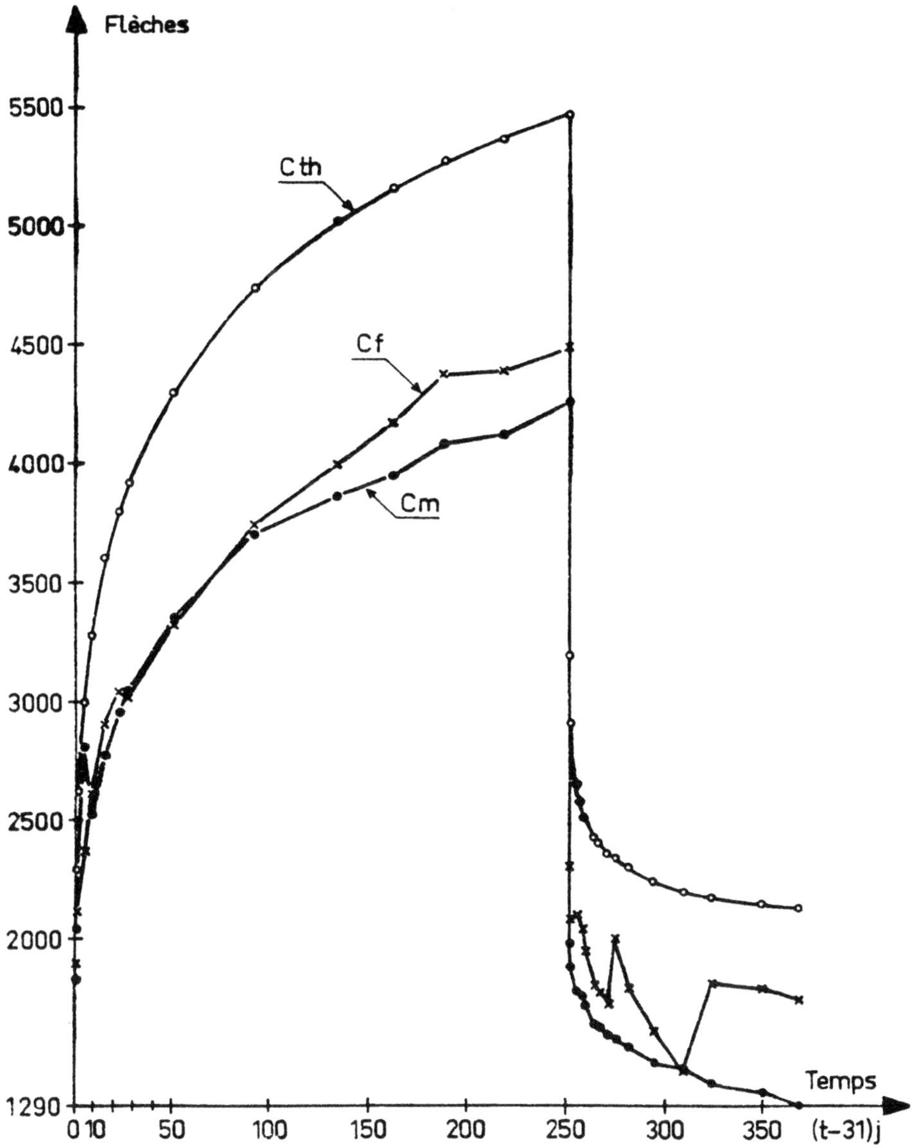
Flèches
5500
5000
4500
4000
3500
3000
2500
2000
1290
Cth
Cf
Cm
Temps
0 10
50
100
150
200
250
300
350
(t-31)j

Figure 2.

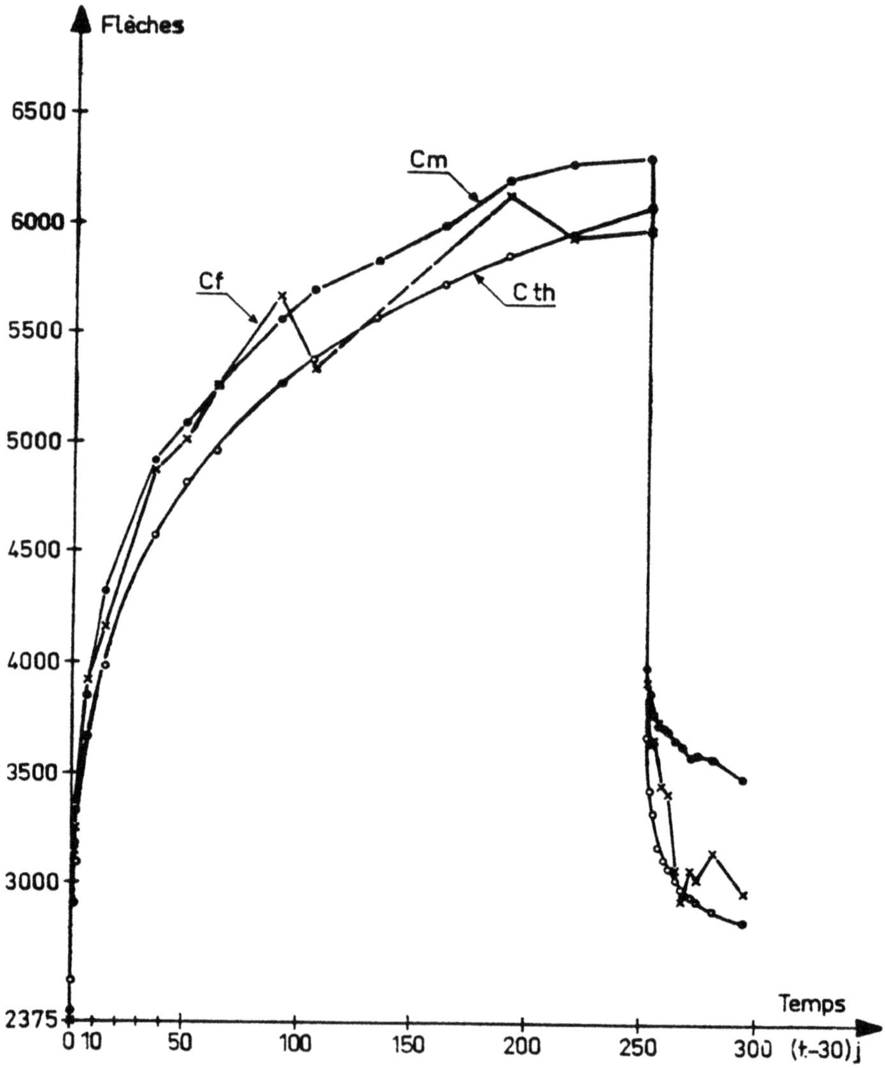
Flèches
6500
6000
5500
5000
4500
4000
3500
3000
2375
Cm
Cf
C th
Temps
0 10
50
100
150
200
250
300 (t-30)j

Figure 3.

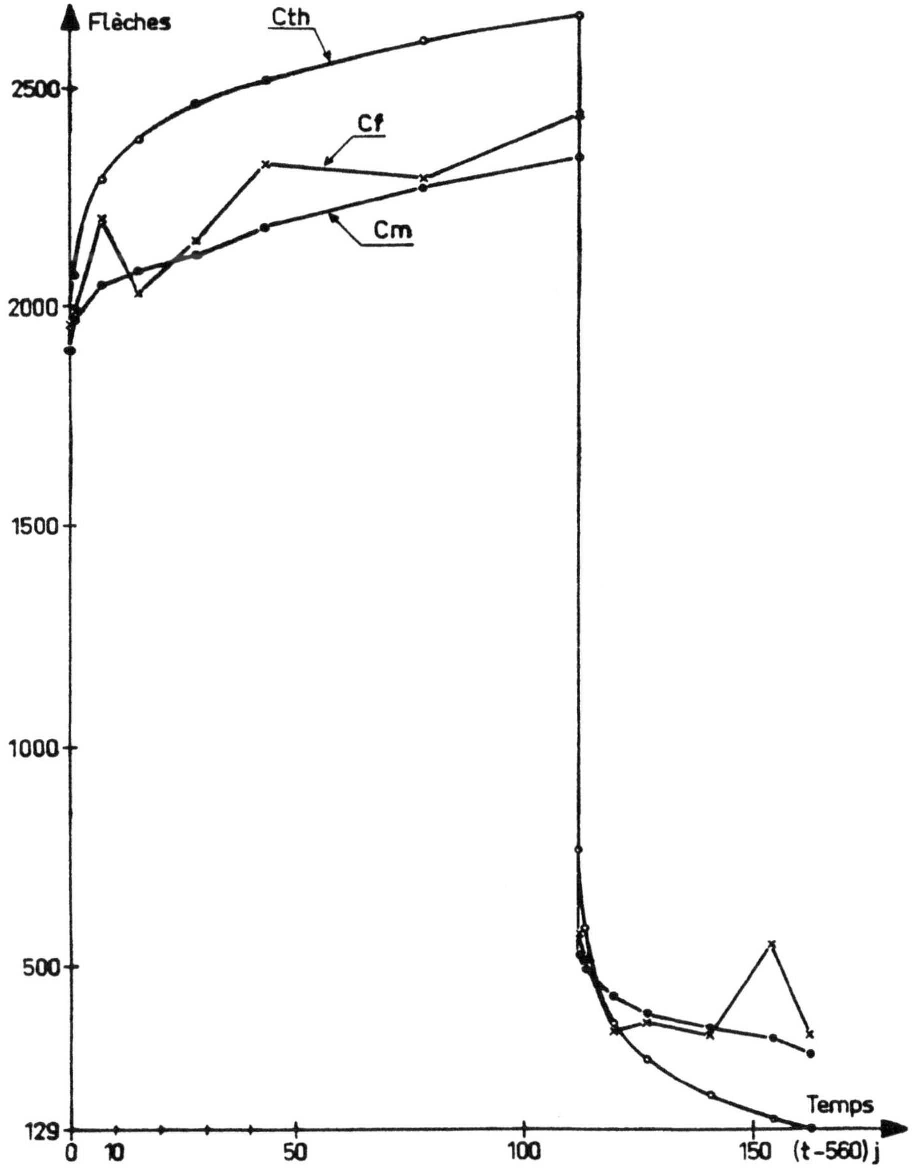
Flèches
Cth
Cf
Cm
2500
2000
1500
1000
500
129
0
10
50
100
150
(t-560) j
Temps

Figure 4.

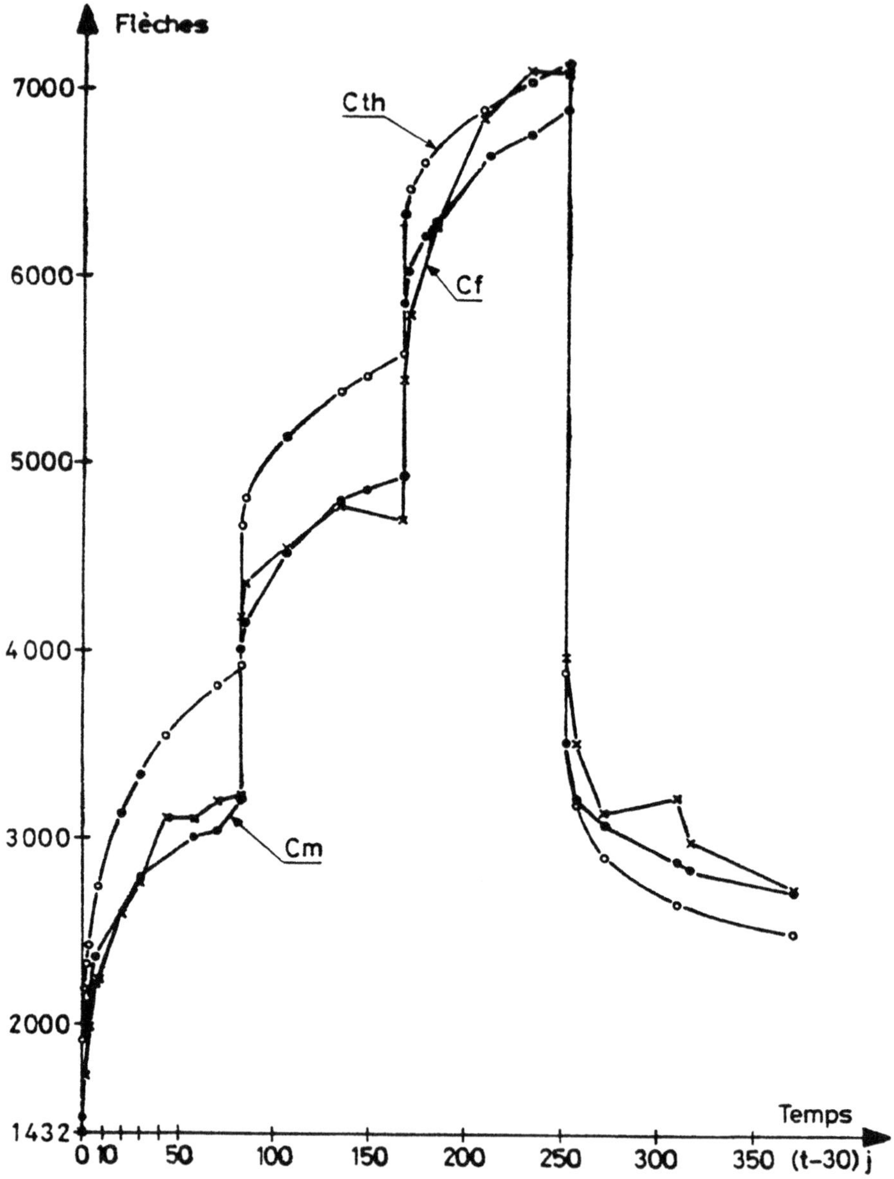

Flèches
7000
6000
5000
4000
3000
2000
1432
Cth
Cf
Cm
Temps
0
10
50
100
150
200
250
300
350
(t-30) j

Figure 5.

FIGURE 6 - TABLE OF LOCAL DEFORMATION (IN 10^{-6}) OF LONGITUDINAL FIBERS OF CENTRAL SECTION OF BEAM EY1 at 400 DAYS

Level of fiber / deformations	+ 100mm upper face	+ 80 mm	+ 40 mm	0 (middle)	- 40 mm	- 80 mm	- 100 mm (lower face)
Linear regression on theory	- 243	-277	-344	-410	-477	-544	- 577
Theoretical (creep + shrinkage)	- 240	-275	-346	-416	-479	-542	- 574
Measures	- 231	-412	-424	-434	-494	-551	- 648
Linear regression on measures	- 302	-333	-395	-456	-518	-580	- 611

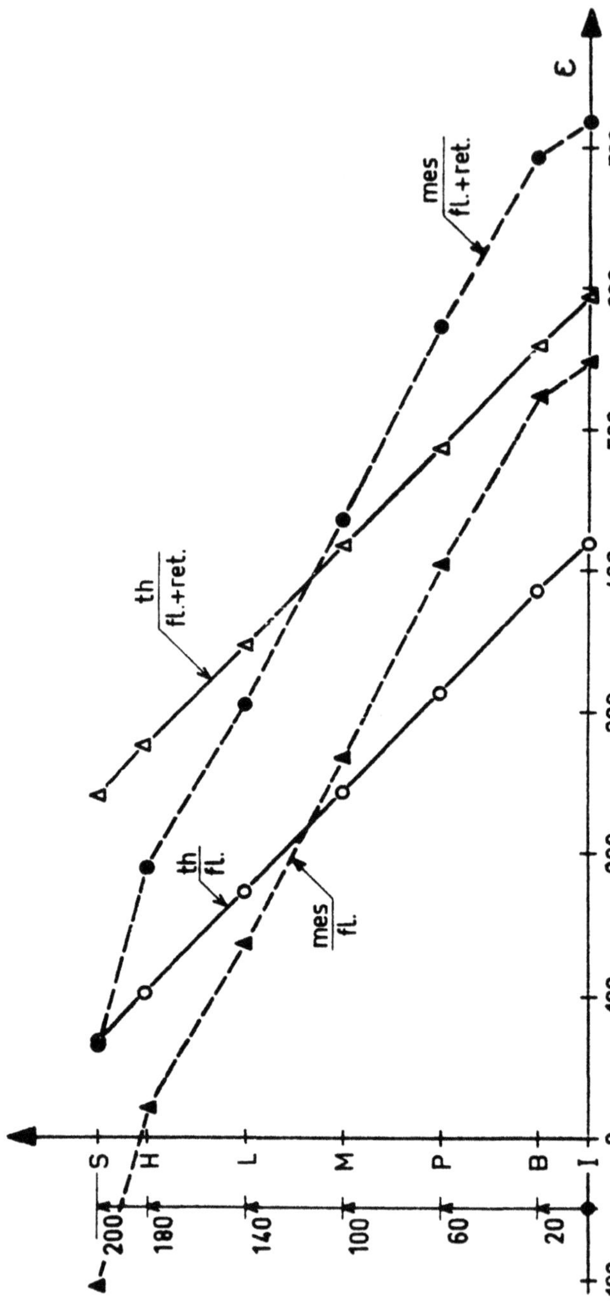
S
H
L
M
P
B
I
200
180
140
100
60
20
100
0
-100
-200
-300
-400
-500
-600
-700
ε
th
fl.+ret.
mes
fl.+ret.
th
fl.
mes
fl.

Figure 7.

EARLY SHRINKAGE OF HARDENING CEMENT PASTES

L. Brüll, K. Komloš

Civil Engineering Research Institute and Institute of Construction and Architecture of the Slovak Academy of Sciences, Bratislava, Czechoslovakia

ABSTRACT

The submitted paper deals with the behaviour of cement pastes in their early period of hardening. The proposed method enables to carry out measurements of volume changes from the first setting phase of the measured sample. The measurement is made in a special device by the contact way with the aid of inductance transducers. The mould of this measuring device has profiled movable fronts which transfer the length changes of the investigated sample to inductance transducers. The method can be applied for measuring volume changes of cement pastes, mortars and concretes.

L'article présenté traite la conduite des pâtes de ciment au jeune âge de leur durcissement. La méthode proposée permet de procéder à la mesure des variations de volume dès la première phase de durcissement de l'éprouvette mesurée. La mesure est réalisée par un appareillage spécial de la manière à contact a l'aide des capteurs à variation d'inductance. Le moule de cet appareillage de mesure a les plaques frontales profilées qui transmettent les variations de longeur de l'éprouvette étudiée aux capteurs à variation d'inductance. Cette méthode peut être utilisée à la mesure des variations de volume des pâtes de ciment, des mortiers et des bétons.

INTRODUCTION

Previously performed investigations [1] showed that the shrinkage of fresh cement pastes, mortars and concretes significantly influences their physico-mechanical properties in the later hardening stage.

Deformations caused by volume changes of the cement paste during setting and hardening have a characteristic course [2] . From the form of the cement paste shrinkage curve two phases of volume changes, following one after the other are obvious. The primary shrinkage takes place before the setting and during it and it is explained by the energy exchange between the solid phase, the liquid phase and its vapours, mainly when the water in capillaries is exposed to underpressure due to evaporation. The secondary shrinkage proceeds during hardening. It is an endogenic phenomenon, connected with physical and chemical processes linked up with cement hydration.

PHASES OF EARLY VOLUME CHANGES

The early course of cement paste volume changes can be divided into four phases (FIG 1) [3] .

First phase (A - B)

Shortly after placing the investigated sample into the mould on its surface a water layer is forming, its thickness being dependent, mainly, on the water-cement ratio. In dependence on the composition of the sample, for a certain time there do not occur any volume change in the sample. The duration of this period is variable and lasts from one to six hours, sometimes even more. It often occurs that in the first phases samples are swelling due to water migration between the grains.

Second phase (B - C)

As soon as the surface water layer evaporates,the sample begins to shrink. It is the phase approximately in accordance with the setting. It is called primary shrinkage. Its maximum value depends mainly on the cement composition and on the water-cement ratio. By increasing the quantity of batched water, this phase is being shortened and the maximum value reached during primary shrinkage is reduced. Therefore it can be stated that the value of primary shrinkage is inversely dependent on the water-cement ratio.

Third phase (C - D)

For several hours after reaching the maximum of primary shrinkage, it is possible to observe very small volume changes only. Two antagonistic phenomena are encountered in the material: at the one hand, shrinkage proceeds, on the other hand, a swelling takes place,the entire result being usually a slight swelling. The course of this phase depends mainly on the gypsum content. For the most of cement types the swelling lasts for ten to twelve hours, however for some cement types it is even shorter.

Fourth phase (D - E)

The fourth phase is the secondary shrinkage. It proceeds after the end of setting. The rate of secondary shrinkage is many times smaller approx. 20 times than the rate of primary shrinkage. It depends on the cement composition, on the duration of the third phase, and on the water-cement ratio.

TESTING METHOD

The principle of the proposed testing method is the measurement of changes in the investigated sample's length. It proceeds from the previous manner of measuring in modified moulds [4, 5, 6]. In the faces of these moulds openings were drilled through which probes were passing that transferred the volume changes of the investigated sample to the shift sensors. This measuring method had the disadvantage that the modified moulds had fixed walls and therefore it was not possible to measure expansive materials in them.

The newly developed testing device for measuring volume changes of cement pastes, mortars and concrete in the early stage of setting and hardening allows the measurement of shrinkage as well as swelling. It consists of a mould with the dimensions of 16 x 4 x 4 cm and fixing clamps for the shift transducers (FIG 2) . The side walls are detechable, this reduces the friction of the sample on the walls of the mould to a minimum value. Before filling the mould with the investigated mixture, the side walls of the mould are secured by bolts simultaneously fixing the faces of the mould. By their structure, the faces of the mould are ensuring by

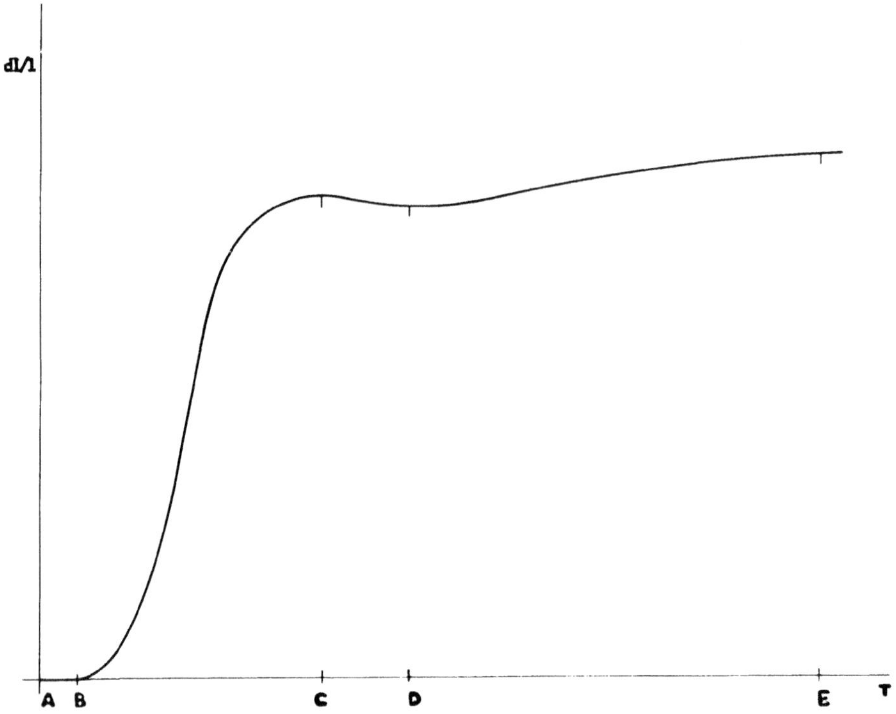

FIG 1
Course of volume changes for cement pastes

their entire area a thoroughout transfer of the investigated sample's volume changes to the shift transducers. The shift transducers are fixed in clamps that can be set into the necessary position during the measurement.

The volume changes of the investigated mixtures were measured by Novotechnik shift transducers, Type F 210, 0-10 mm.

For the automatization of this measuring process, the measuring system Hewlett-Packard, Type HP 3052A, was used. For calculations and the control of the measuring system, the calculator HP 9825A was applied.

The arrangement of the measuring system is shown in FIG 3.

FIG 2
Mould for measuring volume changes

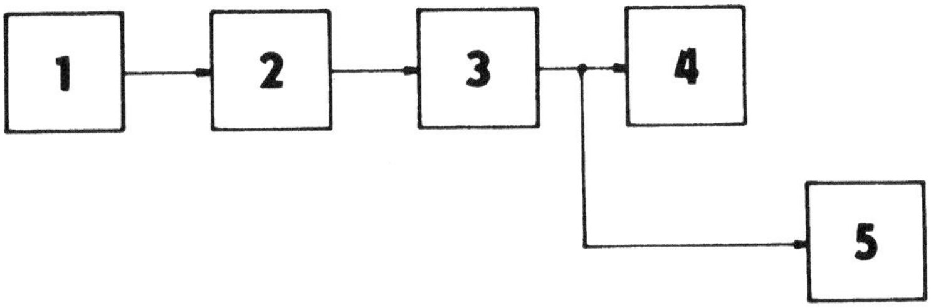

FIG 3
Arrangement of the measuring system
1 - investigated sample, 2 - transducer, 3 - electronic circuits, 4 - digital indicator, 5 - memory

MEASURING PROCEDURS

For the measurement of volume changes of cement pastes in the early stage of setting and hardening, three types of cement were used:

1. SC 70 cement (road cement) ,
2. SPC 325 cement (slag-portland cement) ,
3. PC 400 cement (portland cement) .

The water-cement ration was 0.35 in all cases.

After filling the moulds with cement paste, the shift transducers were connected to the measuring system. Before the beginning of measurement, the side walls were unclamped, this caused also the unclamping of the mould's faces. The volume changes of the cement paste for one cement type were measured always in three moulds simultaneously (FIG 4) .

The measurement duration was 24 hours and the separate values were recorded on a magnetic tape every 30 minutes. The measured values were graphically plotted by the aid of a HP 9871A printer.

MEASURING RESULTS

The measurement results are presented in FIGS 5 to 8.

FIG 5 shows the course of volume changes of a cement paste made of SC 70 cement. In the beginning of the setting, a small swelling occurs, then the primary shrinkage takes place that ends approximately 11 hours after mixing and reaches the maximum value between approx. 1.5 and 3.5×10^{-4}. In the further hours, up to 24 hours after mixing, only slight volume changes are taking place in the investigated samples.

FIG 6 shows the courses of volume changes of a cement paste made of SPC 325 cement. After the first tranquil phase the primary shrinkage takes place that reaches the maximum value of approximately 25.5×10^{-4} nine hours after mixing. Up to 24 hours the investigated sample is slightly swelling and the shrinkage value is reduced to approx. 22.5×10^{-4}.

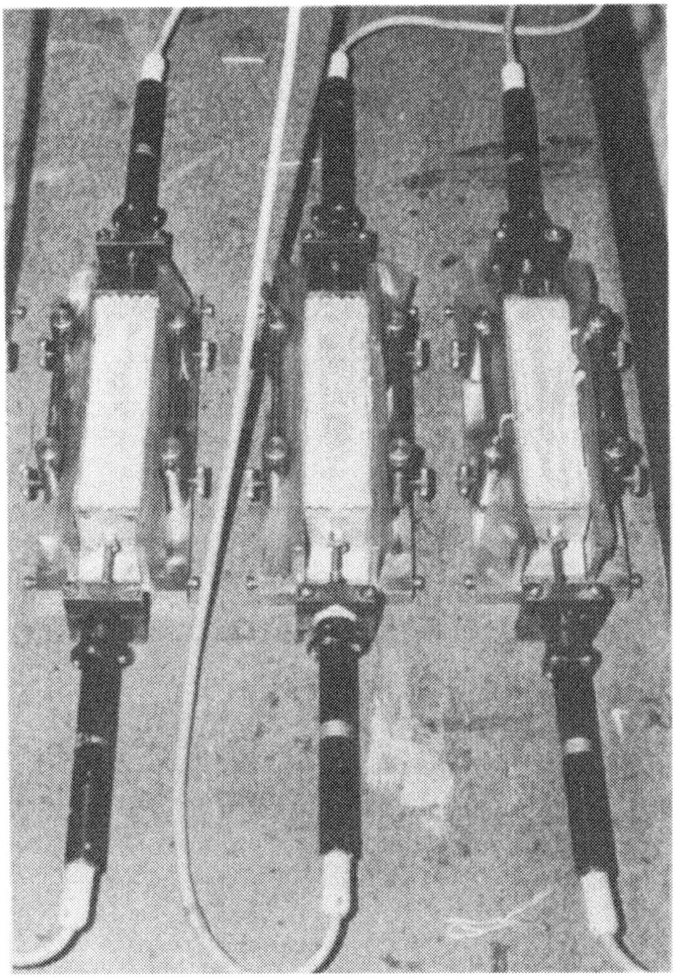

FIG 4
Simultaneous measuring on 3 moulds

FIG 7 shows the courses of volume changes of a cement paste made of PC 400 cement. The maximum value of primary shrinkage is approximately 40 x 10^{-4} fifteen hours after mixing. In the further hours, the sample is slightly swelling and 24 hours after mixing, the full shrinkage has a value of approx. 36 x 10^{-4}.

In FIG 8 a comparison is shown of the mean results of volume changes of cement pastes made of the three given cement types.

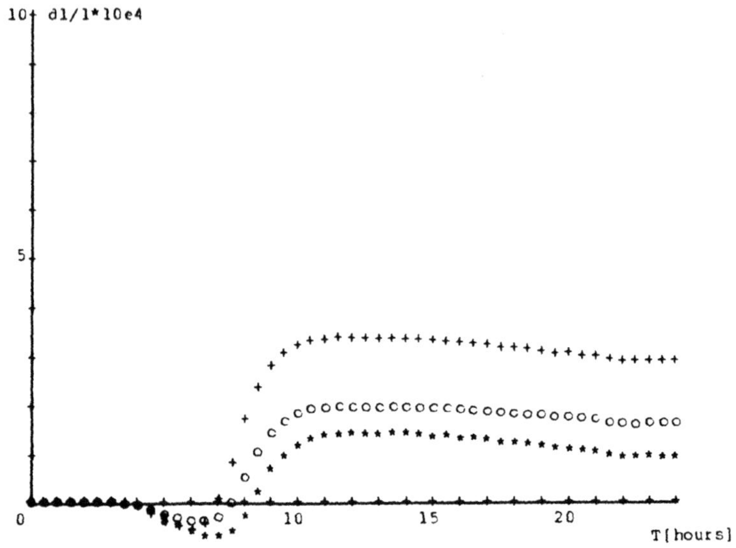

FIG 5
Early shrinkage of cement pastes made of SC 70 cement

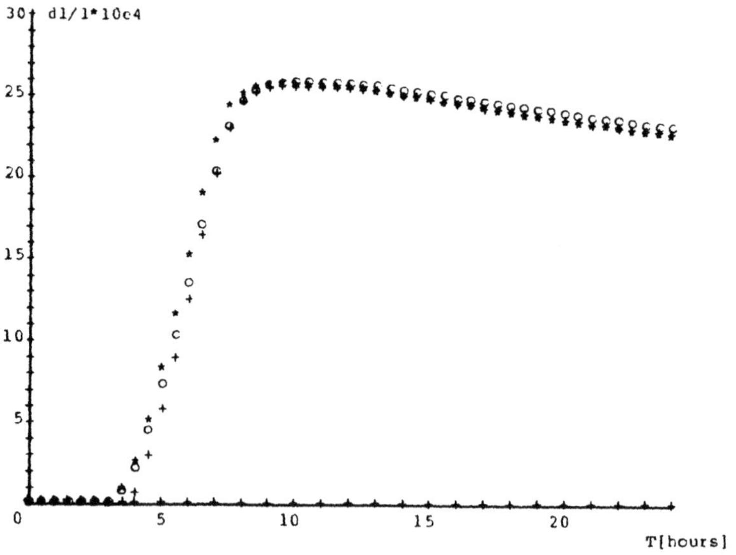

FIG 6
Early shrinkage of cement pastes made of SPC 325 cement

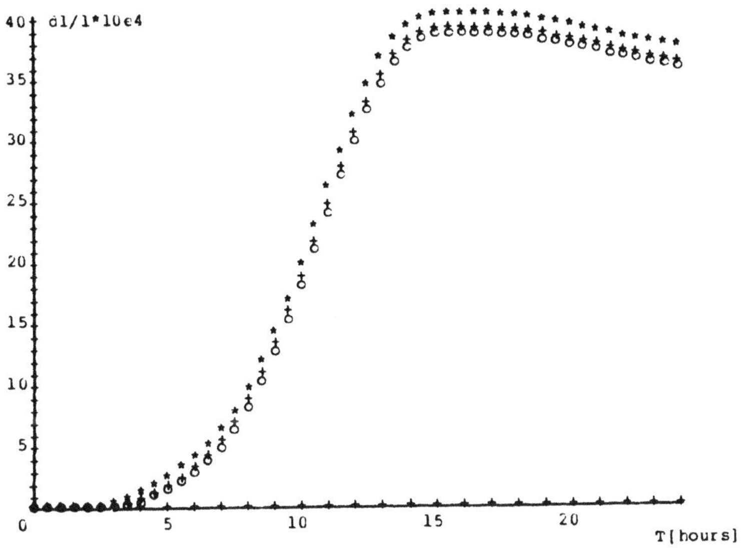

FIG 7
Early shrinkage of cement pastes made of PC 400 cement

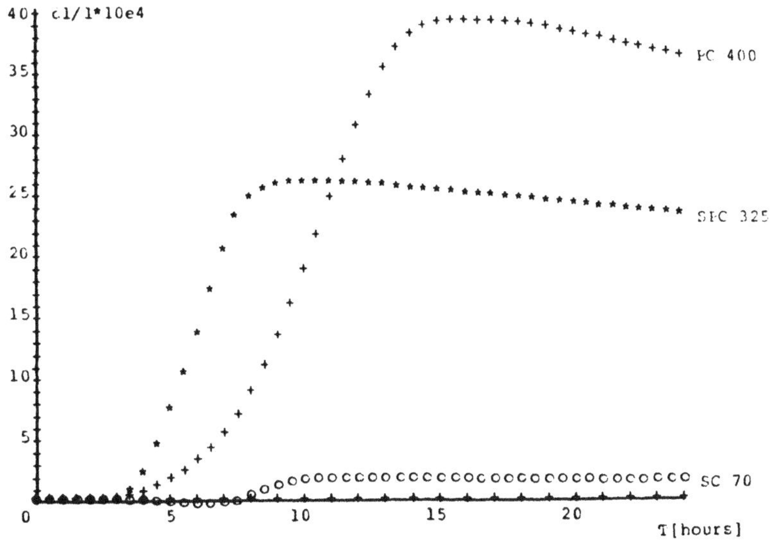

FIG 8
Mean values of early shrinkage of cement pastes made of SC 70, SPC 325, and PC 400 cement

CONCLUSIONS

The investigation results showed that by the selected method it is possible to obtain a high reproducibility of measurements. In comparison with the previous method, this device is far more sensitive,because the friction of the sample on the walls of the mould was reduced to the minimum, and because the transfer of the investigated sample's volume changes is made by the entire faces of the mould.

The given device allows the measurement of volume changes in the early stage of setting and hardening not only of cement pastes, but of mortars and concretes as well, provided that the mould dimensions are modified.

REFERENCES

1. Wittmann, F.H. "Die Bedeutung des Kapillarschwindens für die Technologie der Fertigteilprodukte" Kongress BIBM, Wien (1978) I 23 - I 29.
2. Détriché, C.H. and J. C. Maso. "Premier et second retrait des couches minces de mortier" Matériaux et Constructions 12 (1979) 185-191.
3. Détriché, C.H. "Analyse expérimentale du retrait de couches minces de mortier. Mesure depuis le moulage" Matériaux et Constructions 11 (1978) 247-259.
4. Baron, J. "Fissuration du béton par hydratation localement différé du ciment" Laboratoire Central des Ponts et Chaussées, Paris, Rapport de recherche, No.15 (1971).
5. Brüll, L., K. Komloš and B. Majzlan. "Early shrinkage of cement pastes, mortars and concretes" Materials and Structures 13 (1980) 41-45.
6. Brüll, L. and K. Komloš. "Early shrinkage investigation of cement pastes, mortars and concretes" Stavivo 57 (1979) 306-309.

Part IV
Constitutive Relations for Creep and Shrinkage

NON LINEAR CONSTITUTIVE EQUATIONS FOR CONCRETE

Guillermo J. Creus

Curso de Pós-Graduação em Engenharia Civil
Universidade Federal do Rio Grande do Sul, Porto Alegre, Brasil

ABSTRACT

A representation is introduced for the linear and non linear behaviour of concrete in the presence of time varying humidity, based in the state variables concept.

E proposta uma representação para o comportamento linear e não linear do concreto em presença de umidade variável, baseada no conceito das variáveis de estado.

1. INTRODUCTION

There has been in recent years a considerable advance in the knowledge of the mechanical behaviour of inelastic materials. Practical needs, arising from their use under severe environmental and loading conditions and the continuous growth of computational power are recognized as the main reasons for such advances.

In the case of concrete, progress is particularly difficult due to its complex internal structure and dependence on environmental stimuli (1). In this case, it has been specially helpful the insight derived from observations and speculations at the micro-structural level. For example, some of the papers that first discussed the hygromechanical behaviour of cement paste (2, 3) were inspired by the work of T.C. Powers, in particular its very readable ref. (4). This paper is a good example of a discussion of internal behaviour that helps to structure the macroscopic (continuum) theory.

The continuing growth of knowledge of the internal structure of concrete will certainly spur new, more adequate constitutive theories. In this connection, a crucial point is the interrelation between microstructural and continuum variables. The rigourous approach of mixtures theory may prove valid in the future. Meanwhile, a simpler practical alternative is the use of state variable representations (5, 2) which are also convenient from the point of view of computacional efficiency.

T.C. Powers (4) describes cement paste as a porous adsorbent material with a large specific internal surface area. Changes of humidity at the boundary of an element produce variations in the partial water contents that, because of adsorption forces, result in strain changes. This is shrinkage. Creep is explained by mechanical stimuli activating adsorption or desorption. These mechanisms, that may be responsible for at least part of the delayed deformations of cement paste, shall be used to motivate the basic equations in the following Section 2. In Section 3 the state variable representation is introduced and analysed. In Section 4 the representation is extended to the nonlinear case.

Throughout this paper our emphasis is on basic concepts and qualitative results directly related to the central ideas, namely : (i) consideration of coupled hygromechanical constitutive relations; (ii) use of a state variable representation; (iii) continuity of the linear and non linear viscoelastic regimes. Curve fitting of numerical results and applications may be found in the references cited. Much additional research is still needed to represent quantitatively the main features of the complex concrete behaviour. Important work in this direction has been done by Z.P. Bazant using the formalism of endochronic theory (6).

2. BASIC CONSTITUTIVE RELATIONS

We will restrict our attention to situations characterized by small deformations ε_{ij} (x_j, t) and Cauchy stresses σ_{ij} (x_j, t) where x_j indicates the particle position with respect to a fixed cartesian frame and t is the time. Eventually, to simplify notation, time and location dependence will be not shown explicitly, and a symbolic notation for tensors, i.e. $\underset{\sim}{\varepsilon}$, $\underset{\sim}{\sigma}$ will be used.

We represent with w (x_j, t) the specific water content of the material. As this water may be in different states (i.e. free, adsorbed, etc.) we may eventually define partial specific water contents. The concentration of the water vapor within the solid or in the atmosphere is defined through an adequate measure of the humidity $h(x_j, t)$.

In practical situations, we shall have unhomogeneous distribution of water and humidity and a flux of water within the solid subject to the conservation equation

$$\frac{dw}{dt} + \frac{1}{\rho} q_{i,i} = 0 \tag{1}$$

where ρ is the density of the solid skeleton.

Physical and phenomenological considerations lead to the assumptions that the stress $\underset{\sim}{\sigma}$ and the specific water content w are functions of the histories of strain and humidity

$$\begin{aligned} \underset{\sim}{\sigma}(\underset{\sim}{x}, t) &= \mathfrak{F}\{\underset{\sim}{\varepsilon}(\underset{\sim}{x}, \tau) \overset{t}{\underset{\tau=0}{;}} h(\underset{\sim}{x}, \tau)\} \\ w(\underset{\sim}{x}, t) &= \mathfrak{G}\{\underset{\sim}{\varepsilon}(\underset{\sim}{x}, \tau) \overset{t}{\underset{\tau=0}{;}} h(\underset{\sim}{x}, \tau)\} \end{aligned} \tag{2}$$

where $\mathfrak{F}$ and $\mathfrak{G}$ are functionals of the indicated arguments. The double dependence on $\underset{\sim}{\varepsilon}$ and h introduces the coupling effect observed in real phenomena. The history dependence may come about because of the presence of internal relaxation phenomena and or because of the existence of irreversible processes (i.e. adsorption hysteresis, non linear deformations of the solid skeleton etc.). It is worth noticing that equations (2) are not dependent on the particular picture of internal structure proposed by Powers, but will be adequate also for other creep hypotheses.

For the study of coupled hygromechanical situations with nonhomogeneous water content distribution, a constitutive law for the flux of water is required. We shall assume the simple relation

$$\underset{\sim}{q}(\underset{\sim}{x}, t) = C \operatorname{grad} h(\underset{\sim}{x}, t) \tag{3}$$

where C is a functional of the histories of strain an humidity. This is a rather strong assumption as the actual behaviour may be more complicate. We could think of two kinds of flux, one for the adsorbed phase and one for the free phase. We could in addition to h introduce other potentials that cause flow. More complex situations may be analysed along the lines of Bazant (3) who also considers the flow of solids. Simpler solutions have been used in the solution of practical problems (7).

When the functionals in Eq. (2) are linear, we may use the well known linear integral representation. Thus, for the uniaxial case and using stresses as inputs we have

$$\begin{aligned} \varepsilon (t) &= \int_0^t D (t, \tau) \sigma (\tau) d\tau + \int_0^t A_1 (t, \tau) h^{\cdot} (\tau) d\tau \\ w (t) &= \int_0^t A_2 (t, \tau) \sigma (\tau) d\tau + \int_0^t A_3 (t, \tau) h^{\cdot} (\tau) d\tau \end{aligned} \tag{4}$$

For the case $\sigma^{\cdot} = 0$ the integrals in the second term indicate changes of strain and water content due to humidity changes (i.e. expansion and adsorption isotherms). They do not in general correspond to shrinkage results in concrete, where measurements are made for nonhomogeneous humidity distributions. The first terms indicate the changes in strain and water content due to stress variation in a situation in which the humidity does not change. Again, these are not values generally measured in creep tests, where we have usually nonhomogeneous time-varying distributions of w and h. Comparison with experimental values should be done through the solution of boundary values problems.

3. STATE VARIABLE REPRESENTATION

The notion of state is closely related to the concept of phase space in statistical mechanics. It can also be arrived at from purely phenomenological considerations (5). To this effect, let us consider two sets of identical specimens. We subject the specimens in the first set to the strain and concentration histories $\varepsilon^{(1)}(\tau)$, $h^{(1)}(\tau)$ and the ones in the second set to $\varepsilon^{(2)}(\tau)$, $h^{(2)}(\tau)$ on the time interval (0, t). The two sets of specimens are said to be in the same state at time t if identical future histories of strain (measured with respect to the configuration at time t) and humidity produce identical stress and water content in both sets, for times $\tau > t$.

We now consider the set H of all strain and humidity histories on (0, t). We divide H into subsets such that histories belonging to S_i give rise to the same state at t. Each S_i is then considered as a simple element of a new set Σ. If the number of elements in the set Σ is much less than the number of elements in H then it would be advantageous to work with the elements of Σ rather than with the original histories. It was shown by E.T. Onat in (5) that the set Σ may be converted into a metric space by introducing a physically meaningful definition of distance between its elements.

It follows from the definition of state that the stress and water content are functions of state, namely

$$\underset{\sim}{\sigma} = F\ (\underset{\sim}{S}) \qquad\qquad w = G\ (\underset{\sim}{S}) \tag{5}$$

where $\underset{\sim}{S}$ represents a point in the state space. It follows further that

$$\underset{\sim}{S}^{\cdot} = J\ (\underset{\sim}{S},\ \underset{\sim}{\varepsilon}^{\cdot},\ h^{\cdot}) \tag{6}$$

This equation is called the growth low of the state. It is readily seen that the integration of Eq. (6) for given histories of h(t) and suitable initial conditions would result in a functional dependence of type Eq. (2). Rate equations of the form (5), (6) are more convenient for the application of numerical methods of analysis.

Applying the state representation, it may be shown (2) that Eqs. (5), (6) may be written as

$$\varepsilon = \sum_{\alpha=1}^{j} D_\alpha\, q_\alpha + \sum_{\gamma=j+1}^{k} A_{1\gamma}\, q_\gamma \qquad\qquad w - w_o = \sum_{\delta=k+1}^{m} A_{2\delta}\, q_\delta + \sum_{\nu=m+1}^{n} A_{3\nu}\, q_\nu \tag{7}$$

where the q's are state variables, that may be determined in this case through the solution of linear first order differential equations. For example

$$q_\alpha^{\cdot} + \frac{q_\alpha}{T_\alpha} = Q\ \sigma\ \ (t) \tag{8}$$

To consider the effect of aging, Q is made dependent on an additional internal variable m that characterizes maturity and in general on the histories of humidity and temperature

$$m^{\cdot} + \frac{m}{T_m} = M\ (\theta,\ h) \tag{9}$$

This procedure allows the representation of most of the usual aging criteria for concrete (7). When necessary, several state variables may be used in (9) to better adjust the experimental data.

4. NONLINEAR VISCOELASTIC BEHAVIOUR

For stresses higher than about 1/3 the failure stress, non-linear effects must be considered. To produce a uniform representation of viscoelastic deformations in the linear and nonlinear ranges, the same rate form of Eqs. (5), (6) is mantained. The nonlinearity in the mechanical term (associated to the presence of microcracks in the solid structure) is represented by a nonlinear spring in the Kelvin elements. Using for the nonlinear spring the simplest parabolic form, we obtain again (8)

$$\dot{q}_n + \frac{q_n}{T_n} = Q\,\sigma\ (t) \qquad (10)$$

with a strain depèndent relaxation time

$$T_n = \frac{\eta_n}{E_n\ (1 - \beta_n\ q_n)} \qquad (11)$$

where $E\ (1 - \beta\ q)$ may be considered as a decreasing elastic modulus and β is the nonlinear parameter. Considering the behaviour of the model under a constant stress σ_o we may see that for $\sigma_o \leq E/4\beta$, $\varepsilon(t)$ is bounded while for $\sigma_o > E/4\beta$ a condition similar to failure is reached after a finite time. The stress $E/4\beta$ may be interpreted as the strength under permanent load. The fitting of experimental results may be improved using several elements in series.

There are several ways to extend this model to multiaxial stress situations. We may write

$$N = 1 - \frac{E\ \bar{\varepsilon}\ (1 - \beta\ \bar{\varepsilon})}{\bar{\sigma}} \qquad (12)$$

where $\bar{\sigma}$ and $\bar{\varepsilon}$ are equivalent stress and strain measures obtained as combinations of the basic invariants (9). Then

$$\begin{aligned} \eta\ \dot{e}_{ij} &= \langle N \rangle\ s_{ij} \\ \eta\ \dot{\varepsilon}_{kk} &= \langle N \rangle\ \sigma_{kk} \end{aligned} \qquad (13)$$

where $\langle x \rangle = 1/2\ x + 1/2\ |x|$, and s_{ij}, e_{ij} are the deviatoric components of σ_{ij}, ε_{ij}. Numerical examples using this relation may be found in (9).

Acknowledgment : The support of C.N.Pq. and FINEP is gratefully acknowledged.

5. REFERENCES

1. Bazant, Z.P., "Theory of Creep and Shrinkage in Concrete Structures : A Precis of Recent Developments", Mechanics Today, Vol. 2, (Pergamon, 1975).

2. Creus, G.J. and E.T. Onat, "Mechanics of Porous Adsorbent Materials", Int. J. Eng. Science, 10, (1972), 649-658.

3. Bazant, Z.P., "Thermodynamic Theory of Deformation at Variable Temperature and Humidity", Div. of Struct. Eng. and Struct. Mech., Univ. of Calif., Berkeley, Report 69-11, (1969).

4. Powers, T.C., "Some Observations on the Interpretation of Creep Data", RILEM Bull. 33, (1966), 381-391.

5. Onat, E.T., "The Notion of State and its Implications in Thermodynamics of Inelastic Solids", Proc. IUTAM Symposium on Irreversible Aspects of Continuum Mechanics, H. Parkus and L.I. Sedov (Ed.), (Springer-Verlag, Berlin, 1968).

6. Bazant, Z.P. and A.A. Ashgari, "Constitutive Law for Nonlinear Creep of Concrete", J. Eng. Mech. Div., ASCE, V. EM 103, (1977), 113-124.

7. Pister, K.S., J.H. Argyris and K.J. Willam, "Creep and Shrinkage of Aging Concrete", D. McHenry International Symposium on Concrete and Concrete Structures, ACI, (1977).

8. Helman, H. and G.J. Creus, "A Nonlinear Rheological Model Describing Time Dependent Deformation and Failure", Int. J. Non Linear Mechanics, 10, (1975).

9. Creus, G.J. and A.J. Ferrante, "Analisis No Lineal de Estructuras de Hormigon Armado", Conference on Structural Analysis, Design and Construction in Nuclear Power Plants, Brasil, (1978).

ANALYSES OF SHRINKAGE IN CONCRETE

Jan-Erik Jonasson

Swedish Cement and Concrete Research Institute
Stockholm

ABSTRACT

To get a tool for theoretical studies of drying shrinkage effects in different complex situations, material parameters, determined by fitting of shrinkage data in accordance with constitutive relations based on thermodynamics, are presented. The necessary fundamental data of basic creep and shape of free shrinkage in time are chosen according to some practical formulations, and for a number of shrinkage data sets and for two different creep functions, the obtained material parameters refer to a typical ordinary concrete and a typical cement mortar.

ZUSAMMENFASSUNG

Für die theoretische Behandlung von Schwinden durch Austrocknen in verschieden komplexen Situationen werden Materialparameter gegeben, die durch Anpassung an Schwindwerte gemäss grundlegender, auf Thermodynamik bauende Materialgesetze, erhalten wurden. Die erforderlichen Werte bezüglich Kriechen bei konstanter Temperatur und Feuchtigkeit und der Zeitabhängigkeit von freien Schwinden wurden übereinstimmend mit für die praktische Anwendung vorgesehenen Zusammenhängen gewählt. Eine Anzahl Schwind-Zeit-Kurven sowie zwei verschiedene Kriechfunktionen gelten für einen typischen Zementmörtel und einen typischen Konstruktionsbeton.

1 INTRODUCTION

Two types of mathematical models for time dependent deformations of concrete may be distinguished:

a) constitutive relations for the behaviour of a small element of the material
b) models for prediction of the "over all" material behaviour in simple basic situations.

For many problems in design practice, the latter type of models is quite satisfactory, and most codes for concrete design have formulations of this type.

However, to be able to study the nonlinear strain and stress variations due to time dependent deformations within an element for different complex situations, such as the study of failure due to drying shrinkage when combining two layers of concrete (1,2), the analyses must be based on constitutive relations.

In the last ten years computer methods have made it possible to study structural elements with complicated material properties. Constitutive equations for concrete creep and shrinkage based on thermodynamics (3) has turned out to be able to model most of the complex properties of concrete (4), and an illustrative example showing the qualitative features is presented in (5).

The purpose of this paper is to determine the necessary material parameters to be able to study the behaviour of creep, drying process and shrinkage according to the model in (4) with the extensions of (1,2) when the fundamental behaviour of basic creep and free shrinkage are described by some practical formulations.

In cases where the fundamental data are functions of strength and composition parameters of concrete mix, such as in (6), it is straight-forward to identify the determined model parameters to a similar dependence on strength and composition, but this is out of the scope of this paper.

2 SHAPE OF SHRINKAGE CURVES IN TIME

Recently, (6) a very simple formula for predictions of unhindered shrinkage was shown to give quite close agreement with a great number of data sets, and it is expressed as

$$\varepsilon_s/\varepsilon_u = ((t-t_o)/(3 \cdot \tau_{50}+t-t_o))^{1/2} \qquad (1)$$

in which ε_s = free shrinkage, ε_u = ultimate shrinkage, t = age of concrete (in days), t_o = curing time (in days), and τ_{50} = duration of drying when the shrinkage has reached 50 per cent of the ultimate value. τ_{50} may be called shrinkage half-time and be expressed by

$$\tau_{50} = \tau_d \cdot \gamma_d \text{ with } \gamma_d = (k_s \cdot \frac{d}{0.15})^2 \text{ and } d = 2 \cdot \frac{v}{s} \tag{2}$$

where τ_d = reference shrinkage half-time, k_s = shape factor of the actual cross section (values given in (6)), and v/s = volume to surface ratio of the studied concrete member.

Since one purpose of this study is to keep the "independent parameters" at the utmost minimum, I here choose a reasonable value of τ_d, and I consider ε_u as a quantity to be varied. According to the fitting, see figure 1, of Swedish shrinkage data (7), τ_d = 333 days is here employed as a representative value.

3 BASIC CREEP IN LINEAR RANGE

One reasonable choice of basic creep curves may be according to a function proposed by ACI (8), previously used in (1), of the form

$$C(t,\tau) = \frac{1}{E(\tau)} \cdot (1 + \phi(\tau) \cdot g(t-\tau)) \tag{3}$$

$$\text{with } E(\tau) = E(28) \cdot \left(\frac{\tau}{4+0.85 \cdot \tau}\right)^{1/2}$$

$$\phi(\tau) = \phi(7) \cdot \frac{1.25}{\tau^{0.118}}$$

$$g(t-\tau) = \frac{(t-\tau)^{0.6}}{10+(t-\tau)^{0.6}}$$

in which $C(t,\tau)$ = strain at time t caused by a unit stress sustained since time τ, $E(\tau)$ = elastic (Young's) modulus at age τ, and $\phi(\tau)$ = final creep ratio for loading since time τ. Eq. 3 has the advantage of having only two determining parameters, both well known to designers, namely E(28) and $\phi(7)$.

Another proper choice may be the so called double power law, which have been shown (6) to agree quite closely with most experimental data on basic creep, and it is formulated as

$$C(t,\tau) = \frac{1}{E_o} \cdot (1 + \phi_1 \cdot (\tau^{-m} + \alpha) \cdot (t-\tau)^n) \tag{4}$$

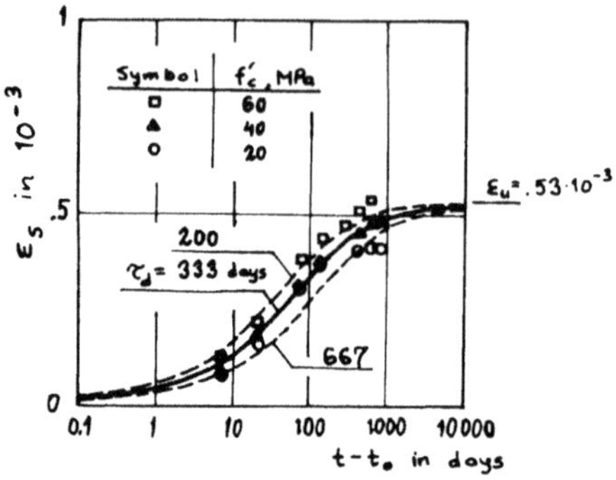

a) Concrete, square prism 0.10x0.10x0.40 m.

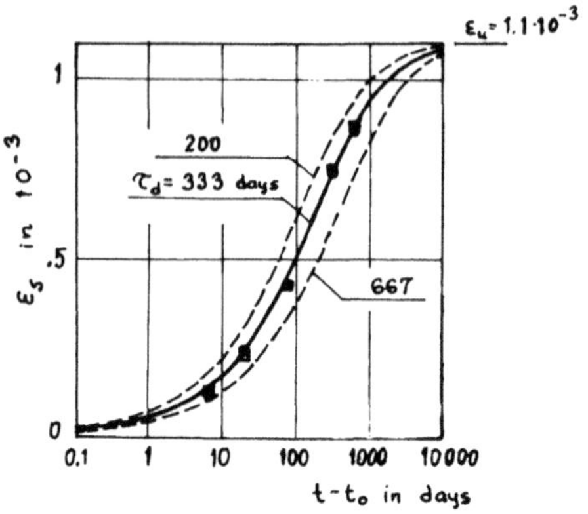

b) Cement mortar, square prism 0.16x0.16x0.60 m.

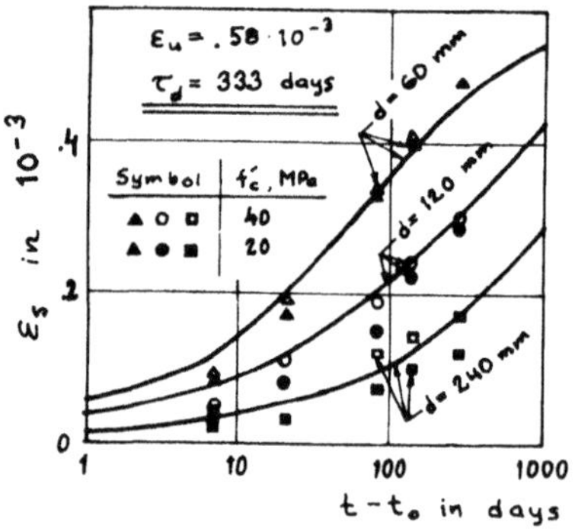

c) Concrete, infinite slabs.

FIG 1. Fitting of Shrinkage Values according to Eq. 1 with Test Data of Ödman (3) (previously used in (1)).

in which E_o, ϕ_1, m, α and n are material parameters.

Here I intend to analyse the cases of a typical ordinary concrete and a typical cement mortar, and one possible way according to the "ACI"-function (=Eq. 3) is to choose basic creep values as

$\phi(7)$ = 2 for concrete; and $\phi(7)$ = 4 for cement mortar.

Using the "DPL-function (=Eq. 4), m = 1/3 may be employed to model most concretes (6), and the remaining parameters are here chosen as:

α = 1/(40·w/c) (6), where w/c = water-cement-ratio. For E(28) = 30 GPa a reasonable choice is w/c = 0.6.

n, ϕ_1 and E_o are determined according to the conditions

i) $\phi_1 = 10^{3n}/(2(28^{-m} + \alpha))$ (6)

ii) $C(t=28.1, \tau=28) = 1/E(28)$ (6)

iii) Agreement between Eq. 3 and Eq. 4 at t ≈ 10000 days (≈ 30 years) when τ = 7 days.

The obtained material parameters are presented in table 1, and the corresponding creep curves are shown in figure 2.

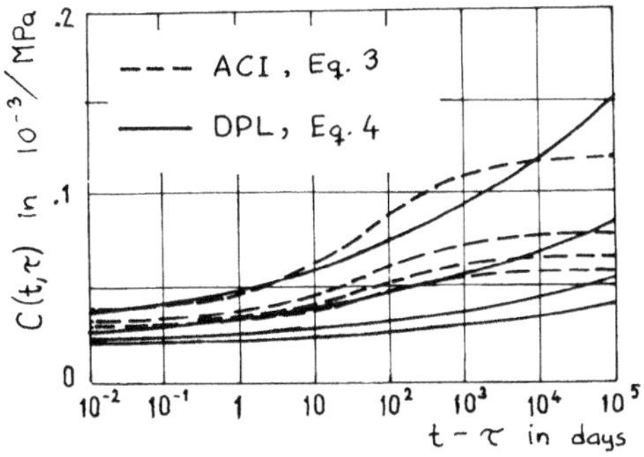

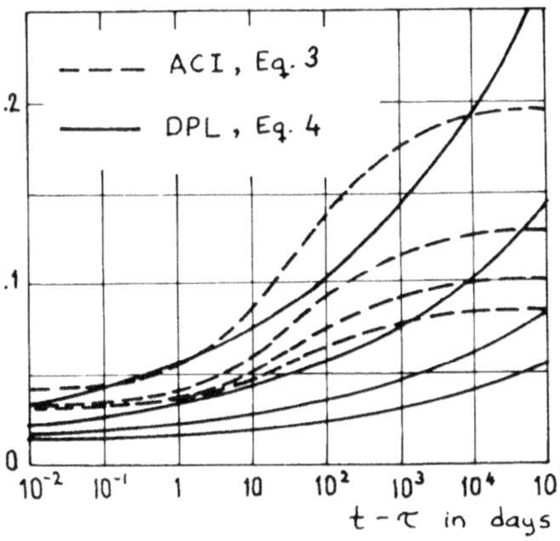

FIG 2. Chosen Basic Creep Curves for Two Typical Cases. The Curves Shown Correspond to τ= 7, 70, 700 and 7000 days, respectively.

Typical case	Creep function according to						
	DPL, Eq. 4					ACI, Eq. 3	
	E_0 GPa	m	n	α	ϕ_1	E(28) GPa	$\phi(7)$
Ordinary concrete	56.7	.333	.125	.0417	3.2	30	2
Cement mortar	120	.333	.15	.0417	10	30	4

TABLE 1. Chosen Parameters for Description of Basic Creep.

4 FITTING OF SHRINKAGE CURVES

For the fitting of shrinkage curves according to the model in (4) with the extensions of (1,2), at least three material parameters are to be determined; 1) C_o = reference drying diffusivity; 2) S_o = strain value modelling the magnitude of the momentaneous shrinkage with respect to the change in pore humidity; and 3) S_1 = = stress value modelling the magnitude of the delayed shrinkage with respect to the change in pore humidity.

However, the running of the Marquardt optimization algorithm could not give stable solutions of especially S_o and S_1, which is physically logical as the relationship between momentaneous and delayed shrinkage cannot be expected to be determined only from fitting of shrinkage curves, especially not when the shape in time is assumed.

One possible way of overcoming this difficulty is to base the fitting on an assumption of the ratio between delayed and momentaneous shrinkage at the ultimate state, i.e. $t \to \infty$, which can be done by the following technique:

1) Calculation of free shrinkage values for various S_1, while $S_o = 0$ and C_o = near the expected value. The resulting ultimate free shrinkage is here denoted ε_1.
2) S_1 is chosen for a certain value of $\varepsilon_1/\varepsilon_u$, and the optimization algorith is running for S_o and C_o.

For an infinite cylinder with the diameter of 0.10 m calculated ε_1-values (C_o=1 · 10^{-10} m^2/s) are shown in figure 3, and the creep curves behind are according to the choice in table 1. The environmental conditions correspond to commonly used indoors conditions, i.e. T = 20°C and h_e = environmental humidity = 0.5.

The optimum fits of step 2 for three ratios of $\varepsilon_1/\varepsilon_u$ are presented in figure 4. The values marked with dots represent "true" shrinkage values (Eq. 1) of equal weight when running the optimization algorithm.

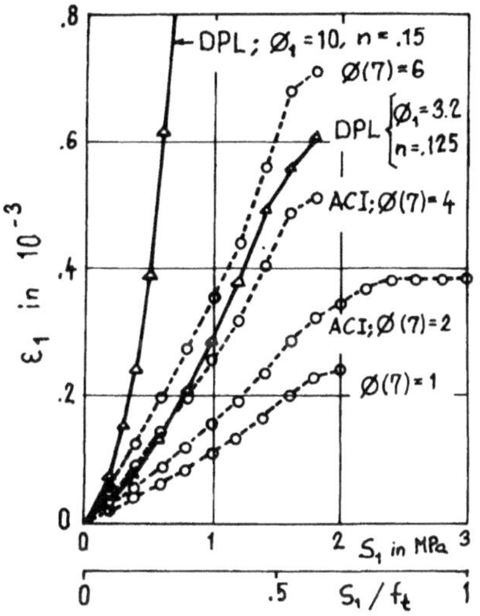

FIG 3. ε_1 as a Function of S_1 for different Creep Curves.

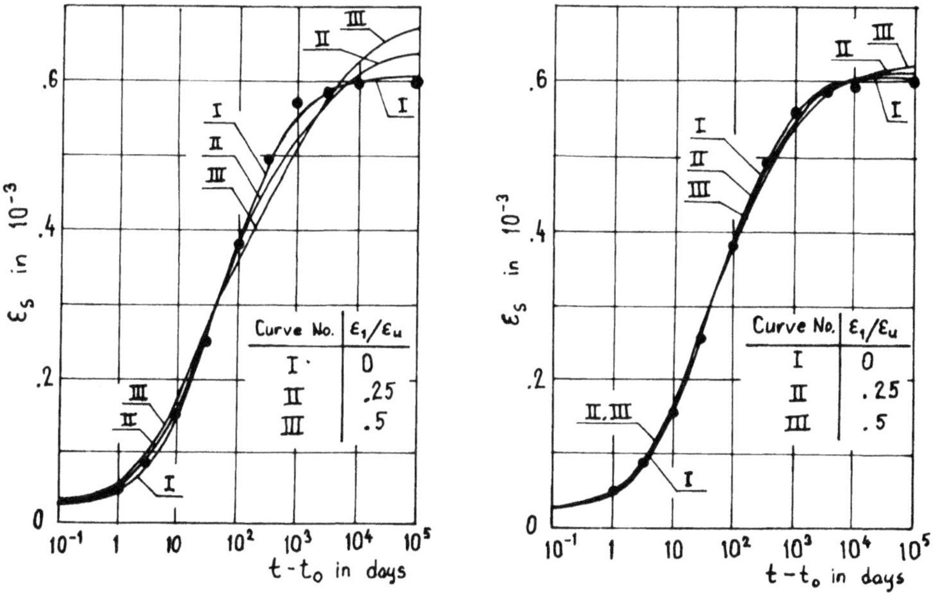

a) DPL, Eq. 4.

b) ACI, Eq. 3.

FIG 4. Fitting of Shrinkage Curves for Different $\varepsilon_1/\varepsilon_u$-values for Ordinary Concrete and $\varepsilon_u = 0.6 \times 10^{-3}$.

As can be seen in figure 4 the best fit seems to appear for $\varepsilon_1/\varepsilon_u = 0$, but this should only be regarded as a pure numerical result and it is not experimentally justified. Besides, one must point out that test results often show a rather scattering picture, see for instance (6), and all the calculated results are within the usually obtained variation of test results. However, one conclusion might be that $\varepsilon_1/\varepsilon_u$ should be rather small, which also is supported by test results of (9, 10), where one possible interpretation is that momentaneous shrinkage represents the major part, up to a level of approximately 80 per cent, of the ultimate shrinkage. So, $\varepsilon_1/\varepsilon_u = 0.25$ is here taken as a representative value for the fitting in the sequel.

In figure 5 the fitted shrinkage curves for different ultimate shrinkage values are shown for the typical cases chosen herein, and the obtained material parameters are presented in table 2.

Typical case		Creep curves according to DPL, Eq. 4			ACI, Eq. 3		
	ε_u 10^{-3}	$C_o \times 10^{10}$ m^2/s	S_o 10^{-3}	S_1 MPa	$C_o \times 10^{10}$ m^2/s	S_o 10^{-3}	S_1 MPa
Ordinary concrete	.2	.864	.199	.24	.810	.190	.33
	.4	1.04	.391	.45	.893	.384	.65
	.6	1.02	.645	.63	.932	.624	.96
	.8	1.06	.860	.77	.940	.833	1.22
	1.0	1.13	1.07	.90	.964	1.05	1.45
	1.2	1.20	1.27	1.02	.996	1.26	1.68
Cement mortar	1.0	1.05	1.00	.40	.952	1.00	.98
	1.5	1.10	1.50	.49	1.00	1.51	1.33
	2.0	1.25	2.25	.55	1.04	2.05	1.64

TABLE 2. Material Parameters (C_o, S_o and S_1) for the Fitting in Figure 5.

All shrinkage data of figure 5 are determined from Eq. 1 with $\tau_d = 333$ days ($\tau_{50} = 49$ days). Variation in τ_d means different drying rates, and as is shown in figure 6 this can consequently be modelled by a change in C_o.

The technique used herein, i.e. only varying three material parameters (C_o, S_o and S_1) and starting with a proper value of $\varepsilon_1/\varepsilon_u$, seems to be able to model drying shrinkage in a reasonable way for different data of free shrinkage and basic creep curves.

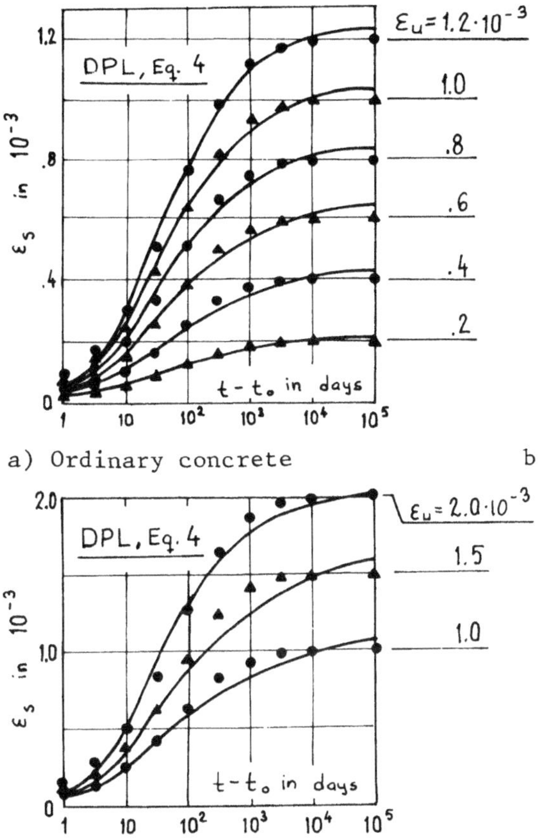

a) Ordinary concrete

b) Ordinary concrete

c) Cement mortar

d) Cement mortar

FIG 5. Fitting of Shrinkage Curves for Different Ultimate Shrinkage Values for $\varepsilon_1/\varepsilon_u = 0.25$.

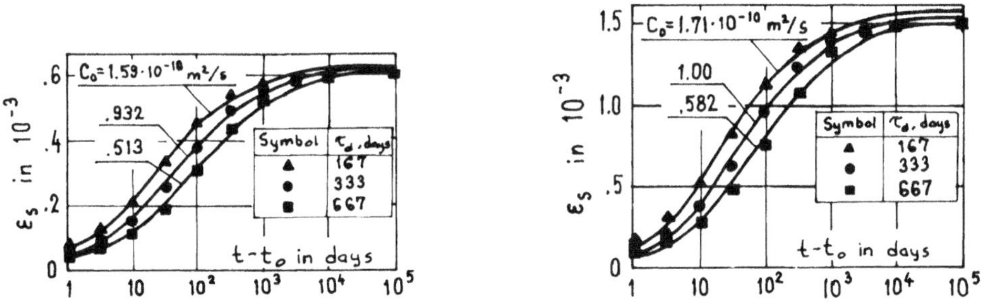

a) Ordinary concrete

b) Cement mortar

FIG 6. Fitting of Shrinkage Curves for Different Drying Rates by Change of C_o. Creep Curves according to ACI, Eq. 3.

So, the obtained parameters are probably a good basis as a tool for theoretical studies of the effects of variable environmental conditions and arbitrary spatial configurations for many "unsolved" problems in concrete design.

REFERENCES

1. Jonasson, J-E. "Computer programs for nonlinear analyes of concrete in view of shrinkage, creep and temperature" (in Swedish with English summary). *Swedish Cement and Concrete Research Institute* 7:77(1977).
2. Jonasson, J-E. "Analyses of creep and shrinkage in concrete and its application to concrete top layers". *Cement and Concrete Research* 8(1978) 441-454.
3. Bazant, Z.P. "Advanced topics in inelasticity and failure of concrete". *Swedish Cement and Concrete Research Institute* (1979).
4. Bazant, Z.P. and T. Wu. "Creep and shrinkage law for concrete at variable humidity". *Journal of the Engineering Mechanics Division*, No EM6 (1974) 1183-1209.
5. Jonasson, J-E. "A theoretical model for analyses of shrinkage and creep of concrete". Contribution at the Nordic Concrete Research Seminar in Copenhagen, March 21 (1980). (To be published by the Danish Engineering Academy, DIAB).
6. Bazant, Z.P. and L. Panula. "Practical prediction of time-dependent deformations of concrete. Part I: Shrinkage, and Part II: Basic creep". *Materials and Structures* 11 (1978) 307-328.
7. Ödman, S.T.A. "Effects of variation in volume, surface area exposed to drying and composition of concrete on shrinkage". *Swedish Cement and Concrete Research Institute*, Reprint No 55(1969).
8. American Concrete Institute, Committee 209, Subcommittee 2. "Prediction of creep, shrinkage and temperature effects in concrete structures". Special publication No 27 (1971) 51-94.
9. Bazant, Z.P., A.A. Asghari and J. Schmidt. "Experimental study of creep of hardened portland cement paste at variable water content". *Materials and Structures* 9 (1976) 279-290.
10. L'Hermite R. "Experiences et théories sur la technologie du béton, deuxième partie, deformation du béton sans charge (retrait et gonflement)". *Annales de l'Institut Technique du Batiment et des Travaux Publics*, No 375 (1979).

COMPARISON OF PREDICTION METHODS FOR CREEP COEFFICIENTS OF STRUCTURAL CONCRETE WITH EXPERIMENTAL DATA

H.S. Müller, H.K. Hilsdorf
Institut für Baustofftechnologie
University of Karlsruhe, W.Germany

ABSTRACT

Various prediction methods for creep of concrete are compared with experimental data and evaluated by statistical methods. Predicted creep functions exhibit average coefficients of variation between 25,o and 39,3 percent. Average errors of a final creep coefficient may range from 23,2 to 61,9 percent depending on the particular method considered. In a detailed evaluation the sources of errors are pointed out.

Nach verschiedenen Vorhersageverfahren ermittelte Kriechverformungen werden mit Versuchsergebnissen verglichen und mit Hilfe eines statistischen Verfahrens beurteilt. Es zeigt sich, daß bei der Vorherbestimmung des zeitlichen Verlaufs von Kriechfunktionen je nach Verfahren mit einem mittleren Variationskoeffizienten zwischen 25,o % und 39,3 % gerechnet werden muß. Für die Endwerte der Kriechfunktionen ergeben sich Fehler zwischen 23,2 % und 61,9 %. Eine differenzierte Auswertung zeigt Fehlerquellen bei den einzelnen Vorhersagemethoden auf.

1 INTRODUCTION

Various methods have been developed by national and international organizations which allow an estimate of instantaneous and time dependent concrete strains. In order to obtain some insight into the reliability of various methods to predict concrete strains an analytical study is under way at our institute in which concrete strains particularly creep strains as predicted by various methods are compared to experimental data. Some of the results will be presented in the following. Furthermore an attempt will be made to analyse the sources of deviations between predictions and experimental data and to make suggestions for improvement.

In this study the following prediction methods will be dealt with:

a) Method proposed by ACI committee 2o9 (ACI), /1/
b) Method proposed by Bazant and Panula (BaP), /2/
c) Method proposed in the CEB/FIP Model Code 1978 (C78), /3/
d) Method of the German prestressed concrete code (DIN), /4/
e) Method proposed in the CEB/FIP recommandations 197o (C7o), /5/
f) Method proposed by the British Concrete Society (BCS), /6/.

2 BRIEF DESCRIPTION OF PREDICTION METHODS

On the basis of the mathematical formulation the prediction methods may be subdivided into two groups. In the first group which includes the methods ACI, C7o and BCS creep coefficients $\varphi(t)$ are obtained from product functions:

$$\varphi(t) = k_1 \ldots k_i \ldots k(t) \tag{1}$$

In eq. (1) k_i are independent coefficients to describe the effect of particular parameters upon concrete creep; k(t) describes development of creep with time.

The BCS method is a simplified version of the C7o method. It allows an estimate of final creep. No continuous function for the development of creep with time is given.

The second group includes the methods BaP, C78 and DIN, in which total creep is described as the sum of individual strain components. In C78 and DIN creep is subdivided into a delayed elastic strain (reversible creep) and into flow (irreversible creep):

$$\varphi(t) = \varphi_{rev}(t) + \varphi_{irr}(t) \tag{2}$$

In eq. (2) irreversible creep may be estimated from the product of several interrelated coefficients. The development of creep with time as well as the effect of age at loading may be estimated from one continuous time function. Reversible creep is independent of age at loading and depends only on time under load. In C78 irreversible creep is subdivided into flow which occurs during the first 24 hours after load application and into flow after 24 hours. In the BaP-method concrete strains are separated into components which are linked to different creep mechanisms. The following creep function is proposed:

$$\Phi(t) = 1/E_o + C_o(t) + C_d(t) - C_p(t) \tag{3}$$

In eq. (3) $1/E_o + C_o(t)$ describe the sum of elastic strain and basic creep, $C_d(t)$ represents drying creep and $C_p(t)$ describes the decrease of creep due to moisture loss. In contrast to all other methods the creep function used in BaP does not approach a final value. For practical purposes a "final creep value" may be estimated from creep after 5o years of loading.

In Table 1 the parameters which are taken into account in the various methods are summarized. From this it follows that the methods differ considerably in the number of parameters which are taken into account. The methods ACI, C7o and BaP can only be applied accurately if concrete composition is known.

3 CHOICE OF EXPERIMENTS

For the evaluation only normal weight structural concretes have been used. Creep strains had to be observed under a constant sustained load at normal environmental conditions over a period of at least one year. In the majority of experiments evaluated the duration of loading exceeded 3 years. The effect of the following experimental parameters has been studied: age at loading $1 \leqslant t_o <$ 33oo days; $45 \leqslant RH \leqslant$ 1oo percent; $7o \leqslant$ size $\leqslant$ 61o mm; 17 N/mm2 < compressive strength < 55 N/mm^2; normal, slow or rapid hardening portland cements. So far 1o2 experiments have been evaluated. They will be described in more detail in a report to be published. A detailed list of references is available now upon request.

In the evaluation creep functions rather than creep coefficients have been presented. Diagrams as shown in Fig. 1 have been prepared for the evaluation with the aid of suitable computer programs.

4 STATISTICAL EVALUATION

The individual experimental data presented on a log time scale have been hand smoothed. For each method a standard error s_i has been determined from the differences between measured and calculated values; these differences have been taken at constant intervals along the log time axes.

$$s_i = \sqrt{\frac{1}{n-1} \cdot \sum_{j=1}^{n} (\Delta\Phi_{ij})^2} \tag{4}$$

In eq. (4) $\Delta\Phi_{ij}$ describes the difference between predicted and experimental strain at time j of experiment i; n presents the number of differences taken for each experiment. A mean creep function value $\bar{\Phi}_i$ has been calculated for each experiment.

$$\bar{\Phi}_i = \frac{1}{n} \cdot \sum_{j=1}^{n} \Phi_{ij} \tag{5}$$

TABLE 1

Required Input Data for Various Prediction Methods

	prediction method					
required input data	ACI	BaP	C78	DIN	BCS	C7o
size and shape of member	x	x	x	x	x	x
age at loading	x	x	x	x	x	x
humidity of environment	x	x	x	x	x	x
compressive strength of concrete	x	x	x	x	x	x
consistency of fresh concrete	x		x	x		
temperature		x	x	x		x+)
type of cement	x	x	x	x	x	x
cement content		x				x
unit weight of concrete	x	x				
water-cement ratio		x				x
sand-aggregate ratio	x	x				
aggregate-cement ratio		x				
air content of concrete	x					
curing conditions (humidity and beginning of drying)		x				

+) During the curing period prior to loading

TABLE 2

Results of Statistical Evaluation

		prediction method					
	N	ACI	BaP	C78	DIN	BCS	C7o
$\bar{V}$(%)	1o2	-	32,o	24,1	25,1	-	23,1
	72	24,8	31,o	24,5	26,1	-	22,5
$\bar{\bar{V}}$(%)	1o2	-	39,3	27,6	28,4	-	25,5
	72	28,3	35,5	28,2	29,1	-	25,o
$\bar{F}$(%)	1o2	-	61,9	27,2	23,2	-	24,2
	72	27,8	53,6	28,5	24,5	31,2	24,9

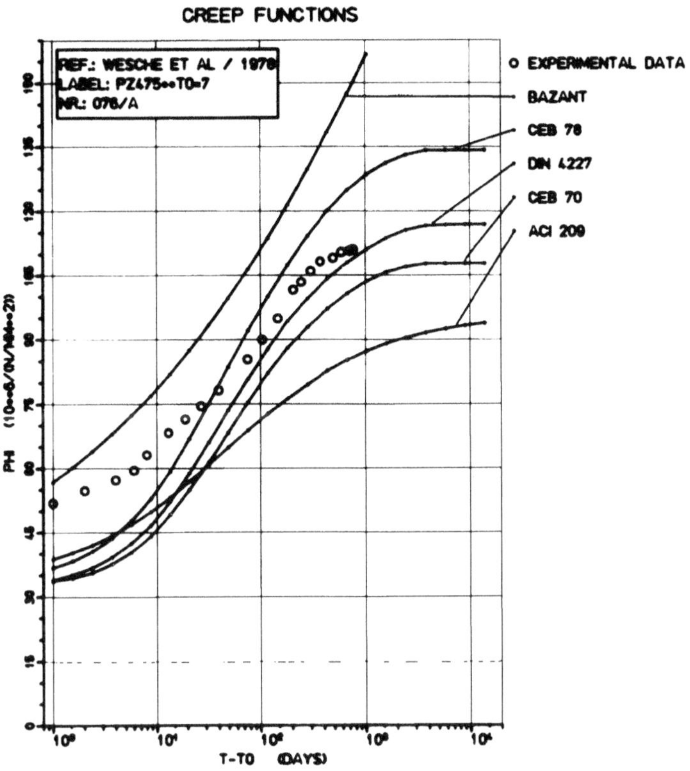

Fig. 1 Creep functions of various prediction methods

In eq. (5) Φ_{ij} describes the strain observed at time j. From this a coefficient of variation V_i can be determined:

$$V_i = \frac{s_i}{\bar{\Phi}_i} \cdot 100\% \tag{6}$$

To evaluate the accuracy of a particular method on the basis of N experiments the following values have been calculated:

$$\bar{V} = \frac{1}{N} \cdot \sum_{i=1}^{N} V_i, \text{ and } \bar{\bar{V}} = \sqrt{\frac{1}{N} \cdot \sum_{i=1}^{N} V_i^2} \tag{7}$$

As a consequence of the procedure described above, the evaluation gives more weight to short and mean durations of loading. However, because of the practical significance of creep after longer periods of loading also the final values of the creep functions obs $\Phi_{i\infty}$

have been estimated using the method of Ross (7). The error F_i may be estimated as follows:

$$F_i = \frac{\text{cal } \Phi_{i\infty} - \text{obs } \Phi_{i\infty}}{\text{obs } \Phi_{i\infty}} \cdot 100\% \qquad (8)$$

Finally, a mean error for each particular prediction method has been calculated from N experiments:

$$\bar{F} = \sqrt{\frac{1}{N} \cdot \sum_{i=1}^{N} F_i^2} \qquad (9)$$

5 RESULTS OF EVALUATION

Table 2 summarizes the results of the evaluation. Because of limitations in the range of applicability of some methods not all experiments could be used for the evaluation of all methods.

The various methods are not equally reliable for various ranges and combinations of parameters. Therefore, in Table 3 average coefficients of variation of creep functions are summarized for an age of loading of less than 7 days (A), for an age of loading more than 6o days (B), for experiments which describe normal conditions of structural concrete (C) and for mass concrete (D).

In Table 4 the effects of specimen size, relative humidity and age of loading as predicted by the various methods are compared to experimental data.

6 DISCUSSION AND CONCLUSIONS

6.1 Table 2 demonstrates that the more complicated procedures are not necessarily the more accurate ones. However, with the exception of the method BaP the difference in accuracy between the various methods is minor.
The simple method BCS gives surprisingly good results.

6.2 According to Table 3 the ACI-method is weak for higher ages at loading and for basic creep. The BaP-method is erroneous for very young concretes.
The methods C78, C7o and DIN show relatively poor results for drying creep but particularly C78 and C7o are in good agreement with creep data for basic creep. Similar tendencies are observed for the average error of final values of the creep functions.

TABLE 3

Results of Detailed Statistical Evaluation

	prediction method	$\bar{V}$ (%)	$\bar{\bar{V}}$ (%)	N	parameters
A	BaP C78 DIN C7o	43,8 24,1 21,2 25,3	58,8 27,2 25,6 27,3	17	$t_o < 7$ d
B	ACI BaP C78 DIN C7o	31,2 27,8 25,8 23,6 24,1	36,7 3o,6 29,1 28,2 26,1	12	$t_o > 6o$ d
C	ACI BaP C78 DIN C7o	25,3 33,7 28,7 29,5 26,3	27,4 37,3 32,o 31,8 28,o	28	7 d $\leqslant t_o \leqslant$ 6o d 2oo mm $\leqslant \varnothing \leqslant$ 6oo mm drying creep
D	ACI BaP C78 DIN C7o	32,8 21,7 14,9 2o,o 17,o	37,3 22,8 18,4 24,o 18,2	14	$t_o \geqslant 7$ d basic creep

6.3 Table 4 clearly demonstrates that compared to experimental data the methods BaP, C78, DIN underestimate the effect of member size whereas the methods ACI and C7o in many cases overestimate the influence of this parameter. The effect of rel. humidity is underestimated by all methods. The effect of age at loading is overestimated or underestimated by the various methods depending on the particular range considered.

6.4 The errors in the prediction of creep functions are large no matter which method is taken into account. Therefore, some consideration should be given to the inherent scatter of experimental data under controlled laboratory conditions. In our analysis creep functions of concrete specimens made from one batch as well as creep functions of similar concrete specimens made from different batches and fabricated at different times and in different laboratories - but exposed to creep under identical conditions - are compared to each other. The following coefficients of variation have been estimated:

one batch: $\bar{\bar{V}}$ = 5,o %
different batches: $\bar{\bar{V}}$ = 8,5 %

TABLE 4

Effect of Member Size, Relative Humidity and Age at Loading on Creep of Concrete -Experimental Data and Predictions-

		test result	ACI	BaP	C78	DIN	BCS	C7o	Ref.
effect of member size	$\frac{\varphi_\infty(\emptyset = 152\ mm)}{\varphi_\infty(\emptyset = 61o\ mm)}$	1,24	1,43	o,91	1,11	1,13	1,29	1,44	8
	$\frac{\varphi_\infty(\emptyset = 7o\ mm)}{\varphi_\infty(\emptyset = 2oo\ mm)}$	1,2o	1,24	1,o1	1,o5	1,o3	1,o	1,18	9
	$\frac{\varphi_\infty(\emptyset = 1oo\ mm)}{\varphi_\infty(\emptyset = 6oo\ mm)}$	1,24	1,59	1,o3	1,1o	1,12	1,3o	1,57	1o
effect of rel.humidity	$\frac{\varphi_\infty(H = 5o\ \%)}{\varphi_\infty(H = 1oo\ \%)}$	3,45	-	1,37	2,75	3,o4	3,65	2,87	11
	$\frac{\varphi_\infty(H = 5o\ \%)}{\varphi_\infty(H = 1oo\ \%)}$	3,12	1,56	2,34	2,77	3,o8	3,65	2,87	12
	$\frac{\varphi_\infty(H = 5o\ \%)}{\varphi_\infty(H = 99\ \%)}$	4,22	1,54	1,98	2,7o	2,89	2,47	2,73	13
effect of age at loading	$\frac{\varphi_\infty(t_o = 1\ d)}{\varphi_\infty(t_o = 28\ d)}$	1,26	-	2,o4	1,95	1,81	-	2,43	14
	$\frac{\varphi_\infty(t_o = 3\ d)}{\varphi_\infty(t_o = 28\ d)}$	1,22	-	1,62	1,57	1,47	-	1,6o	14
	$\frac{\varphi_\infty(t_o = 21\ d)}{\varphi_\infty(t_o = 365\ d)}$	1,92	-	2,3o	2,92	2,78	2,61	2,17	13
	$\frac{\varphi_\infty(t_o = 28\ d)}{\varphi_\infty(t_o = 329o\ d)}$	6,62	1,83	2,96	6,46	5,76	-	2,71	14

Apparently, there is quite some room for improvement of the prediction methods.

6.5 Such improvement may be accomplished by improved optimization of the methods including the relations to predict elastic strains. Furthermore, experimental values for the elastic strains rather than predicted values may be used. On the basis of 5o experiments the value $\bar{\bar{V}}$ for the creep function according to DIN decreased from 27,1% to 19,5% when obs E has been used instead of cal E. Similar tendencies have been reported in /2/.

6.6 Considering the large errors in the prediction of creep functions for creep sensitive structures an analysis should be carried out both for lower and for upper bound creep functions.

7 REFERENCES

/1/ ACI Committee 2o9, subcommittee II, "Prediction of creep, shrinkage and temperature effects", 2. Draft October 1978

/2/ Z.P. Bazant, L. Panula "Practical prediction of time-dependent deformations of concrete", Part I: Shrinkage, Part II: Basic creep, Materials and Structures, Vol. 11, No. 65, 1978, Part III: Drying creep, Part IV: Temperature effect on basic creep, Materials and Structures, Vol. 11, No. 66, 1978, Part V: Temperature effect on drying creep, Part VI: Cyclic creep, nonlinearity and statistical scatter, Materials and Structures, Vol. 12, No. 69, 1979.

/3/ CEB/FIB Model Code, Paris, London, Berlin 1978

/4/ DIN 4227, Teil 1, Spannbeton, Ausgabe Dezember 1979

/5/ CEB/FIP International recommendations for the design and construction of concrete structures, Paris, London, 197o

/6/ British Concrete Society, Movements working party, "A simplified method for estimating the elastic modulus and creep of normal weight concrete", Cement and Concrete Association Training Centre, TDH 7376, June 1978

/7/ A.D. Ross "Concrete creep data", Structural Engineer, Vol. 15, No. 8 (1937)

/8/ T.C.Hansen, A.H.Mattock "Influence of size and shape of member on the shrinkage and creep of concrete", American Concrete Institute, Journal, Vol. 63, February 1966

/9/ M.Mamillan "Les mesures de déformations sur ouvrages en béton précontraint", Supplément aux Annales de l'Institut Technique du Bâtiment et des Travaux Publics, No. 2o7/2o8, 1965, pp. 346-354.

/1o/ G.Weil "Influence des dimensions et des contraintes sur le retrait et le fluage du béton", RILEM Bulletin, No.3, July 1959.

/11/ M.M.Mamillan "Evolution du fluage et des propriétés du béton", Supplément aux Annales de l'Institut Technique du Bâtiment et des Travaux Publics, No. 154, Octobre 196o

/12/ G.E.Troxell, I.M. Raphael, R.E. Davis "Long Time creep and shrinkage tests of plain and reinforced concrete", Proceedings ASTM, Vol. 58, 1958, pp. 11o1-112o.

/13/ R.L'Hermite, M.Mamillan, C.Lefèvre "Nouveaux Résultats de Recherche sur la Déformation et la Rupture du Béton", Annales de l'Institut Technique du Bâtiment et des Travaux Publics, No. 2o7/2o8, 1965, pp. 325-345.

/14/ K.Wesche, J.Schrage, W. vom Berg "Versuche zum Einfluß des Belastungsalters auf das Kriechen von Beton, DAfStb Heft 295, Berlin 1978

ON THE PREDICTION OF CREEP FUNCTIONS FOR CONCRETE

Lauge Fuglsang Nielsen

Building Materials Laboratory
Technical University of Denmark
Copenhagen

ABSTRACT

A composite theoretical method is demonstrated which predicts the rheological parameters applying to concrete - including the Young's modulus and the shape of the creep developing function. Only three parameters control the creep function. These are the Young's modulus, E_C, the delayed elastic coefficient, α_C, and the flow modulus, K. They are all three related in a simple way to concrete composition. That is water- and aggregates/cement ratios, W/C and A/C. The resulting creep function is dependent on the time of very first loading. Thus, the phenomenon of consolidation is included. The prediction method is consistent with methods of creep analysis previously developed by the author. These are the so-called Modified- or Improved Dischinger Methods.

ZUSAMMENFASSUNG

Die rheologischen Koeffizienten für Beton werden vorausgesagt auf Grundlage der Komposittheorie; der Elastizitätsmodul sowie die Form der Kriechfunktion sind hier einbegriffen. Die Kriechfunktion ist nur durch drei Koeffizienten festgelegt: Elastizitätsmodul, E_C, verzögert-elastischer Koeffizient, α_C, und Fliss-modul, K. Alle drei lassen sich auf einfache Weise von der Betonzusammensetzung ableiten, d.h. vom Wasser-Zement-Verhältnis (W/Z) und Zuschlag-Zement-Verhältnis (A/Z). Die erreichte Kriechfunktion ist abhängig vom Zeitpunkt der allerersten Belastung des Betons, das Phänomen der Konsolidation ist somit einbegriffen. Die aufgestellte Methode ist in Übereinstimmung mit den früher vom Verfasser entwickelten Kriechberechnungsverfahren (Modified- oder Improved Dischinger Methods).

INTRODUCTION

One way of establishing a creep prediction method is to consider concrete as being a composite made of *varying* volume fractions of water, cement, gel (hydration products) and aggregates - each component having basicly *fixed* mechanical properties (1-3). Water, cement and aggregates behave practically according to very simple stress strain relations. At the present time however a number of hypothetical assumptions are needed when characterizing the properties of the gel phase. In a recent theory developed by the present author (4) such assumptions have been avoided by considering concrete as being a three phase composite made of *fixed* volume fractions of water, basic paste and aggregates. The basic paste which has *varying* mechanical properties is defined to be that part of the water-cement system (cement paste) which will hydrate 100 % corresponding to an effective water-cement ratio (by weigth) of W/C = 0.38 (5). The mechanical properties of the basic paste are those which may be deduced from experiments directly on concrete (or cement paste). The aggregates are hereby considered as being linear elastic while the water (in excess of W/C = 0.38) is considered to posess only a negligable stiffness.

Practically we may consider concrete to behave like a linear viscoelastic material as long as the stress applied is less than about 40 % of the concrete strength (6). This concept implies that also the basic paste is linear viscoelastic. It may also be deduced that the constitutive equation of basic paste must be simpler than the one applying to concrete. As concrete behaves practically as a Burger material with a "rapidly working" Kelvin component (7,8) it is then justified as to assume a Pseudo Maxwell behavior of the basic paste. That is, the mechanical model is a Burger model where the delayed elastic strain develops allmost instantaneously, see figure 1. The creep function of a Pseudo Maxwell material (PM material) may be written as follows.

$$c(t,\tau) = \frac{1}{E(\tau)} \{1 + \alpha(\tau)\cdot H(t,\tau)\} + v(t,\tau) \tag{1}$$

where

$$v(t,\tau) = \int_{\tau}^{t} \frac{1}{\eta(\theta)}\, d\theta = F(t) - F(\tau) \quad ; \quad (\dot{F} = \frac{dF}{dt} = \frac{1}{\eta(t)}) \tag{2}$$

Time in general and time at load application are denoted by t and τ respectively. The Young's modulus is $E(t)$. The developing function, $H(t,\tau)$, of the delayed elastic strain has a value of 0 at $t = \tau$. For increasing t the final value of 1 is obtained very rapidly such that the delayed elastic strain may be expressed as a fraction of the corresponding elastic strain, that is by $\alpha(\tau)/E(\tau)$. The flow function, $F(t)$, expresses the viscous part, $v(t,\tau)$, of the creep function. The viscosity is denoted by $\eta(t)$.

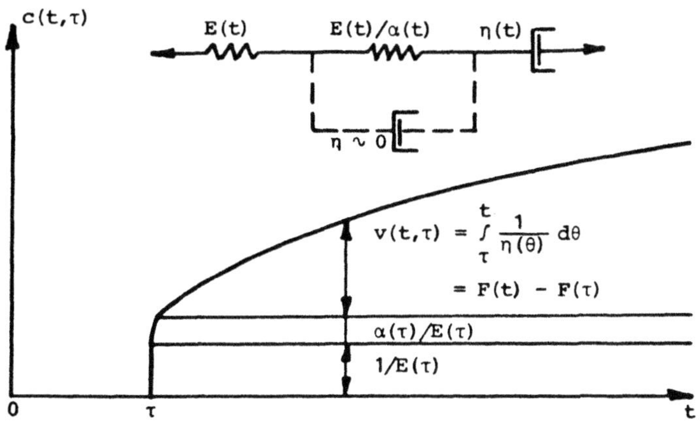

Figure 1. Model and creep function of a Pseudo Maxwell material.

The above mentioned informations on the mechanical behavior of water, basic paste and aggregates have been utilized in a composite theoretical analysis (4) of a number of concrete- and cement paste test results as to deduce quantitatively the rheological parameters of basic paste. The results obtained - and the composite theory - were then used to establish prediction formula for the creep parameters of concrete with *hard* aggregates. These formula are presented in the following sections, and an example is given as to demonstrate their applicability.

PREDICTION METHOD

Informations needed

Concrete composition - is characterized by the following volume parameters. The water- and aggregates/cement ratios (by weigth) are denoted by W/C and A/C respectively.

$$A_w = \begin{cases} \left(\frac{0.38}{W/C}\right)^{1.3} & \text{when } W/C \geq 0.38 \\ 1 & \text{when } W/C < 0.38 \end{cases} \quad (3)$$

$$c_a = \begin{cases} \frac{38 \cdot A/C}{100 \cdot W/C + 38 \cdot A/C + 32} & \text{when } W/C \geq 0.38 \\ \frac{38 \cdot A/C - 84 \cdot W/C + 32}{100 \cdot W/C + 38 \cdot A/C + 32} & \text{when } W/C < 0.38 \end{cases} \quad (4)$$

$$A_a = \frac{1 - c_a}{1 + c_a} \quad (5)$$

The capillary pores of the cement paste are considered by A_w. The exponent, 1.3, applies for concretes of normal *consistency*. At decreasing quality of consistency the exponent will increase in magnitude. The volume fraction of aggregates in the concrete is denoted by c_a.

Degree of hydration - is denoted by g. At high relative humidities it may be given the following time dependence when $t > 0.2$ days.

$$g(t) = 1 - 0.5\left(\frac{3}{\rho t}\right)^{0.2\cdot\kappa} \quad ; \ (\kappa \simeq 1) \tag{6}$$

where ρ depends on temperature, T (^{o}C), and type of cement used, that is normally- (NPC) or rapidly hardening (RPC) Portland cement.

$$\rho_{NPC} = \left(\frac{T + 10}{30}\right)^{2.5} \quad ; \quad \rho_{RPC} = \left(\frac{T + 15}{30}\right)^{2.5} \tag{7}$$

The κ-parameter varies somewhat with W/C (4). However, for any practical purpose (W/C = 0.25 - 0.65) a value of 1 is very good.

Surface parameter - is denoted by R. It is defined as follows.

$$R = \frac{2\cdot V}{S} \quad \text{(in cm)} \tag{8}$$

where the volume and the surface of the structural element considered are given by V (cm^3) and S (cm^2) respectively.

Relative humidity - is denoted as follows.

RH_m in the central parts of the structure
RH_b at the boundary of the structure (curing humidity) (9)

Aggregates elasticity - is characterized by the Young's modulus, E_a. The theory presumes hard aggregates. That is

$$\frac{E_a\,(MPa)}{3.2\cdot 10^4\,(gA_w)} > 1.5 - 2 \tag{10}$$

When the paste-aggregates *bond is defective*, E_a should be replaced by an effective quantity, $E_{a,eff} = \gamma\cdot E_a$, where the efficiency parameter, γ = effective surface/total surface of the aggregates.

Creep functions for concrete

For moisture cured concrete at room temperature we predict,

$$c(t,\tau) = \frac{1}{E_c(\tau)}\left\{1 + \alpha_c\left(1 - \left(\frac{\tau}{t}\right)^2\right)\right\} + \frac{1}{K}\left\{\log_e\left(\frac{t}{\tau}\right) + \sqrt{\frac{7}{t_o}}\left(\frac{t_o}{\tau}\right)^2\left(1 - \left(\frac{\tau}{t}\right)^2\right)\right\} \ /MPa \tag{11}$$

where time in general and time at load application are denoted by t (days) and τ (days) respectively. The time at which the concrete is *first loaded* is denoted by t_o (days). Other input data in eq. 11 are the following.

The *calculational Young's modulus* - is given by

$$E_c = E\cdot(gA_w)^{0.2\cdot c_a} \quad ; \quad (c_a < 0.9,\ gA_w > 0.1) \tag{12}$$

where the *dynamic Young's modulus*, E, is determined by

$$E = 3.2\cdot 10^4 (gA_w) \frac{A_a(gA_w) + N}{(gA_w) + A_a N}\ \text{MPa} \quad ; \quad (N = \frac{E_a(\text{MPa})}{3.2\cdot 10^4}) \tag{13}$$

The calculational Young's modulus represents elastic strain and the very "fast" part of the delayed elastic strain. Its magnitude is less than the dynamic Young's modulus and greater than the static Young's modulus, see (4).

The *calculational delayed elastic coefficient* - is given by

$$\alpha_c = 1 - c_a \tag{14}$$

The *flow modulus* - is given by

$$K = 5.7\cdot 10^4 \frac{A_w}{A_a}\ \text{MPa} \tag{15}$$

The first term in eq. 11 represents strain which is basicly elastic. The second term describes viscous strain as it corresponds to $v(t,\tau)$ in figure 1 and eqs. 1 and 2. It is noticed that the so-called *consolidation strain* (8) is included. That is a strain which is caused exclusively by load variations relatively close to the time, t_o, when first loading the concrete. The phenomenon of consolidation is discussed in more details in (4,8).

Experimentally determined creep functions are so-called *virgin creep functions*, $c_V(t,\tau)$. That is, they refer to concretes which have never been loaded until $t = \tau$. Obviously virgin creep functions are obtained by eq. 11 introducing $\tau = t_o$. The result is,

$$c_v(t,t_o) = \frac{1}{E_c(t_o)}\{1 + \alpha_c(1 - (\frac{t_o}{t})^2)\} + \frac{1}{K}\{\log_e(\frac{t}{t_o}) + \sqrt{\frac{7}{t_o}}\,(1 - (\frac{t_o}{t})^2)\}\ /\text{MPa} \tag{16}$$

Modifications as to consider different curing conditions

Creep functions for concrete exposed to constant curing conditions (RH_b,T) different from moisture curing at room temperature are determined by eqs. 11 and 16 introducing the following modifications.

Viscous strain - is modified by replacing the flow modulus, K, and the quantity, 7, in the developing function with the following quantities.

$$K \rightarrow K/(\gamma_T \cdot \gamma_H) \quad ; \quad 7 \rightarrow 7/\rho \tag{17}$$

where temperature, T, and relative humidity, RH, are considered by ρ as given in eq. 7 - and the shift factors

$$\gamma_T = \rho^{0.7} \quad ; \quad \gamma_H = RH_m + 5\,\frac{|RH_m - RH_b|}{1 + R/50} \tag{18}$$

where the relative humidities, RH_m and RH_b, and the surface parameter, R, are defined in eqs. 9 and 8 respectively.

Elastic- and delayed elastic strains - are modified by replacing the Young's modulus, E_c, and the exponent, 2, in the developing function with the following quantities.

$$E_c \rightarrow E_c\,\frac{2 + RH_b + (2 + RH_m)\cdot R/50}{3\cdot(1 + R/50)}\;\frac{220 - T}{200} \tag{19}$$

$$2 \rightarrow 2\cdot\gamma_T\cdot\gamma_H \tag{20}$$

Static and dynamic Young's moduli transform in a way similar to E_c.

The modifications, 17 - 20, are estimated to apply practically when $RH_{b,m} > 0.2$ and $-10\ ^oC < T < 100\ ^oC$.

Preconditioning

The flow term (viscous strain) in eq. 11 is the result of a rate process defined by the following flow rate at moisture curing and room temperature, see eqs. 2 and 11.

$$\frac{dF}{dt} = \frac{1}{K}\left\{\frac{1}{t} + \frac{2}{7}\left(\frac{7}{t_o}\right)^{3/2} \cdot \left(\frac{t_o}{t}\right)^3\right\}\ /\text{MPa/day} \tag{21}$$

Deviations from the reference curing conditions are considered using shift factors as indicated in eqs. 17 and 18.

When preconditioning (for example steam curing) has raised the degree of hydration to g_s over a period of s days we may establish fictively the reference conditions for t > s by introducing a time shift according to

$$(t,t_o,\tau) \rightarrow (t,t_o,\tau) + \left(\frac{0.1}{\rho(1 - g_s)^5} - s\right)\ \text{days} \tag{22}$$

which is derived by eq. 6 with ρ corresponding to the temperature applying for $t > s$. The time shift means that we consider the concrete to be x days older than what it actually is. x is the correction term in eq. 21.

The time shift is introduced consequently in the right side of eqs. 11 and 16 - with the exception that E_c developes according to the actual degree of hydration. For most practical purposes where the central parts of the concrete dry slowly we may, however, also here shift the time according to eq. 22. That is, we may use a degree of hydration which develops approximately according to eq. 6 with shifted time.

EXAMPLE

The prediction method presented has been applied to a concrete having W/C = 0.55, A/C = 5.5 and $E_a = 7.5 \cdot 10^4$ MPa. The curing conditions are constant such that the temperature is $T \equiv 20$ °C. The concrete is used in a structural element the surface parameter of which is R = 30 cm. It is assumed that the internal relative humidity does not vary much from $RH_m = 1$ - and that the hydration developes approximately as described by eq. 6.

Some results of the prediction of creep functions are shown in figure 2. It is here illustrated how time at loading (t_o) and relative humidity (RH_m) influence the composition of virgin creep functions 55 years after the load application.

Some codes define the creep function in the following way.

$$c(t,\tau) = \frac{1}{E_{st}(\tau)} (1 + \phi(t,\tau)) \tag{23}$$

where E_{st} is the static Young's modulus and ϕ is the so-called creep factor. Knowing the creep function (eventually from experiments) the creep factor is determined by

$$\phi(t,\tau) = E_{st} c(t,\tau) - 1 \tag{24}$$

The static Young's modulus is a very "softly" defined quantity. A number of definitions exist. Formula are often suggested which predict E_{st} from the concrete strength. The uncertainty on E_{st} is very unfortunate because it influences very much the creep factor. It is easily derived from eq. 24 that

$$\frac{\Delta\phi}{\phi} = \frac{\phi' - \phi}{\phi} = \left(\frac{E'_{st}}{E_{st}} - 1\right)\left(1 + \frac{1}{\phi}\right) \tag{25}$$

where (E'_{st}, ϕ') and (E_{st}, ϕ) are associated pairs of quantities describing a given creep function. Eq. 25 tells us that two "choices" of E_{st} which differ 20 % (which is relevant) give $\Delta\phi/\phi > 40$ %

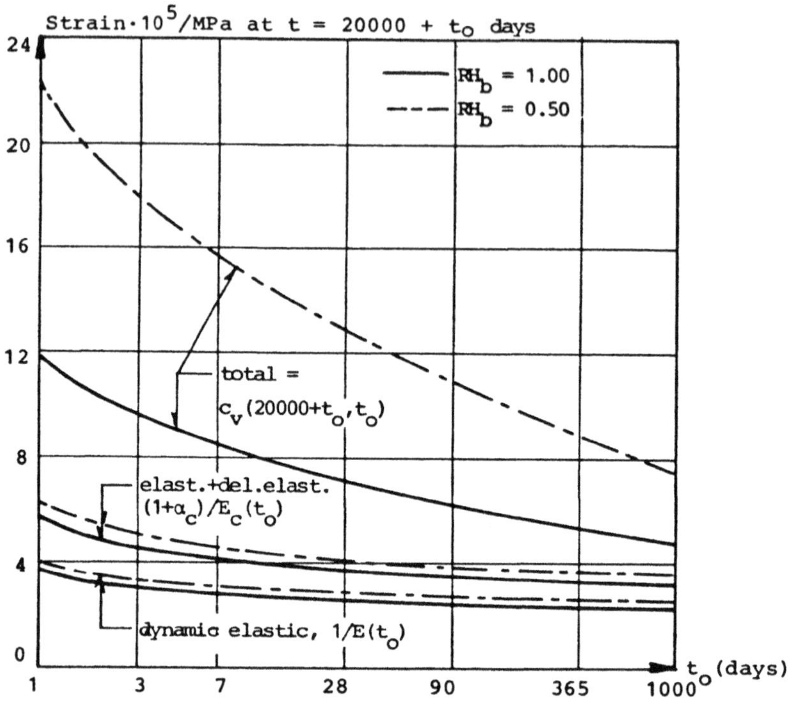

Figure 2. Strain components developed in concrete, constantly loaded over a period of 55 years. The load is applied at $t = t_o$. Thus, total strain = $c_v(20000 + t_o, t_o)$. Concrete composition and curing conditions are given in the text.

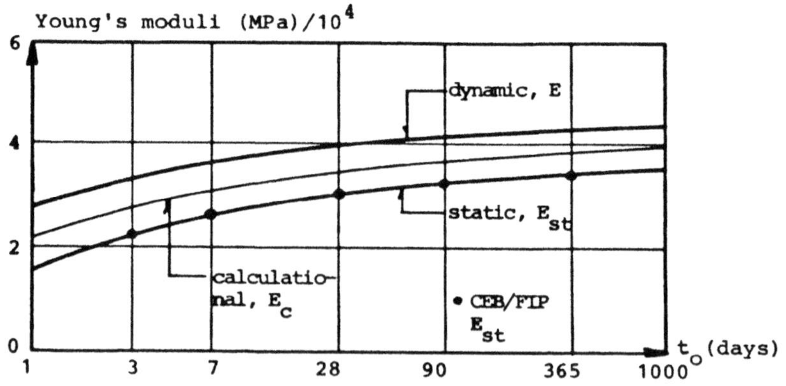

Figure 3. Different Young's moduli applying to the concrete defined in the text.

when $\phi < 1$. This feature is of course very disturbing when evaluating or comparing different codes and methods on the basis of eqs. 23 and 24.

The British Code for Concrete, CP110:1972 relates E_{st} to the dynamic Young's modulus, E, as follows.

$$E_{st} = 1.25(E - 1.5 \cdot 10^4) \text{ MPa} \tag{26}$$

For the concrete previously defined the Young's moduli, E, E_c and E_{st} have been calculated from eqs. 13, 12 and eqs. 13, 26 respectively. The results are shown in figure 3. It is interesting to notice that E_{st} determined in this way practically coincide with values obtained by the CEB/FIP Model Code for Concrete Structures: 1978 when a 28 days cylinder strength of $f_c(28) = 31$ MPa is *assumed*. The CEB/FIP code predicts empirically E_{st} by the following expression.

$$E_{st} = 9500 \cdot \sqrt[3]{f_c(t)} \text{ MPa} \tag{27}$$

The relative strengths, $f_c(t)/f_c(28)$, are given by 0.4, 0.65, 1, 1.2 and 1.35 for the ages 3, 7, 28, 90 and 365 days respectively when NPC concretes are considered.

Figure 4 shows some creep factors as they apply to the concrete previously defined when a static Young's modulus according to eq. 26 has been assumed. Also shown in the figure are creep factors suggested by the British Concrete Society, BCS:1978. The general trends of the two groups of creep factors are seen to be similar. However, the graphs predicted by the present method are more concentrated than those given by the BCS method. At high loading ages the BCS method predicts significantly lower values.

Several reasons may be given as to explain these differences: One reason may be that the BCS method is an average method which covers a broad class of concretes without considering the specific

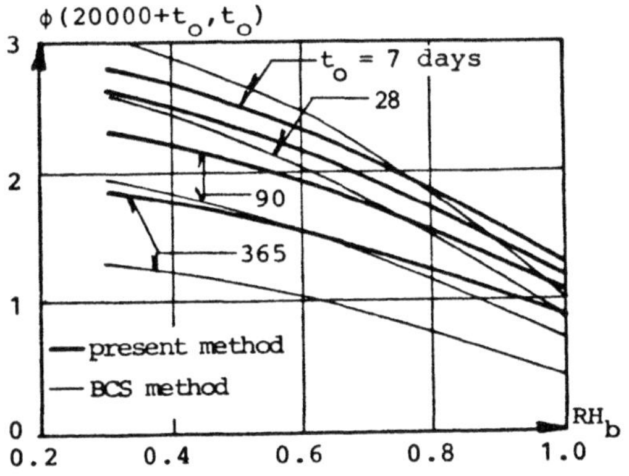

Figure 4. Creep factors as predicted for the concrete defined in the text. The BCS method does not consider details on concrete composition.

concrete composition. Another reason may be that the BCS method is associated with an E_{st}-estimate which is completely different from the expression 26, used in the present method. As previously indicated this feature may influence considerably the creep factors - especially the low ones. Finally, when real long time creep is considered, there are to day an insufficient number of test data. That is, the question on extrapolation becomes important when methods are compared. The present method considers creep as a process which never stops. Especially at high loading ages this concept favours higher creep factors.

FINAL REMARKS

Many current codes on creep of concrete are inconsistent in the way they suggest that creep functions can be constructed by multiplying a quantity, $1 + \phi(t,\tau)$, which is often evaluated precisely and very detailed, by another quantity (the elastic component) which is determined rather crudely by estimates based on strength data. Besides being un-logical this approach presumes indirectly that any strain component in concrete are strength dependent - and furthermore that they are equally strength dependent. These presumtions are by no means obvious - especially when normally loaded structures are considered.

The present work is an attempt to establish a more consistent approach to predict deformational behavior of concrete. On the basis of rheology and composite theory the present method predicts the total creep function. Only three rheological parameters are needed for the prediction. These are the Young's modulus, E_c, the flow modulus, K, and the delayed elastic coefficient, α_c. They are all three related in a very simple way to concrete composition, that is W/C and A/C. Different curing conditions are considered by modifying, essentially only E_c and K.

Finally, it should be mentioned that the prediction method presented is consistent with methods of creep analysis previously suggested by the present author (7-9). These methods may be used, for example, when conversions from creep- to relaxation functions are needed.

LITTERATURE

1. Hansen, Torben C. *Proceedings* 31(1960), Swedish Cement and Concrete Institute, Royal Inst. of Technology, Stockholm.

2. Nielsen, L. Fuglsang. "Strain Properties of Cement Paste and Concrete". *Part of a thesis*. Structural Laboratory, Technical University of Denmark, Copenhagen, 1967.

3. Bazant, Z.P. "Viscoelasticity of solidifying porous Materi-

als". *Research Report* 5(1977), Cement och Betong Institutet, Stockholm.

4. Nielsen, L. Fuglsang. "On the Prediction of Rheological Parameters for Concrete". "Nordic Concrete Research Seminar on Shrinkage-, Creep- and Temperature Deformations in Concrete Structures", 21. march 1980, Copenhagen. *Proceedings, DIA-LOG* 1/80(1980), DIA-B, Techn. Univ. Denmark.

5. Powers, T.C. and T.L. Brownyard. *ACI Journ., Proc.* 41(1946-1947), 101, 249, 469, 549, 669, 845, 865, 933, 971.

6. McHenry, D. *ASTM Proc.* 43(1943), 1069.

7. Nielsen, L. Fuglsang. *Beton und Stahlbetonbau* 65(1970), 272.

8. Nielsen, L. Fuglsang. *Cement and Concrete Res.* 7(1977), 149.

9. Nielsen, L. Fuglsang. "The Improved Dischinger Method as related to other Methods and practical Applicability". ACI Fall Convention in Houston, Texas oct.-nov. 1978. *Proceedings*.

Part V
Observations on Structures

CREEP AND SHRINKAGE BEHAVIOR OF TALL BUILDINGS AND LONG SPAN BRIDGES

Henry G. Russell

Director
Structural Development Dept.
Portland Cement Association
Skokie, Illinois, U.S.A.

K. Nam Shiu

Structural Engineer
Structural Development Dept.
Portland Cement Association
Skokie, Illinois, U.S.A.

ABSTRACT

Measured strains on concrete columns of two high-rise buildings and on three segments of one long span post-tensioned concrete box girder bridge are reported. For each structure, standard creep and shrinkage measurements were made on samples of concrete used in the structure.

RESUME

Les deformations mesurées sur des colonnes de béton de deux batiments élevés et sur trois voussoirs évides d'un pont précontraint de longue portée sont reportées. Des essais standards de retrait et de fluage ont été effectués sur des echantillons de béton utilisé dans chacune de ces structures.

HIGHLIGHTS

In recent years, the Construction Technology Laboratories of the Portland Cement Association has undertaken a series of projects concerned with creep and shrinkage of reinforced and prestressed concrete structures. All projects have included field instrumentation on actual structures and laboratory tests on samples of concrete used in the instrumented structures.

Instrumented structures have included both high-rise

buildings and long-span bridges. Concrete columns in three major buildings were instrumented during construction to measure time-dependent shortening. Two post-tensioned concrete box girder bridges were also instrumented to measure strains caused by instantaneous shortening, creep, and shrinkage. One bridge was cast-in-place and the other was precast.

The objective of this paper is to summarize the strain measurement data obtained on two buildings and one concrete bridge. For each structure, samples of concrete were obtained during construction. Creep and shrinkage data as measured on the samples are reported. Tests have been started at ages ranging from 28 to 730 days and have continued for as long as five years.

Evaluation of creep and shrinkage is an important consideration in design of tall concrete buildings and long-span bridges. Field data given this paper show the amount of deformation that can be expected in large structures. Laboratory data indicate basic properties of concretes used in actual structures.

BACKGROUND

The use of concrete in very tall buildings and in long span bridges has increased rapidly within North America during the past decade. Consequently, several projects have been undertaken to determine long-term deformations in some of these structures. All projects included field instrumentation on actual structures and laboratory tests on samples of concrete used in the instrumented structures.

FIELD MEASUREMENTS

Strains on structures were measured using internal strain meters and surface mounted mechanical strain gages.[1] Internal strain meters were installed prior to concrete placement. Mechanical strain gage points were installed on the concrete surface as soon as possible after formwork removal. Wherever possible, readings were taken during construction and subsequent occupancy of the structure. Measured strains included the combined effects of instantaneous shortening, creep, and shrinkage.

Instrumented structures included two high-rise buildings and a long-span bridge. Two major concrete buildings located in Chicago, were instrumented during construction. The buildings were Lake Point Tower and Water Tower Place, currently the world's tallest concrete building. The box girder bridge was Kishwaukee River Bridge near Rockford, Illinois.

Lake Point Tower

In August 1966, the Portland Cement Association began a field investigation at Lake Point Tower. The purpose of the investigation was first, to measure deformations of columns in an unusually tall structure during and after construction; second, to relate field measurements to laboratory data; and third, to develop methods for predicting time-dependent deformations and differential shortening between columns and core walls.

Lake Point Tower has a total height above grade of 197 m (645 ft). In plan, the building has three wings. Each wing is 20 m (65 ft) wide and extends 36 m (117 ft) from the center of the building. A photograph of the building is shown in Fig. 1. Other features of the building are described in more detail elsewhere.[2] Data on column shortening were obtained on a regular basis from August 1966 to December 1967 and have been reported previously.[2]

In 1978, additional measurements were taken on two columns at the first floor level. Measured total shortening of two 1.02 m (40 in.) diameter circular columns at the first floor level are shown in Fig. 2. The two columns had nominally identical loading histories and concrete properties. Differences between data for each column are well within the scatter for measured field data.

Water Tower Place

Water Tower Place, shown in Fig. 3, is a 76-story reinforced concrete building. Its record height of 262 m (859 ft) exceeds that of any reinforced concrete building previously constructed. The 13-story lower part of the structure occupies a city block 65 by 162 m (214 by 531 ft). The 63-story tower, located at one corner of the base structure measures 29 by 67 m (94 by 221 ft) in plan.

Sizes and reinforcement of columns and walls vary continuously throughout the height of the structure. Column concrete strengths vary from 62.1 MPa (9000 psi) at the lowest levels to 27.6 MPa (4000 psi) at the top of the building. Further details of the structure are given elsewhere.[3]

Fig. 1 Lake Point Tower

Fig. 3 Water Tower Place

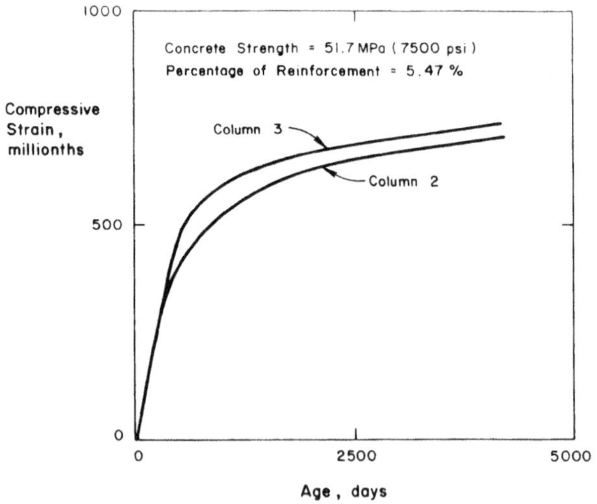

Fig. 2 Measured Strains on Columns at Lake Point Tower

Water Tower Place was selected for instrumentation to extend the work started at Lake Point Tower. This building was chosen because of the different sizes and layout of columns. In addition, Water Tower Place had higher design strength concrete in its lower columns.

Field measurements were made at six levels in Water Tower Place. Instrumented levels were selected to measure shortening in each of the five different concrete strengths. Two levels were located where 62.1 MPa (9000 psi) concrete was used in columns.

For each strain reading, corresponding air and concrete surface temperatures were also measured. All gage readings were corrected for temperature changes of concrete and reference standard bar by adjusting to a column temperature of 23C (73F). The coefficient of thermal expansion for each concrete was measured in the laboratory. Measured strains for all instrumented columns and walls in the building are shown in Fig. 4. Strains measured at all stories except levels 73-74 have approximately the same range of values. Strains in walls were less than those measured in the columns. For levels 14-15, 44-45, and 57-58, readings ceased when construction prevented access to gages. A more detailed description of this project has been given elsewhere.[3-5]

Kishwaukee River Bridge

Kishwaukee River Bridge consists of twin single-cell prestressed concrete box girders. Each bridge has five spans measuring 52 m (170 ft) for the end spans and 76 m (250 ft) for three interior spans. Total length of each structure is approximately 357 m (1170 ft).

The bridge is a constant depth precast concrete post-tensioned box girder. Longitudinal post-tensioning was applied to resist positive and negative moments. Transverse post-tensioning was provided in the roadway.

Each span was constructed by cantilevering from each main pier.[6] Longitudinal strains were measured on three segments of a main span cantilever. A photograph of the bridge and location of instrumented segments are shown in Fig. 5. Measured strains are shown in Fig. 6.

LABORATORY MEASUREMENTS

Coinciding with field measurements, laboratory tests were conducted on 152x305 mm (6x12 in.) cylinders made from the

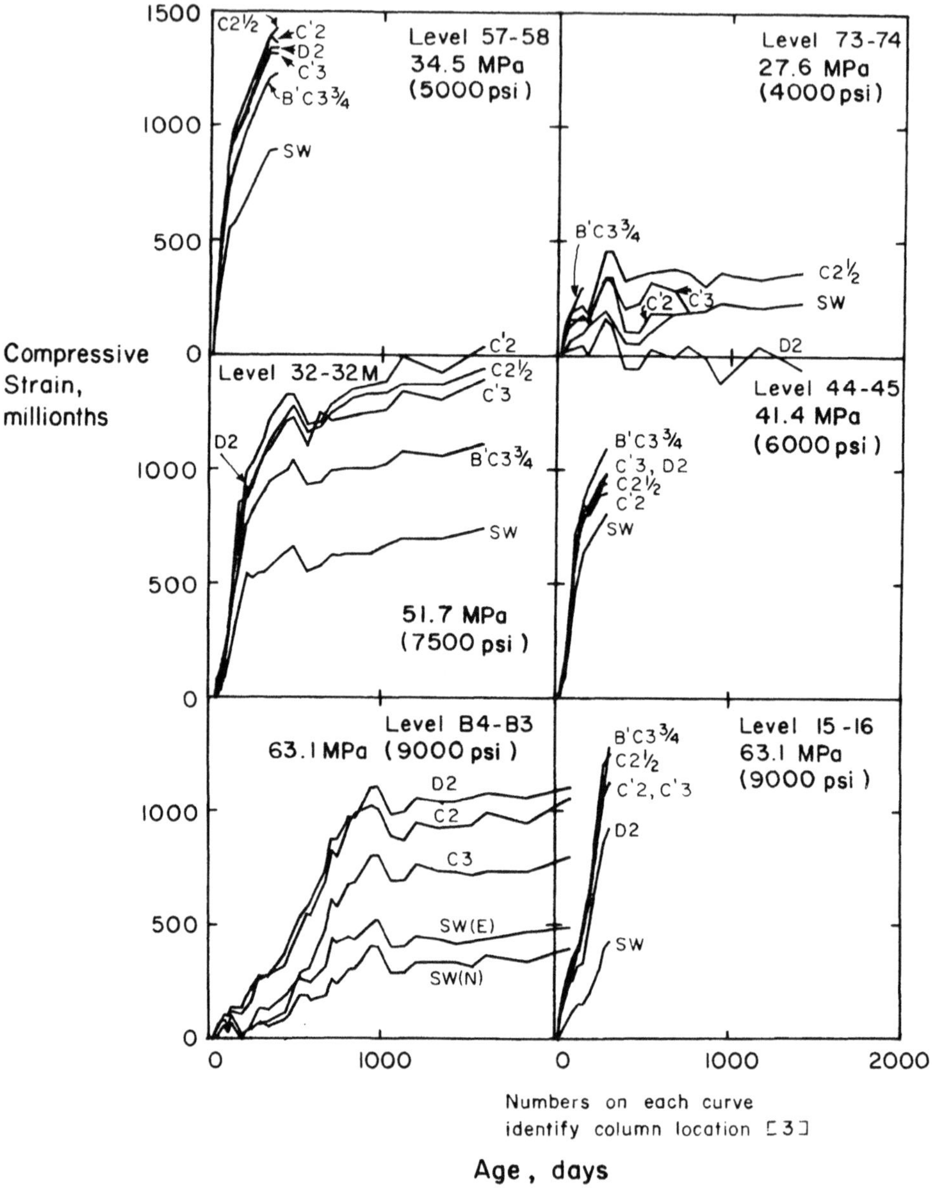

Fig. 4 Measured Vertical Shortening of Columns and Walls at Water Tower Place

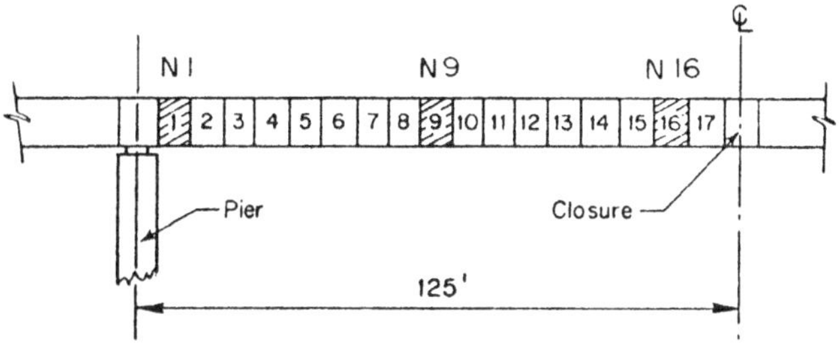

Fig. 5 Location of Segments on Kishwaukee River Bridge

same concrete used to cast the instrumented structures. Compressive strength, elastic modulus, coefficient of thermal expansion, creep, and shrinkage at various ages were measured. Only creep and shrinkage data are reported in this paper. Further details of laboratory measurements and other results are given elsewhere. [2,5] All creep and shrinkage data were obtained at 23C (73F) and 50% relative humidity except for outdoor specimens.

Creep

For each concrete sample, creep tests were started at ages ranging from 28 to 730 days. Cylinders were loaded in creep frames in accordance with procedures of ASTM Designation: C512. [7] Creep data for the three structures are shown in Fig. 7.

Shrinkage

Shrinkage strains on one set of cylinders for each concrete were measured from age 7 days. Data for the three structures are shown in Fig. 8.

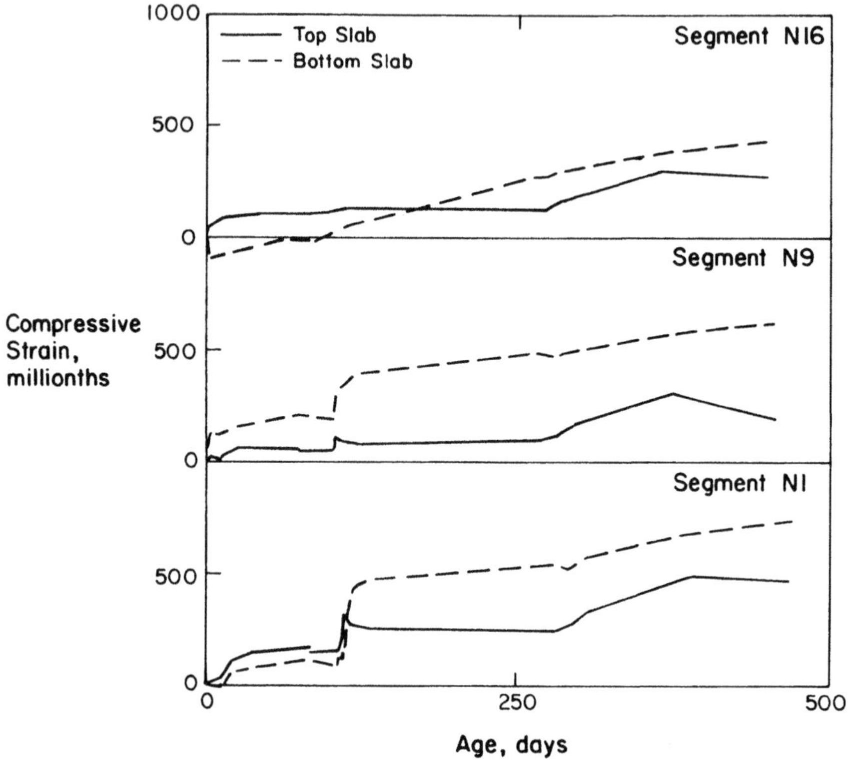

Fig. 6 Measured Longitudinal Strains on Kishwaukee River Bridge

CONCLUSIONS

This paper summarizes results from measurements on three unique concrete structures. Data indicate that axial strains in columns of a high-rise building were as high as 1500 millionths after 4 years. In the precast bridge, strains up to 750 millionths after 1-1/2 years were measured.

The data also show that not all columns and walls at the same level shortened the same amount. Consequently, differential movements between adjacent columns can result. Furthermore, columns in lower levels of a building shortened less than columns at a higher level.

Data from the precast bridge showed that changes in strain gradient occurred as the cantilever was segmentally

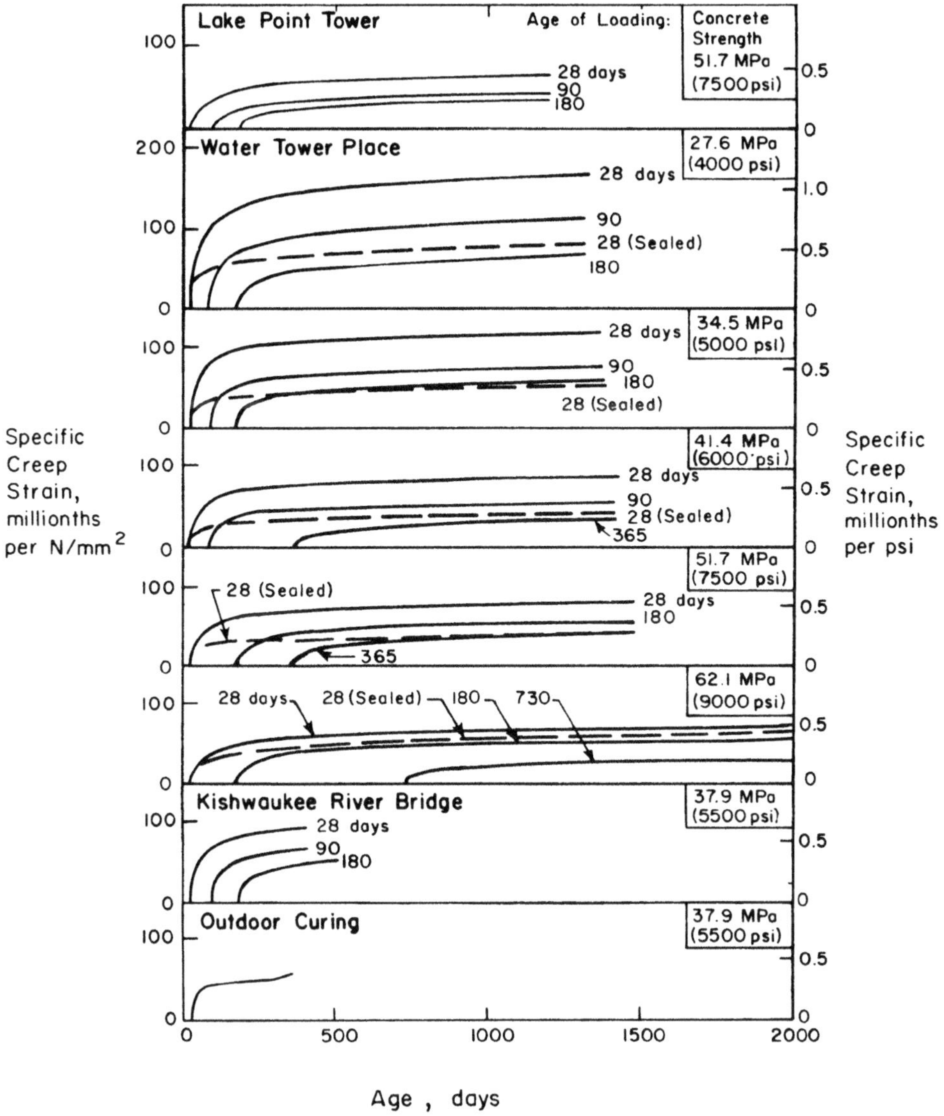

Fig. 7 Creep Data

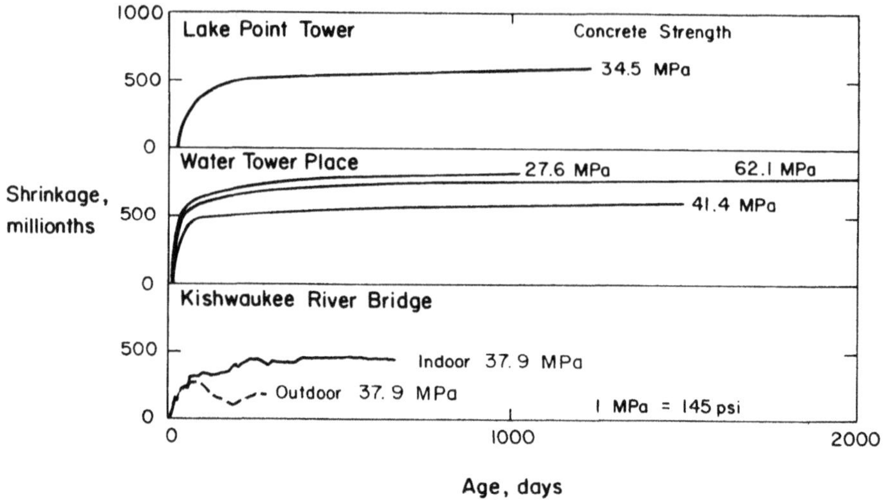

Fig. 8 Shrinkage Data

erected. Large strain gradients can result in large deflections at the end of cantilevers.

It is evident from these investigations that time-dependent deformations of large concrete structures are significant and must be considered during design and construction.

ACKNOWLEDGMENTS

Information given in this paper was obtained from several research investigations conducted in the Engineering Development Division of the Portland Cement Association Construction Technology Laboratories under W. G. Corley, Divisional Director. Sponsors of the research were National Science Foundation, Chicago Committee on High-Rise Buildings, Concrete Reinforcing Steel Institute, Material Service Corp., State of Illinois, Federal Highway Administration, and Portland Cement Association. Any opinions, findings, and conclusions expressed in this publication are those of the author and do not necessarily reflect official views of sponsors who have not necessarily reviewed the paper. This report does not constitute a standard, specification, or regulation.

REFERENCES

1. Hanson, N. W., Hsu, T. T. C., Kurvits, O. A., and Mattock, A. H., "Facilities and Test Methods of PCA Structural Laboratory--Improvements 1960-65," Journal of the PCA Research and Development Laboratories, Vol. 3, No. 2, May 1961, pp. 27-31, Vol. 7, No. 1, January 1965, pp. 2-9, and Vol. 7, No. 2, May 1965, pp. 24-38. Also Portland Cement Association Development Department Bulletin D91.

2. Pfeifer, D. W., Magura, D. D., Russell, H. G. and Corley, W. G., "Time Dependent Deformations in a 70-Story Structure." Symposium on Designing for Effects of Creep, Shrinkage and Temperature in Concrete Structures, ACI Special Publication SP-27, American Concrete Institute, Detroit, Michigan, 1971, pp. 159-185.

3. Russell, H. G. and Corley, W.G., "Time-Dependent Behavior of Columns in Water Tower Place," ACI Publication SP55, Douglas McHenry International Symposium on Concrete and Concrete Structures, Paper SP55-14. Also Portland Cement Association, Research and Development Bulletin RD052.01B.

4. Helgason, Th., Russell, H.G., Corley, W.G., and Hognestad, E., "Time-Dependent Behavior of Columns in the World's Tallest Reinforced Concrete Building," Preliminary Report of the Inter-Association Colloquium on Behavior in Service of Concrete Structures, Liege, Belgium, June 1975, 13 pp.

5. Russell, H.G. and Corley, W.G., "Long-Term Deformations of the World's Tallest Concrete Building," Preliminary Report of the International Symposium on "Testing In-Situ of Concrete Structures," RILEM, Budapest, Hungary, Vol. II, September 1977, pp. 152-163.

6. Anon, "Truss Launches Quickly in U.S. Box Girder Debut," Engineering News Record, October 19, 1978, pp. 45-48.

7. Standard Method of Test for Creep of Concrete in Compression, C512-69, American Society for Testing and Materials, Philadelphia, Pa., 4 pp.

ANALYSIS OF RESTRAINED CONCRETE STRUCTURES FOR CREEP AND SHRINKAGE

Charles H. Thornton, Ph.D., P.E., President
I. Paul Lew, P.E., R.A., Senior Associate
Lev Zetlin Associates, Inc., New York, New York, USA

ABSTRACT: Analysis of restrained concrete structures for creep and shrinkage. Examples of the detrimental effects of these internal strains are illustrated. A method of analyzing and designing for internal strains is presented in three case studies. The case studies are 1) a long span girder; 2) a deep beam; 3) a ring slab system. Each case is analyzed by computer and finite element techniques. The results are compared to the physical condition and were found in agreement.

СОДЕРЖАНИЕ: Расчет защемленных железобетонных конструкций на усадку и ползучесть. Примеры остаточного эффекта от действия внутренних усилий. Методика расчета конструкций по внутренним усилиям для трех основных случаев: 1. Балка большого пролета. 2. Высокая балка. 3. Кольцевое железобетонное перекрытие. Расчеты произведены на ЭВМ по методу конечных элементов. Результаты согласуются с эксперементальными данными.

KEYWORDS

Bridges (cantilever)

Buildings

Concrete (reinforced and prestressed)

Creep

Columns

Deformations

Field Tests

Laboratory Tests

Shrinkage

Strains

Time-dependence

INTRODUCTION

The structural engineer is often confronted with the analysis of concrete structures that have experienced internal strain deformation due to the volumetric changes caused by creep, shrinkage and thermal effects. The magnitude of stresses created by these internal strains when a member is fully restrained is so large that permanent damage to the structure and/or the restraining elements usually results.

The structural designer typically details a structure to minimize the built-in restraint to the movements caused by the internal straining of concrete. This is accomplished by providing adequate expansion joints in the structure and the use of sliding bearings and/or low friction supports. The provision of releases to allow for internal movements creates a new set of conditions that have structural consequences. The most significant of these consequences is a significant reduction in the stiffness of the structure to resist lateral or horizontal loadings. In the case of long-span structures supported at a limited number of points, the lateral resistance of the building may be so greatly reduced that the designer may elect to design the structure to accept the internal strains and not provide the releases mentioned above. Another factor that must be accounted for in providing the releases in the structure to allow internal movement is the significant cost of the details.

In every day practice, the engineer must weigh the benefits of releasing the structure against the benefits of not releasing the structure and designing for internal strain. To accomplish this, the structural designer must be able to analyze the structure for the stresses and movements that exist if the external restraint to internal strains is not relieved. These internal strains become more critical in the case of prestressed and posttensioned concrete structures because the components have axial compression forces acting upon them and the creep movements are greatly magnified due to these axial forces.

For the case of the evaluation of an existing structure, the engineer does not have the option of releasing an existing structure from internal strain. He must evaluate the existing condition for its adequacy and serviceability. The engineer must also evaluate the effect of the internal strains on non-structural elements in the building which are connected to the structural elements.

This paper is concerned with the analysis of internal strains in restrained concrete structures and evaluation of stresses that result from this analysis.

BACKGROUND ON DETERMINATION OF MAGNITUDE OF INTERNAL STRAINS

The determination of the amount of creep and shrinkage strain has been presented by many authors. The American Concrete Institute Committee 209 1 has presented techniques for the prediction of creep, shrinkage and temperature strains in concrete.

This paper is further expanded upon in a paper by Branson and Christiason 2 .

Zetlin, Thornton and Lew, 3 showed in their paper that internal stresses are induced by internal time dependent non-linear strains resulting from shrinkage and temperature changes in concrete which vary throughout the depth of the member.

These references provide a time history relationship of the internal concrete strains considering the following parameters:

1. Curing.
2. Age of the concrete at the time of loading.
3. Relative humidity.
4. Thickness of the structure.
5. Slump of the concrete mix.
6. Cement content of the concrete mix.
7. Percent fines.
8. Air content.

By properly using these factors, an approximation of the magnitude of internal strain can be made. Once the magnitude of these internal strains is established, an analysis of the effects of these internal strains on the structure can be undertaken.

EXAMPLES OF THE EFFECTS OF INTERNAL STRAINS IN CONCRETE

When the restraint to internal strains is large, i.e., over 25% of the internal strains are not released, the structure usually exhibits significant distress. In the course of the authors' investigations, numerous examples of the action of internal straining of concrete have been observed.

In one example (see Figure 1) a posttensioned structure was stressed between two large concrete tower shear walls. Under the action of shrinkage, creep of the posttensioning, and thermal strains, the slab pulled away from the wall creating a continuous crack up to one inch wide (see Figure 2). In this condition, the shear strength and load carrying capacity of the slab were greatly impaired.

Another typical condition found by the authors is the lack of recognition by designers of the need to detail to allow for internal strains. This situation led to failure of a concrete bracket support for a precast beam (see Figure 3). In the most extreme case, both ends of the precast beam are welded to their respective support brackets. In this case, under the action of internal strains, the brackets are pulled away from the columns. The resulting condition caused a large crack at the face of the column and a potential shear failure of the bracket or beam.

Another example of the damaging potential of internal straining of concrete was found in the case of a long-span concrete girder which was supported by and integrated with two concrete shear walls (see Figure 4). The internal straining of the concrete girder was resisted by the shear walls causing a tensile force in the concrete beam. The resulting tension force reduced the shear strength of the beam, causing a series of diagonal cracks in the beam (see Figure 5).

A final example of the potential distress caused by the internal straining of concrete is a posttensioned slab which was supported on and restrained by a series of foundation and free-standing bearing walls. The tensile stresses resulting from the restraint completely negated the expected compressive stress due to the posttesion forces. The slab cracked extensively up to 1/8-inch in width (see Figure 6). This case will be studied in more detail in the next section.

ANALYSIS OF RESTRAINED STRUCTURES FOR INTERNAL STRAINS

Through the use of high speed digital computer simulation of the structure, it is possible to analyze for the effects of internal strains in concrete for complex structural configurations. In order to perform the analysis, the engineer must exercise two areas of engineering judgement, as follows:

1. The determination of the magnitude and distribution of internal straining that is anticipated.

2. The modelling of the boundary conditions for the particular structure analyzed.

The total amount of creep and shrinkage strains must be determined as well as that portion of the strain that occurs prior to the introduction of the restraints. For example, the contractor will normally use a pouring sequence that will allow a substantial portion of the shrinkage to occur before the restraining elements are placed. The ACI Committee 209 [1] provides equations for predicting the time history for a given total amount of shrinkage or creep strain.

For creep, the recommended equation is:

$$C_t = \frac{t^{0.6}}{10 + t^{0.6}} Cu \qquad \text{(Eq. 1)}$$

where t = time in days

Cu = ultimate creep ratio = $\frac{\text{ultimate creep strain}}{\text{initial elastic strain}}$

C_t = creep ratio at time t

For shrinkage, the recommended ACI Committee 209 [1] equation for moist cured concrete older than seven (7) days is:

$$(e_{sh}) = \frac{t}{35 + t} (e_{sh})u \qquad \text{(Eq. 2)}$$

where t = time in days

$(e_{sh})u$ = ultimate shrinkage strain

(e_{sh}) = shrinkage strain at any time

Typical values for shrinkage and creep are $(e_{sh})u$ = 800×10^{-6} in/in for shrinkage strain and Cu = 2.0 for the creep ratio. Time history graphs for the values of creep and shrinkage based on equations 1 and 2, are shown in Figures 7 and 8.

The residual stresses were obtained using the following residual stress factor developed by Mattock [4]:

$$R = (1 - e^{-(Cu)}) / Cu \qquad \text{(Eq. 3)}$$

where R = residual stress factor

Cu = ratio of total creep strain to elastic strain

e = natural log

Internal straining of concrete due to creep and shrinkage cannot be completely analyzed without including the effects of temperature changes. By including shrinkage, creep and temperature strains a complete picture of the environmental influences on concrete is obtained.

CASE STUDIES OF THE EFFECTS OF CREEP, SHRINKAGE AND TEMPERATURE

Case Study No. 1

A sixteen-foot deep girder was used to clear span 164'-0" between two shear walls. The shear walls were 14'8" deep and their clear height from the underside of the girder to the ground was 38'-0". The girder was poured integral with the wall (see Figure 4).

The structure was posttensioned in a sequential manner that allowed the effective posttensioning force to increase closer to the support (see Figure 9). Axial compressive stresses in the girder ranged between 426 psi at the support and 66 psi at the midspan due to posttensioning forces.

The resulting frame was analyzed with two base conditions (fixed and pinned). The range of degree of fixity at the base was interpolated between these points. The structure was analyzed for the following internal strains:

a.	Shrinkage	(e_{sh})	=	$400 \times 10^{-6} \frac{in}{in}$
b.	Creep ratio	C	=	1.5
c.	Temperature	T	=	-75°F. change

The cumulative unrestrained internal movement of the beams between the shear walls that would result from these internal strains is 1.6 inches.

The cumulative effect of the shrinkage, creep and temperature strains was to induce an eccentric tensile force in the girder due to the restraints of the shear walls. The magnitude of axial tension stress induced from this restraint was 144 psi, excluding eccentricities. Superimposing these stresses on the stresses caused by the posttensioning forces (see Figure 10) indicates that the effective posttensioning force is substantially reduced. At midspan, the net tensile stress resulting from the combined effects of posttensioning, creep, shrinkage and temperature and dead load was 78 psi at the centroid of the beam, whereas the original design anticipated a compressive stress.

The reduction of the effective axial force also reduced the beneficial effect of the compressive axial force on the shear strength of the beam, as specified in ACI Code 318-77 in Equations 11-11 and 11-13. Alternatively, the ACI Code in Section 11.4.2.2 allows the use of an allowable principle tensile stress of $4\sqrt{f'c}$. In this case the significance of the axial stress is readily apparent, --- precompression reduces the principle tensile stress and tension increases the principle tensile stress. The shear strength of the beam was significantly reduced from that of an unrestrained posttensioned beam. This reduced shear strength is visually apparent in the form of diagonal cracks found in the girder near the support (see Figure 5). The ultimate effect of the creep, shrinkage and thermal strain of the concrete was to reduce its available concrete shear strength at the face of the support from 1600 kips if unrestrained to 1400 kips if restrained and from 1400 kips at the quarter point if unrestrained to 1000 kips if restrained.

Case Study No. 2

The design of an arena (see Figure 11) shed light on the effects of internal straining of the concrete in exposed, deep beams subjected to creep, shrinkage and thermal effects. The concrete wall of the arena is used as a deep beam to support the trusses that span the roof of the arena. The 13-foot-high wall is a continuous three span beam with an inner span of 56'-0" and the exterior spans of 42'-0". The column supports are 30 inches in diameter and the clear height of the column to the floor below is only 11 feet (see Figure 12). Therefore, the short columns offer large restraint to shrinkage and thermal movement of the deep beam.

The condition was analyzed for a shrinkage strain of 400×10^{-6} in/in and a temperature change of $\pm$ 75°F. The resultant unrestrained movement between end column would be 1.36 inches due to these internal strains. Three potential end conditions were analyzed for the exterior columns:

1. Column fixed at beam and base.

2. Column fixed at beam and pinned at base.

3. Slide bearing at the top of the column.

The conditions were analyzed through the use of the ICES STRUDL computer program by inputting the internal strain in the beam. The column stiffness was corrected to account for the increased stiffness due to the large difference in the imaginary height of the column to the center of the beam and the actual clear height of the column. Table 1 shows the moment and shears that result from the internal straining and wind loading which can be summarized as follows:

Condition 1. With the column fixed at both ends, high moments are found in the end columns and high shear forces are found due to internal straining. Relatively uniform wind moments are found in all columns. The horizontal component of deflection due to wind was the smallest of all conditions.

Condition 2. With the column fixed at one end and pinned at the other end resulted in a significant reduction in the magnitude of moments at the column ends due to internal strain, but resulted in increased moments due to wind at the fixed end of the interior columns. The deflection due to wind was increased.

Condition 3. By allowing the end of the exterior columns to slide or move freely, no moments resulted in the end columns due to internal straining. However, the moments in the interior columns due to wind load were the highest. Deflection due to wind was the greatest of the three conditions.

As a result of this study, it was determined that Condition 2 provided the best compromise solution to providing adequate lateral load resistance in the structure without allowing excessive internal stresses in the system due to excessive restraint.

Case Study No. 3

This case study exemplifies the type and depth of analysis that is required to reviewing the effects of external restraint to internal volumetric changes in concrete. This structure is comprised of an arena perimeter concourse level posttensioned concrete slab (see Figure 13). The planar slab is restrained by support walls eccentric to the line of the posttensioned slab and located along the longitudinal edges of the slab. The slab was 7½-inches thick and posttensioned to 26 k/ft or about 290 psi compressive stress. The edge wall restraints were located in the corners of the concourse slab ring and the slab had abrupt changes of width along its length that were potential areas for stress concentrations.

The slab was analyzed utilizing finite element computer techniques. The slab was modeled as a membrane utilizing principles of plane stress from the theory of elasticity. The slab was modeled as a quarter structure by utilizing the double symmetry of the perimeter ring around both principal axis (see Figure 13).

The following inernal strain conditions were considered:

1. A creep ratio of 2 with respect to elastic strain as a result of the posttensioned system.

2. A shrinkage strain of 450×10^{-6} in/in.

3. A temperature change of $\pm$ 15°F. This small temperature change was used since the slab was contained within an enclosed space.

The analytical results of the analysis are depicted on the isobar diagrams of the X and Y stresses in Figure 14. Stresses along the long axis of the slab ranged up to 1500 psi tensile stress. With tensile stresses of this magnitude, it becomes apparent that the initial compressive stresses due to the posttensioning of 290 psi compression are totally negated by the tensile stresses caused by the internal straining of concrete. As a result, excessive cracking occurred in the structure (see Figure 6). The cracks have been repaired with epoxy injection techniques and load tests were undertaken for the structure to determine the adequacy of the structure in its present condition.

Although it is not always possible to eliminate all restraint to the internal straining of concrete, it is normally standard practice to minimize this restraint. This is particularly important for a posttensioned structure since the ends of the members will experience large movements due to internal strains. When shear walls are used for lateral resistance in a high-rise building, they should be placed as close to the geometric center of the structure as possible in the direction of their lateral stiffness. An example of this approach was used on the Hyatt Regency Hotel in Minneapolis, Minnesota, USA (see Figure 15). For this structure (see Figure 16), the structure utilizes a central box-like core to resist wind loadings on each wing of the long, narrow building. By positioning the end walls only in the transverse direction, they provide minimal resistance to internal straining of the ends of the slab along its longitudinal axis.

CONCLUSIONS

This paper has shown through the use of several examples and through analysis that the stresses due to the internal straining of the concrete are significant and cannot be ignored by the engineer. Failure to recognize these effects has led to extensive damage and failure of structures. A compromise solution must be developed in many cases since in the process of releasing a structure's restraints, the lateral resistance of the structure to resist wind and seismic forces can be significantly reduced. This paper has presented some analytical techniques to determine the stresses in restrained structures. The effect of these stresses must be evaluated by the structural engineer in the design of a new structure or the evaluation of an existing building in order to determine the adequacy of the structural system.

REFERENCES

1. ACI Committee 209, Subcommittee 2. "Prediction of Creep, Shrinkage and Temperature Effects in Concrete Structures", (ACI Publication SP-27, Detroit, 1971)

2. D.E. Branson and M.L. Christiason. "Time Dependent Concrete Properties Related to Design-Strength and Elastic Properties, Creep and Shrinkage" (ACI Publication SP-27, Detroit, 1971)

3. L. Zetlin, C.H. Thornton, I. Paul Lew. "An Approach to the Problem of Internal Straining of Concrete" (ACI Publication SP-27, Detroit, 1971)

4. A.H. Mattock. "Precast Prestressed Concrete Bridges: 5, Creep and Shrinkage Studies", Journal PCA Research and Development Laboratories, V.3, No. 2 (May 1961) pp. 32-66

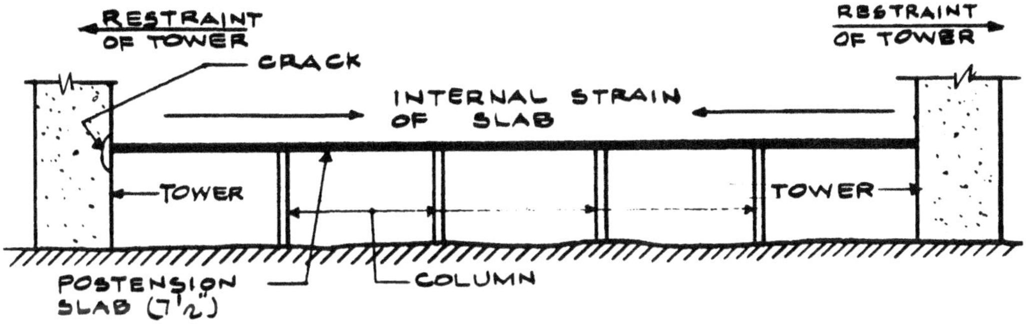

FIG. 1 RESTRAINT OF INTERNAL STRAIN OF SLAB

FIG. 2 CRACK IN WALL

FIG. 3 CRACK AT END OF BEAM

	COLUMNS				WIND DEFLECTION
	END		INTERIOR		
DEGREE OF FIXITY OF END COLUMNS	MOMENTS		MOMENTS		cm. (INCHES)
	DUE TO INTERNAL STRAIN	DUE TO WIND	DUE TO INTERNAL STRAIN	DUE TO WIND	
FIXED	276 tm. (2000 K Ft.)	13 tm. (94 K Ft.)	64 tm. (462 K Ft.)	13 tm. (94 K Ft.)	.13 cm. (.05")
PINNED ONE END	138 tm. (1000 K Ft.)	11 tm. (79 K Ft.)	64 tm. (462 K Ft.)	15 tm. (109 K Ft.)	.20 cm. (.08")
FREE	0	0	64 tm. (462 K Ft.)	26 tm. (188 K Ft.)	.25 cm. (.10")

TABLE I: WALL BEAM MOMENTS & DEFLECTIONS

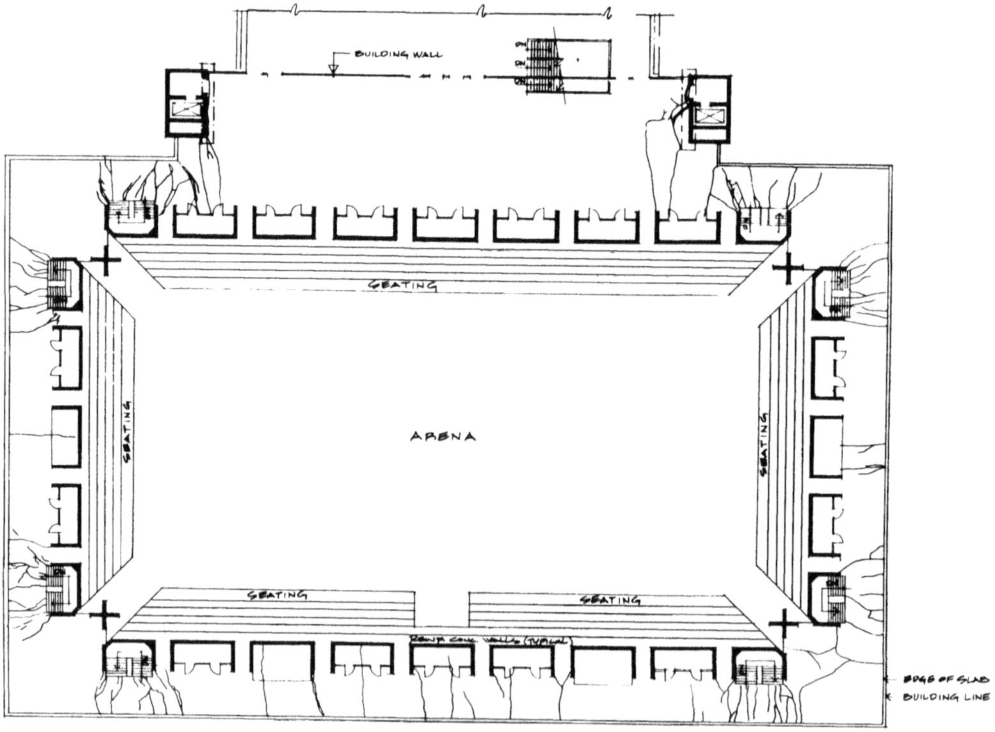

FIG. 6 PLAN OF CRACKS IN SLAB

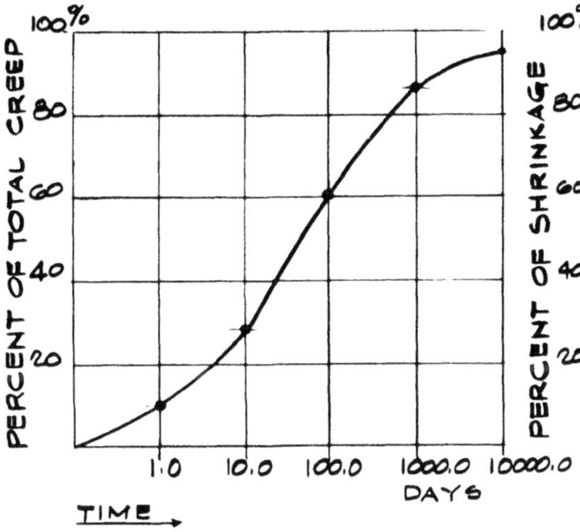

FIG. 7 CREEP GRAPH

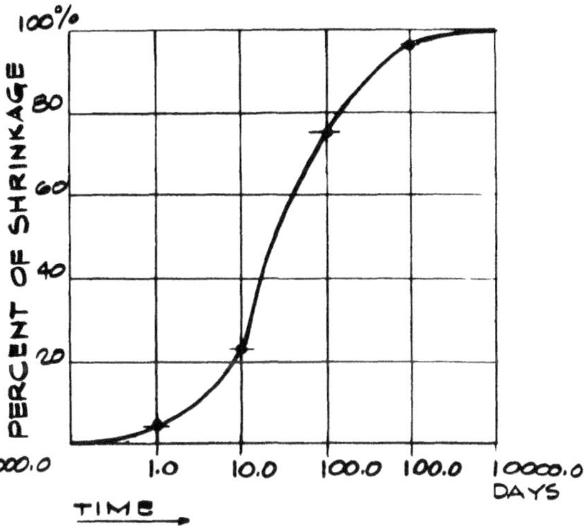

FIG. 8 SHRINKAGE GRAPH

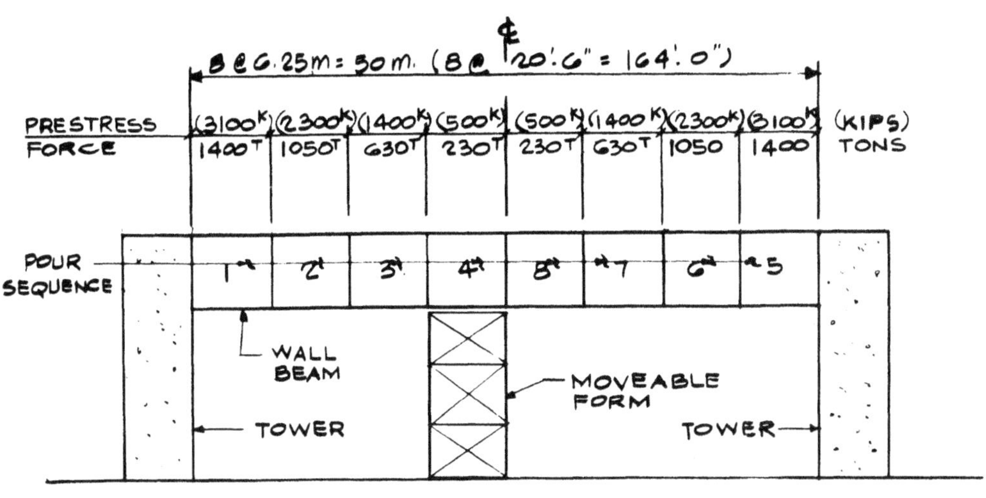

FIG. 9 WALL BEAM POUR SEQUENCE

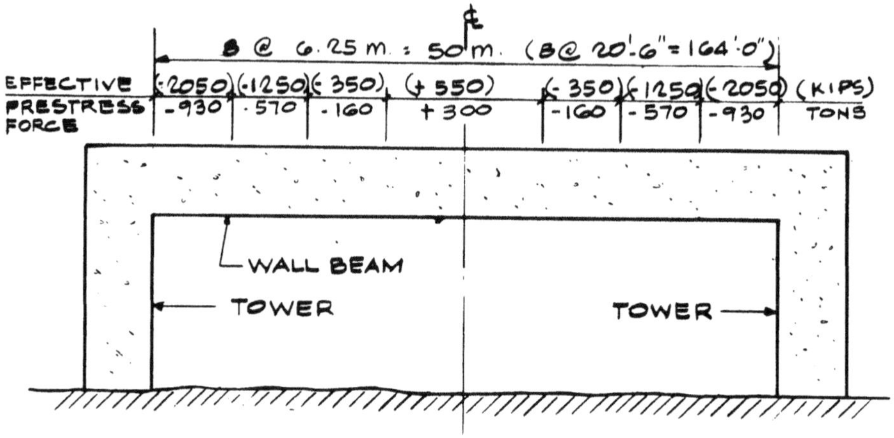

FIG. 10 EFFECTIVE PRESTRESS FORCE

FIG. 11 WEST POINT ATHLETIC FACILITIES

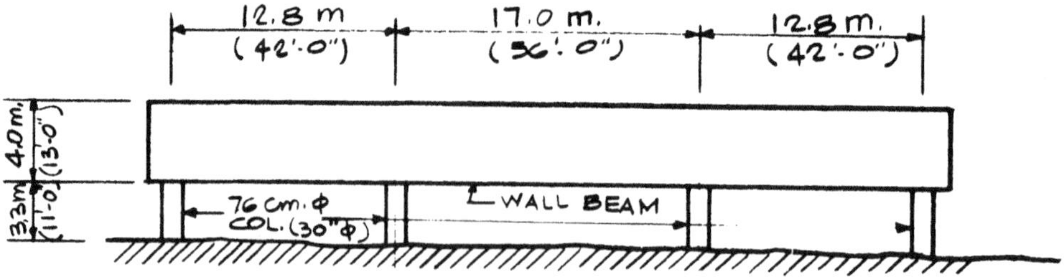

FIG. 12 ELEVATION - WALL BEAM

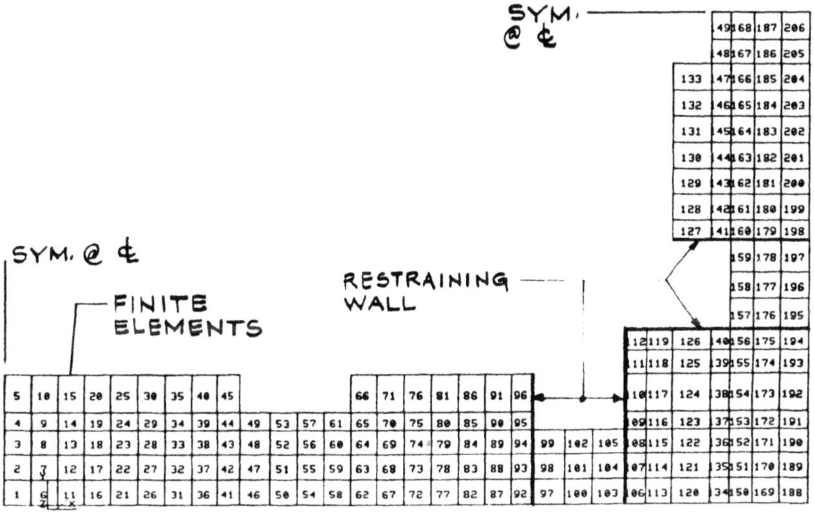

FIG. 13 COMPUTER MODEL: SLAB

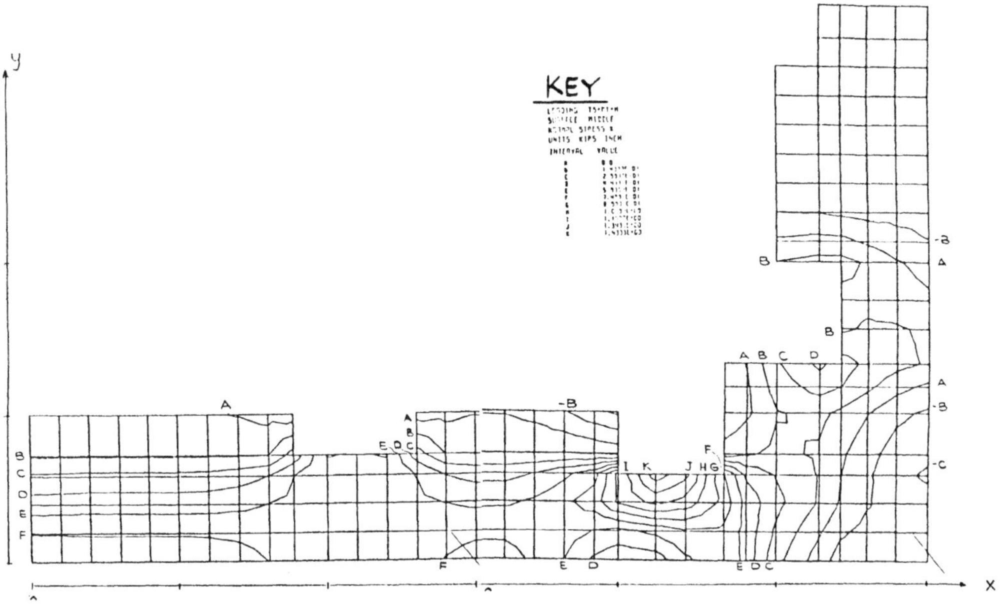

FIG. 14 STRESS CONTOURS OF SLAB

FIG. 15 MINNEAPOLIS HOTEL

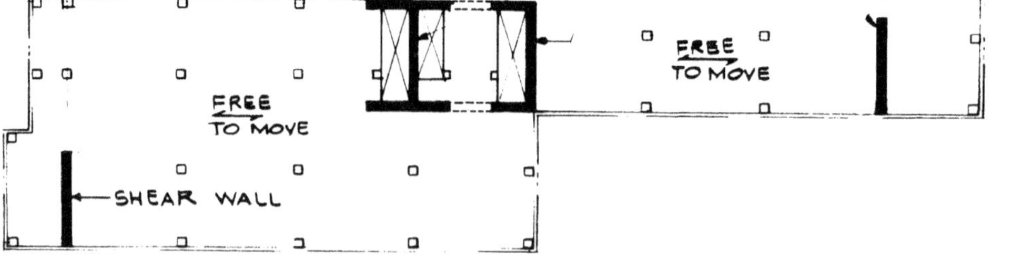

FIG. 16 TYPICAL TOWER FLOOR PLAN

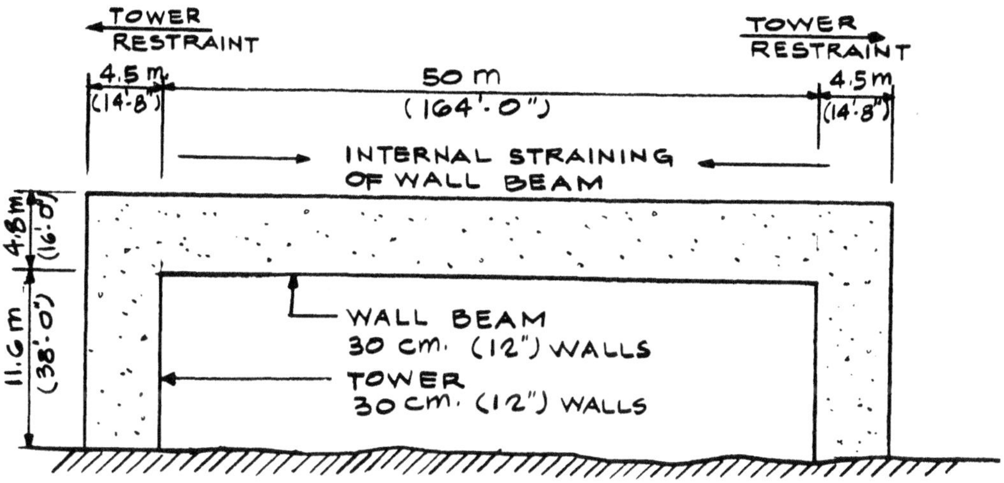

FIG. 4

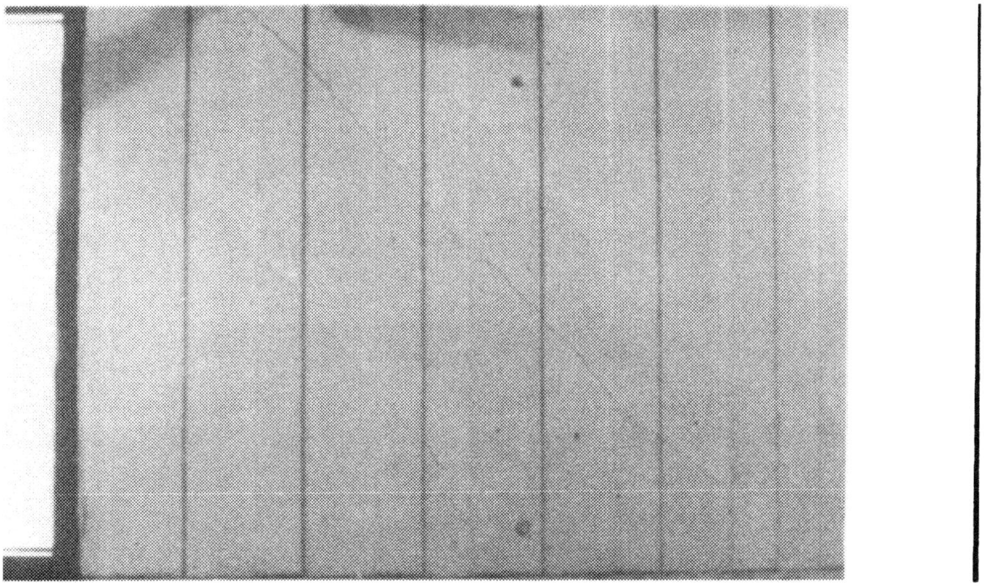

FIG. 5 DIAGONAL WALL BEAM CRACKS

CREEP OBSERVATION OF PRESTRESSED CONCRETE BRIDGES

Tibor JÁVOR, Doc.Ing. CSc.

Research Institute of Civil Engineering /VÚIS/,
Bratislava, Czechoslovakia

ABSTRACT

The creep and shrinkage observation was made by various prestressed bridges. The results of long-term observations were shown by cantilever method erected bridges and one cast in place frame bridge. The strains were measured by checoslovakian vibro-wire embedded gauges. The results showed, that the Dischinger theory is acceptable for generally calculation of the creep on this type of structures.

- . - . - . -

In the case of a structure with difficult statical solution or with a high economy which is to be designed several times in the future it is desirable to check the statical function of the bridge completed and to examine particular stages of stresses in the course of the bridge construction. The problem of the behaviour of structures and considering the influence of the duration of load application is a problem that is drawing more and more the interest of researches.Within the framework of the experimental analysis of prestressed concrete bridges in which the static construction diagram varies with time, a long-term experimental study of shrinkage and creep was undertaken by various Research Laboratories in Europe like Research Institute of Civil Engineering /VÚIS/, Bratislava, LCPC Paris, LNEC Lisboa, EIBS Berlin, SOJUZDORNII Moscou etc., this study included the instrumentation of several bridge structures at the time of their construction and the time-

-measurement of certain variable parameters such as local deformation, deflection, temperature, and so forth, which can improve our knowledge concerning the real behaviour of these structures.

Follow-up observation of local deformations,level measurement and prestressing values during construction can contribute to better knowledge concerning the stress conditions of the structure when it becomes statically indeterminate. Temperature and relative humidity variations, among other factors, greatly modify the experimental values. It was necessary to include concrete thermal and shrinkage speciments which give the distribution of temperature and humidity influences.

The behaviour of prestressed concrete bridges under traffic is the object of following checking:
- in the case of quality deficiencies during the construction,
- in the case of longterm changes of stresses owing to large creep, longterm losses of prestressing and settlements of foundations,
- in the case of danger from overloading or fatigue of the structure.

The purpose of testing bridges in situ is to verify their correct static and dynamic behaviour and quality of work, either subjecting the structure in question to a test load of proper magnitude, or investigating it over a long time period.

The methods for measuring and checking of statical function of bridges are divided according to the kind of values measured as follows:
- methods for measuring of deflections and deviations,
- methods for measuring of strains,
- methods for determinations of various material constants of the structure examine.

All kinds of measuring referred to were carried out with the aim to abanded traditional ways and to establish a high automation of the values measured being read by means of their registration and of measurement mechanisation.

1. Investigation methods of state of stress of bridges

When checking correct statical behaviour of bridges, the investigation of state of stress is very im -

portant both during construction work and immediately after its finishing for check loading tests or long time after the complетation of the structures. Observa - tion of state of stress is made indirectly by evalua - tion of material constants and measuring of relative deformations-strains in the structure. Mutual correlation of material constants and strains is in fact very difficult especially in the case of concrete structu - res. In essence for considerable simplification of evaluation operation we consider the application of Hook's law. It is desirable especially for long-termed observation to take into consideration influences of temperature and humidity changes, of shrinkage and creep of concrete, as well as changes of the concrete modul of elasticity of the bridge examined. From the point of view of the evaluation of the behaviour of bridges it is necessary to study the material constants, especially the changes of elasticity modul of concrete and the course of shrinkage and creep of concrete. These observations can be made best and in the simplest way by observation of laboratory samples, prepared during the concreting of the structure. For this purpose we prepare in essence two kinds of prismatic speciments, the first ones for the systematic laboratory investigation under constant humidity and temperature, if possible in climatic boxes, and the others for the observation direct in situ near the bridge or placed in the hollow cross section of the bridge.

Material constants of finished bridges are generally tested by various non-destructive methods: sclerometric, resonance, sonic and ultrasonic ones. It is advantageous to investigate the development of the Young modul in any cross section of the bridge deck, most often by ultrasound. We can namely investigate the development of the elasticity modul by ultrasound from the beginning of setting and hardening of concrete till the any moment of bridge under the traffic.

The temperature differences at the construction can be observed by two methods on the one hand with normal mercury thermometers and on the other hand with vibro-wire thermogauges embedded direct in characteristic places of the bearing part of the bridge. The temperature changes and changes of air humidity influence considerable not only the deformation state of the whole structure, but its particular cross section too. It appears, that in the case of assembling brief temperature changes must be considered already in the design of the structure but at the same time for proportioning can be considered by a third smaller temperature changes, than they are in reality in the atmosphere for a short time period.

Relative deformations are measured with surface strain gauges, most frequently with strain gauges embedded in concrete and the deflections or deformation displacements of the structure by geodetic methods.Mechanical strain gauges used in situ have relatively lower response and higher measurement errors because of their very simple operation. Deformeters generally enable to measure in a high deformation range and in the case of bridges they serve especially for checking e.g. of joint opening in beams, crack patterns etc.

Because of the importance of traffic safety on bridges, we have analyséd, in a complex survey, the behaviour of bridges in the following stages:
- visual inspections giving a complete image of possible failires such as cracks, bad insulation, bad pavement, deficiences of dilatations and bearings;
- analysis of material including the determination of young moduls, checking of grouting, corrosion analysis of reinforcements by non-destructives tests;
- repeated loading tests on bridges in order to receive information on fatigue characteristics and actual bearing capacity of the structure;
- longterm tensometric and geodetic observations which give information about deformation and state of stress for particular cross sections of the bridge.

The behaviour of prestressed concrete bridges in service is to be checked in the case of:
- quality failures arisen during the construction;
- long-term changes in stress as a consequence of large creep of concrete, long-term losses of prestress and settlements of foundations, and
- danger of oveloading or fatigue of the structure.

In the case of long duration tests, observations are read remotely, from the data logers located in the bridge, by means of telex or by phone directly at the institute, which can be often at a distance of more than hundred km from the site. After dialing the phone number of the site /of the data loger/ the acoustic pic-ups /vibrowire gauges/ are connected gradually and their resonances are read and evaluated remotely, i.e. transformed with the help of a computer, from frequency into strain, and from strain into stress. As for short static loading tests, the same method was also used in the case of the loading test on the new bridge over the Danube in Bratislava, where the strains measured with resistance strain gauges were read automatically in the measuring vehicle, and transmitted by telex to the computer at our institute, and there immediately proces - sed. Within a minute the final stresses were sent back by telex to the site in order that the loading of the bridge might be continued.

2. Deformation state determination during prestressing

At the Research Institute of Civil Engineering in Bratislava we have done experimental research on prestressed concrete bridges since 1954. We have examined various types of concrete bridges. In the case of observation of the state of stress for prestressed,cantilever concreted or bridges assembled of precast box elements we determine the state of strain usually after prestressing or concreting, it means after particular working operations. In Fig. 1 there is shown the course of strains of a cantilever assembled highway /S/ bridge determined in this way. The increase of relative deformations of all investigated box elements assembled by cantilevering of a road bridges had the same character. The largest deformations occured in the upper fibre of the inner web of the cellular cross section of a pre - cast box element. The deformations of the inner web reached as much as the double value in comparison with the strains of the similar fibre of external web. We used enbedded vibro-wire gauges.

The cantilever bridges with prestressed precast concrete elements had already larger deformations du - ring prestressing in comparison with the monolytical concreted cantilever bridges. However one has to note, that the desired prestress in the highway bridge "S" /Sirník, span 60.00 m/ was - 147.3 kp/cm^2 /the measured prestress - 130.7 kp/cm^2/ and in the monolytical - cast in place bridge "N" /Nové Mesto, span 70.00 m/ was - 83.0 kp/cm^2 /the measured - 84.9 kp/cm^2/, which is smaller as in the bridge with precast elements "S". In comparison, the monolytical concreted bridge "KO" /Kolárovo, span 61.61 m, Fig. 2/ made with the better quality of concrete, the maximum deflection in 4 years was only 4.5 cm, and the strains, measured u - sing akustical strain gauges, was 25 to 30 % smaller as in bridge "N". In the bridge "KO" the advantage of the metal joint was observed comparing it with the reinforced concrete joint in the center of the span in bridge

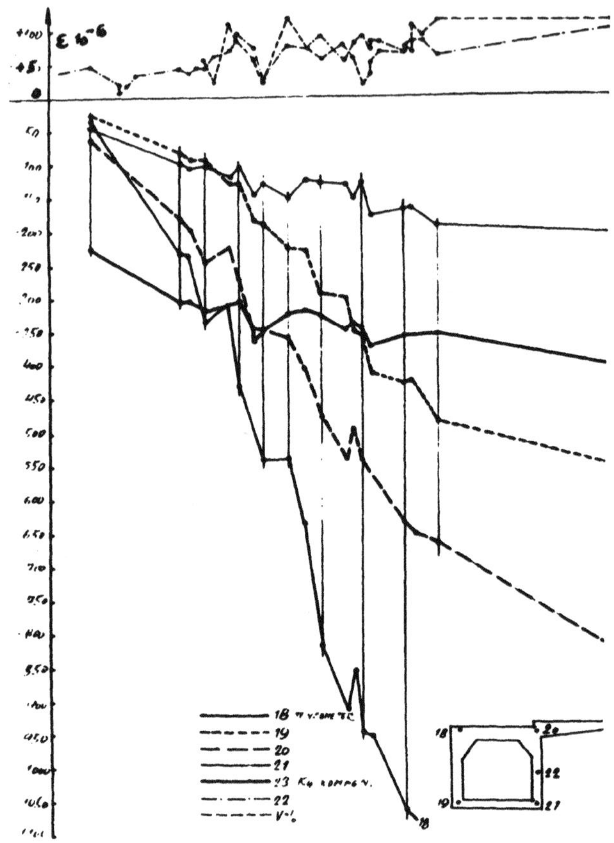

Fig. 1. The course of strains of the first segmental element during the construction of the cantilever prestressed concrete highway bridge "Sirník"

"N". The measured strains after prestressing is shown in Fig. 3.

An other interesting observation was that in the cantilever highway bridge "S" with precast elements,the strains after three years were 100 % higher as in the cantilever monolytical concreted bridge "KO". The de - formations in bridge "S" also stabilized in 4 years, since the quality of the concrete is not bed. In the cantilever bridge "S" the maximum strain was - 850x10^{-6} after 3 years, in the cantilever railway bridge "M" /Margecany, same span, Fig. 4/ the maximum strain was - 800x10^{-6} after 2.5 years /50 % of the later is shrinkage/.

Fig. 2. The cast in place cantilever prestressed concrete bridge "Kolárovo"

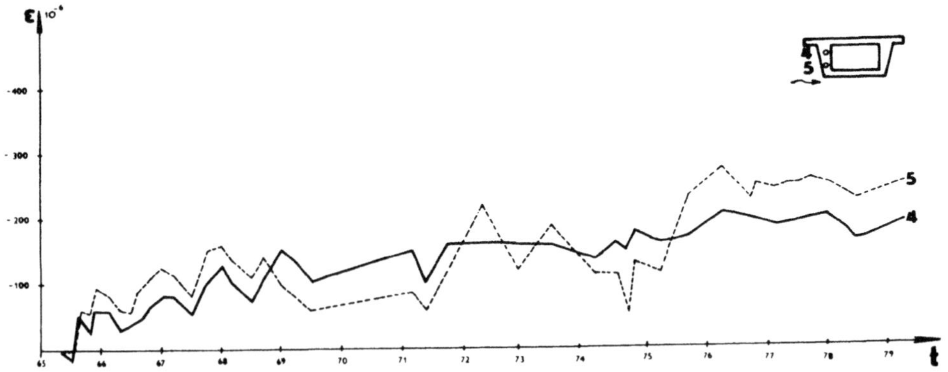

Fig. 3. The course of strains at the fixed end of the cantilever prestressed concrete bridge "Kolárovo"

3. Long-term observation of bridges in situ

Observations on prestressed concrete bridges and their behaviour versus time allow to obtain some date on real behaviour of bridges in traffic or on real state of stress of structures during service. The latter is an important datum for designing of further similar type bridges. Especially for prestressed concrete structures often arises a question concerning maximum creep which can be expected and its influence on state of stress of structure. Unclear is also the whole complex of question concerning the influence of the temperature and moisture changes, shrinkage effects, prestressing losses, possibility of vertical and horizontal foundation displacement or foundation footing and so on. Special attention should be paid to changes of Young-s modulus in various seasons because of variability of

Fig. 4. The cantilever prestressed concrete railway bridge "Margecany"

deformation caused by temperature and moisture changes. We can investigate and examine all the influences referred to in three groups of observation:

- the course of the material constants of the construction,
- long-term strain gauge investigation of deformations,
- long-term geodetic observation of vertical displacements.

Only in the basis of observation made from this point of view it is possible to make a deeper analysis of the results of these investigation compare mutually the relative deformations and deflections measured and transform the measured deformations into strain and stresses so that they may be a suitable basis for checking the static bridge constructions similar to the bridges in long-term observation.

On 12 prestressed concrete bridges of larger spans /continuous precast prestressed concrete structures,rigid frame bridges, monolytical-cast in place concreted structures, box girder bridges, cantilever structures/

observations have been made for 10 to 20 years. During the construction of the bridges referred to in each of these structures examined acoustical vibro-wire strain gauges in number of 50 to 300 were embedded. At the same time thermometers and geodetical marks for measuring vertical displacements of structures were set. In the bridge structures concrete compensation beams permit - ting to determine the direct action of temperature changes and in surrounding air moisture as well as the effects of shrinkage and creep in time are cast in.All strain gauge measurements are automatically remote read from a vehicle which enable to check bridges every 3 months. The evaluation is made with a computer accor - ding to the author's own programme. The deformations determined in this way are transformed into stresses and compared to theoretical assumptions.

The evaluation of the measured values consists,besides computation of real deformations, also in the separation of deformations occasioned by stress only and finally in the transformation of these deformations into stresses.The basic main conditions of the classic transformation of deformations into stresses are:

- homogeneity and isotropy of the material in the zone of the strain gauge,
- independence of the elasticity coefficient /Young modulus/ of the magnitude and duration of loading,
- independence of spontaneous volumetric changes such as shrinkage and swelling due to stress acting in the concrete.

In the case of simple statically determinate structures we can assume that the stress at any "n" time is given by

$$\nu'_n = \frac{{}^{n}\varepsilon - \sum_{\substack{i=0\\ j=0}}^{\substack{i=0\\ j=0}} \nu'_j \, \varepsilon_{\frac{i+j}{2}}^{n-\frac{i+j}{2}}}{{}^{1}\varepsilon_{\frac{m+n}{2}}^{m-\frac{m+2}{2}}}$$

where $i = 0,a,b,\dots,k,\ell,$
$j = a,b,c,\dots,\ell,n$

$'\varepsilon^{n-\frac{m+n}{2}}_{\frac{m+n}{2}}$ = the measured deformation from unit load for concrete at the $\frac{m+n}{2}$ age and during loading period $n - \frac{m+n}{2}$,

σ'_j = the unit stress at the j time and

n_ε = the calculated measure deformation in the direction of the strain gauge axis in the course of n period.

The resulting strain is obtained by superposition of partial stress components. If this procedure of deformation transofmation into stresses is applied, e.g. to a frame bridge structure we obtain results suitable only to the span center and not to the frame cross-beam /Fig. 5/ restrain of the statically underterminate structure, where a more complex computation method would be desirable. The resulting stresses after 12 years determined in this way in the characteristic cross sections of the frame bridge and the comparison of the measured stresses with the values theoretically computed are shown in Table 1. It is always reasonable to check the tensometric investigation with geometrical measurements, so as to read sufficient mutual control of both long-term methods of bridge observations.

Table 1

Place of the cross section	Theoretical value kp/cm^2	Stresses determined from the measured deformations in kp/cm^2	Differences between the theory and the transformed stresses in %
The middle of span:			
above	-43.40	-34.04	21.6
below	-76.70	-88.34	15.2
The third of span:			
above	-49.30	-39.45	20.2
below	-75.10	-27.08	63.6
In the restraint:			
above	-31.50	-34.29	8.8
below	-57.00	-7.41	87.0

Fig. 5. The frame bridge structure

The process of change of deformations stabilized in the bridges constructed with good quality of con - crete /Youngs Modulus E = 350 000 kp/cm^2/ after three years. The total strain /creep and shrinkage/ measured after 10 years was - 200x10^{-6} in a monolytical concreted framework prestressed concrete structure /Highway bridge "K" - Kotešová, span 63.40 m/. The maximum deflection after 10 years was 5.57 cm at the center of the span. 4.87 cm of the above was found after the first three months the bridge was in use. The deflections are shown in Fig. 6.

In the second framework structure /Highway bridge "B" - Bytča with the same span of 63.40 m/ but of worse quality of concrete /Youngs Modulus E -250 000 kp/cm^2/ the maximum deflection was 11.77 cm after three years and is not stabilized. 8.5 cm of the 11.77 cm was found after the first 6 months of traffic on the bridge. The influence of temperature change between summer winter on the deflection oscilations was ± 1 cm. However it is possible that these large deformations were partially caused by foundation movements and therefore recommendations were made to connect the bottom of the founda-

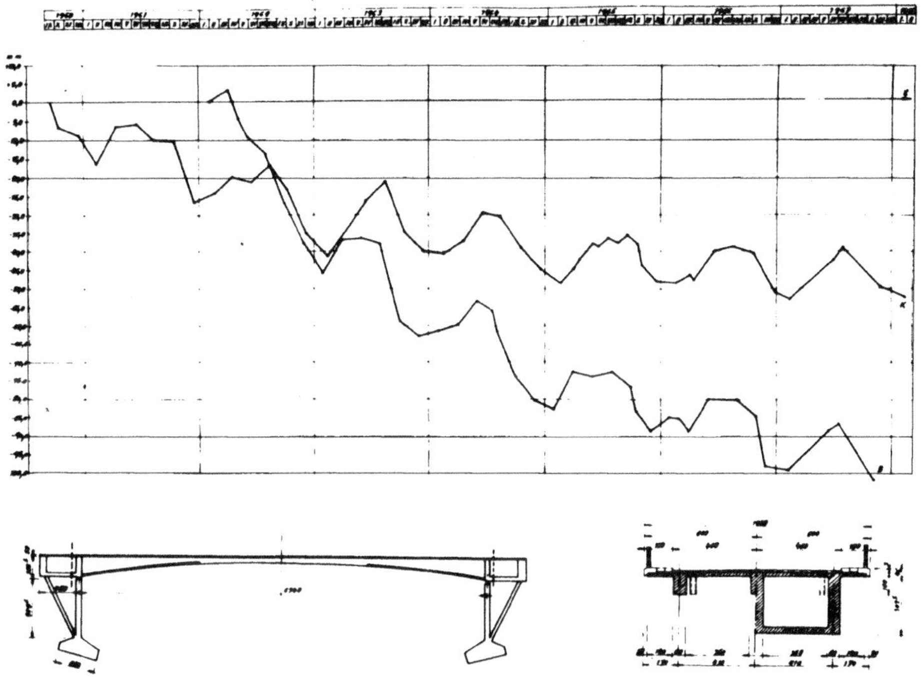

Fig. 6. The deflections of the framework bridge structure "Kotešová" with good concrete quality and "Bytča" with worse quality of concrete

tions with prestressed concrete bars in this type of structure. The course of deflections at the center of span of both bridges are shown in Fig.6.

In this case of long-term tensometric investigation of deformation state of bridge structure in the observation of a frame bridge erected cast in place with a considerably varying cross section and with skew tie, the course of the strains in the centre of the span and at the support of the frame bridge following 15 years observations is shown in Fig. 7. The deformations show, that the creep of this bridge stabilizes after 4 years and the deformation state in the following period oscillates owing to temperature and humidity changes of the atmosphere. The zero state for long-term observation of this bridge is taken as the moment after the ending of the loading test.

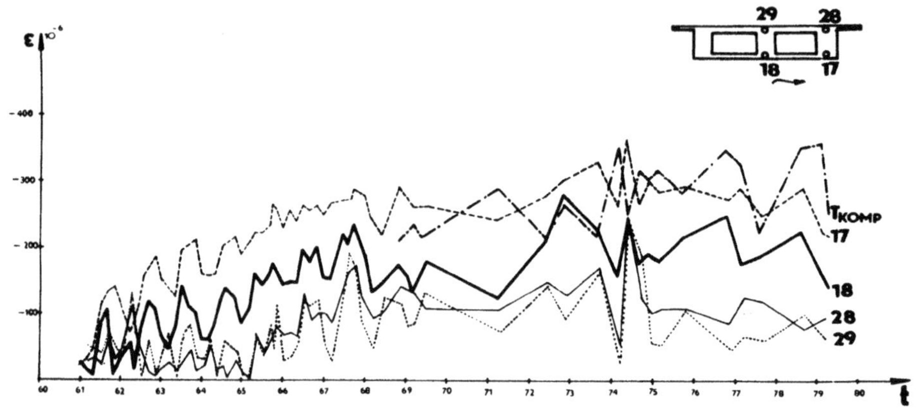

Fig. 7. The course strains of the framework bridge structure "Kotešová"

On several of these bridges loading tests in situ for total loading tests in situ for total loading following 8 to 10 years were repeated and hereby fatigue characteristics of these bridges after a long-term traffic were ascertained. In Fig. 8 are shown the results of observations and the behaviour of one by cantilever method erected prestressed concrete highway bridge versus time in strains $/10^{-6}/$. The long-term deformations of this bridge are examined by geometrical and tensometric measurements. The course of strains was measured by vibro-wire gauges for the 1st precasted box-element of this bridge. The creep and shrinkage settled after 4 years and in the next period take place only displacements of the structure due the temperature changes in the course of the year.

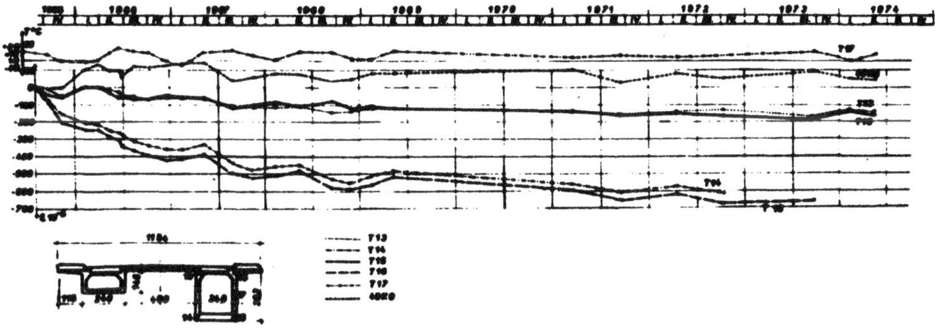

Fig. 8.The results of strain observation of the canti-
lever precasted prestressed concrete highway
bridge "Sirník"

In Fig. 9. are shown the creep observations and strains at the support of the main span of the by can- lever method erected prestressed concrete railway bridge /Markecany/. The course of strains way measured by 80 vibro-wire gauges and the deflections geometrica- lly. The bridge with 60 m main span was erected from box-girder precasted elements. This bridge was checked with statical and dynamical load tests twice, first just after the opening of the traffic and second after 10 years. The differences are not very high. The stres- ses by the repeated load tests are higher about 10-15 % but the quality of concrete after 10 years is also bet- ter.

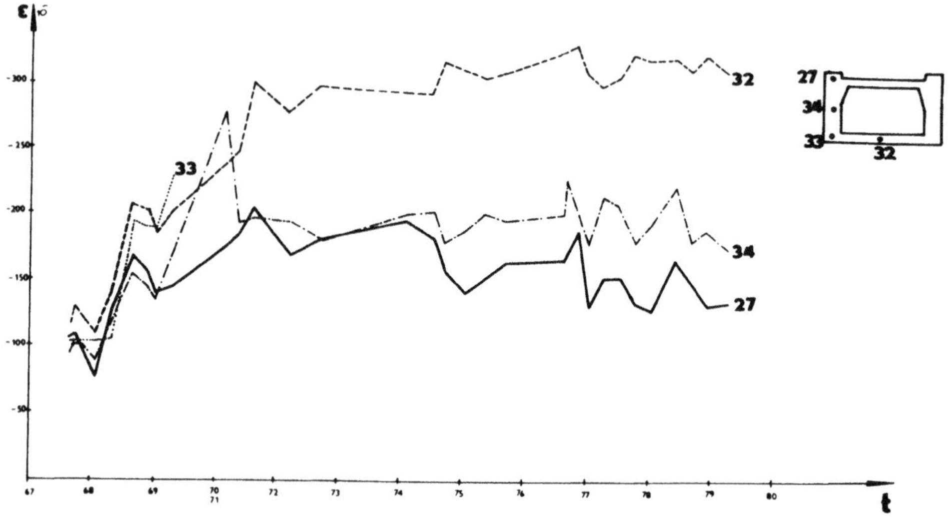

Fig. 9. The creep observation of the cantilever prestressed precasted concrete railway bridge "Markecany"

Conclusions

The creep of the by cantilever method erected bridges generally settled after 4 years. The situation of the prestressed concrete frame bridges was more complicated, because they are very sensitive to the moving of foundations. The highest stress calculated by the designer of the cantilever precasted highway bridge Sirník at the end of erection is -14.73 MPa /measured -13.07 MPa/ and at the casted in place cantilever bridge -8.3 MPa /measured -8.49 MPa/. The creep coefficient for 5 years old concrete of the highway bridge was 1.78 /calculated after Dishinger theory 2.07/ and the creep coefficient of the railway bridge after 5 years was measured 2.06 /equal with theoretical results obtained by classic Dishinger theory/.

Part VI
Structural Creep Analysis

PREDICTION OF SHRINKAGE STRESSES AND DEFORMATIONS IN CONCRETE

R. Iding

Senior Structural Analyst, Wiss, Janney, Elstner & Associates
Emeryville, California

B. Bresler

Principal, Wiss, Janney, Elstner & Associates
Emeryville, California
Professor Emeritus, Department of Civil Engineering
University of California, Berkeley

ABSTRACT An analytical method is described by which shrinkage stresses and deformations in concrete members are predicted. Nonuniform, time-dependent free shrinkage in structural concrete members is modeled by a diffusion equation. The moisture dependence of diffusivity and surface factor are considered. A finite element method is used to solve the nonlinear shrinkage diffusion equation. Inelastic properties and creep characteristics of concrete are considered in calculating the effect of nonuniform shrinkage on stress, deformation, and cracking. Predictions of shrinkage for various sizes of prism are compared to experimental data with good agreement. The effects of ambient relative humidity, surface conditions, and the material characteristics of concrete on shrinkage are discussed.

RESUME Cet article contient la description d'une méthode analytique permettant de prévoir les contraintes et déformations de retrait dans les éléments (structuraux) de béton. Le libre retrait non uniforme est fonction du temps dans les éléments structuraux de béton et répresenté par une équation de diffusion. L'influence de l'humidité sur la diffusivité et le facteur de surface sont pris en considération. On utilise une méthode d'éléments finis pour résoudre l'équation non linéaire de diffusion du retrait. Les propriétes inélastiques et les caractéristiques de fluage de béton sont prises en compte en calculant l'effet du retrait non uniforme sur les contraintes, les déformations, la fissuration. Les prévisions de retrait pour des tailles variées de prisme de béton sont comparées aux données expérimentales et s'accordent avec elles. Les effets de la humidité relative ambiante, des conditions de la surface et les caractéristiques matérielles du béton sur le retrait sont discutés.

INTRODUCTION

The effects of drying shrinkage, creep, and cracking must be properly considered in the design of reinforced concrete structures. These three factors, along with loading and environmental history, can significantly affect the redistribution of stresses with time in a concrete structure. Early studies (1-3) of concrete behavior considered these three phenomena to be independent processes. Pickett (4) considered shrinkage stress, but neglected the effects of cracking and creep. Most methods currently used to calculate deformation due to drying shrinkage are based on empirical equations derived from experimental data (5,6). The range of validity of such predictions of behavior is, however, limited due to the exclusion of the effects of stress history and cracking.

Integrated approaches to the nonlinear analysis of these three phenomena have been developed in several studies (7-9) in each of which differing degrees of simplification were used to model the complex physical processes. The studies reported here show that good agreement between analytical predictions and experimental data can be obtained using relatively simple models of material behavior and the finite element method of analysis.

FREE SHRINKAGE STRAIN ANALYSIS

Free shrinkage is the contraction that an element of concrete would undergo while drying if there were no constraint on its movement. Such shrinkage differs from apparent shrinkage, the experimentally measurable movement within a specimen which includes not only free shrinkage strain, but stress-related strain as well. The mathematical model described here rests on two basic assumptions:

(1) that a change in free shrinkage strain is directly proportional to a change in moisture content, and (2) that moisture flow within concrete obeys the moisture diffusion equation (10). The combination of these two assumptions leads to the shrinkage diffusion equation:

$$K_s \nabla^2 S + Q_s = \partial S/\partial t \tag{1}$$

where the field variable $S(x,y,z,t)$ represents external shrinkage strain, $Q_s(x,y,z,t)$ external shrinkage flow into the body, and K_s shrinkage diffusivity. Equivalently:

Rate at which shrinkage flows from elements adjacent to a node (point)	+	Rate at which shrinkage enters elements adjacent to a node (point)	=	Rate at which shrinkage is stored in elements adjacent to a node	(2)

i.e., shrinkage in an element must equal shrinkage flow into that element from adjacent elements and, where an element is on the boundary, from adjacent and external sources. The boundary condition used to describe the drying behavior of a concrete surface is:

$$\frac{\partial S}{\partial n} = \frac{f}{K_s}(S_\infty - S) \tag{3}$$

where $\partial S/\partial n$ is the shrinkage gradient at the drying surface, f is a surface factor, and S_∞ is the ultimate free shrinkage strain, i.e., the amount of shrinkage that an element of concrete would undergo if it were allowed to attain a steady-state equilibrium with surrounding moisture conditions and if its movement were not restrained by adjacent elements. Ultimate shrinkage is both a material property which depends on concrete mix proportions and a forcing function based on the relative humidity of the surrounding environment.

Shrinkage diffusivity K_s is a material property of concrete and can be defined as the rate at which shrinkage flows through concrete when there exists a unit shrinkage gradient. Diffusivity is highly shrinkage dependent (moisture dependent) and temperature dependent. Surface factor f characterizes the rate of surface evaporation and is both a material property and a description of surface conditions. Given that the moisture dependence of the surface factor resembles that of diffusivity, Pickett (4) postulated that:

$$f(S) = 1.67\, K_s(S) \tag{4}$$

The finite element method can be used to discretize the shrinkage diffusion equation, resulting in the following matrix equation:

$$\underline{K}\ \underline{S} + \underline{Q} = \underline{V}\ \underline{\dot{S}} \qquad (5)$$

where $\underline{K}$ represents the shrinkage diffusivity matrix, $\underline{V}$ the shrinkage velocity matrix, $\underline{Q}$ the external shrinkage flow vector, $\underline{S}$ the nodal shrinkage strain, and $\underline{\dot{S}}$ the nodal shrinkage strain rate of change. Equation (5) is solved by a step-by-step time integration. If material properties are shrinkage dependent, a nonlinear iterative technique is used within each time step to determine the shrinkage distribution $\underline{S}$ which satisfies the matrix balance equation. A computer program (SHRINC) in which these techniques have been implemented is described in Reference 11.

SHRINKAGE STRESS ANALYSIS

The time history of nonuniform free shrinkage strain serves as input data for the structural analysis by which apparent shrinkage movement and shrinkage stress are determined. The free shrinkage strain at each point in the structure is subtracted from apparent mechanical strain to find the portion of the total strain that gives rise to stress. The structural analysis then proceeds following well-established methods for the analysis of nonlinear structures, such as the direct stiffness method of matrix structural theory whereby the structure is discretized into appropriate finite elements. A computer program, SASHTEC (12), has been developed for this purpose. Time-dependent and inelastic properties of concrete--such as creep, cracking, and crushing--are accounted for. Numerous other computer programs could be modified to accept shrinkage history input (13,14).

SIZE EFFECT ON SHRINKAGE DEVELOPMENT

Numerous empirical approximations have been proposed to account for the fact that large concrete specimens exhibit less shrinkage than do smaller ones. The approach developed in this paper is also an empirical approximation, but is based on the laws of physics (the diffusion equation) which governs the process of nonuniform shrinkage, and hence models the size effect in a more rational way. In order to establish the accuracy of this model, experimental data reported for concrete specimens of various sizes were compared to results from analytical simulations of the tests obtained using the computer programs SHRINC and SASHTEC. One set of data was taken from experiments by Keeton (15) in which a large number of 3-inch diameter by 9-inch long (7.62 x 22.86 cm) and 6-inch diameter by 18-inch long (15.24 x 45.72 cm) cylindrical specimens were dried

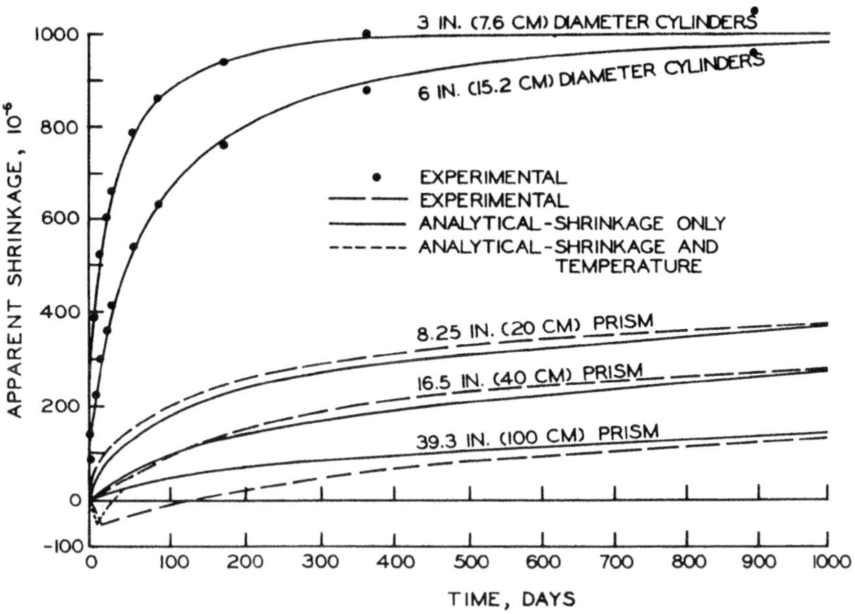

FIGURE 1 EXPERIMENTAL AND ANALYTICAL SHRINKAGE HISTORIES FOR SPECIMENS AT 50% RELATIVE HUMIDITY

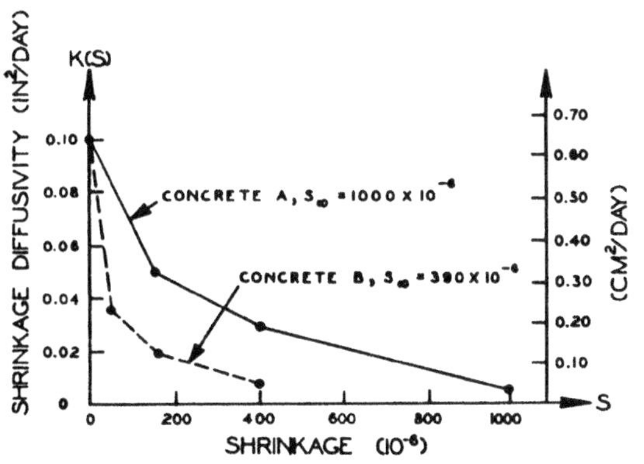

FIGURE 2 SHRINKAGE DIFFUSIVITY FOR CONCRETE

tor periods up to several years in a carefully controlled 50% R.H., and 68°F (20°C) environment. A second set of data was from tests by Mamillan (16) on square prismatic specimens ranging in size from 8.25 to 39.3 inches (20 cm to 100 cm) prepared from a single batch of concrete and dried at 55% R.H. and 68°F (20°C) for up to three years. Shrinkage histories for both sets of data are shown in Figure 1. Since data on properties of the concretes used in these tests were incomplete, most moisture flow and mechanical properties used in the analyses were best estimates. Estimates for shrinkage diffusivity K_s and ultimate shrinkage S_∞ are shown in Figure 2. Other material properties used in the analyses were reported in Reference 7.

Free shrinkage distribution history for each specimen was calculated using the computer program SHRINC and the finite element meshes shown in Figure 3. Time steps were varied throughout the simulated drying period; the initial one-day steps were gradually increased to one-hundred-day steps after about three hundred days of drying. Apparent shrinkage values, i.e. the observed average shrinkage of each cylinder or prism, were calculated from the free shrinkage values using the computer program SASHTEC and compared to the experimental data shown in Figure 1. Correlation for all sizes of specimen was excellent except during the early stages of drying of the largest specimens. The latter discrepancy could have been caused by volume changes due to undissipated heat of hydration which would have affected observed strains; a thermal analysis using the computer program FIRES-T3 (17) was performed to determine free thermal expansion, the results of which (see Figure 4) were then included in the stress analysis. The shrinkage history including thermal effects is shown in Figure 1 and accounts for the expansion observed in the drying tests. The excellent correlation obtained above where a single set of material properties was used for all prisms (or cylinders) suggests that the proposed analytical method

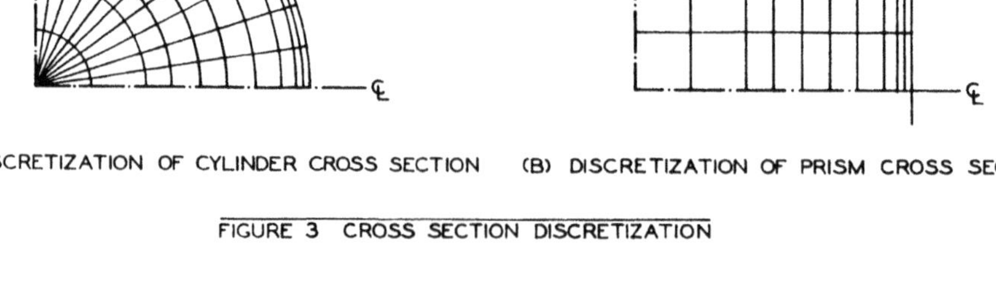

(A) DISCRETIZATION OF CYLINDER CROSS SECTION (B) DISCRETIZATION OF PRISM CROSS SECTION

FIGURE 3 CROSS SECTION DISCRETIZATION

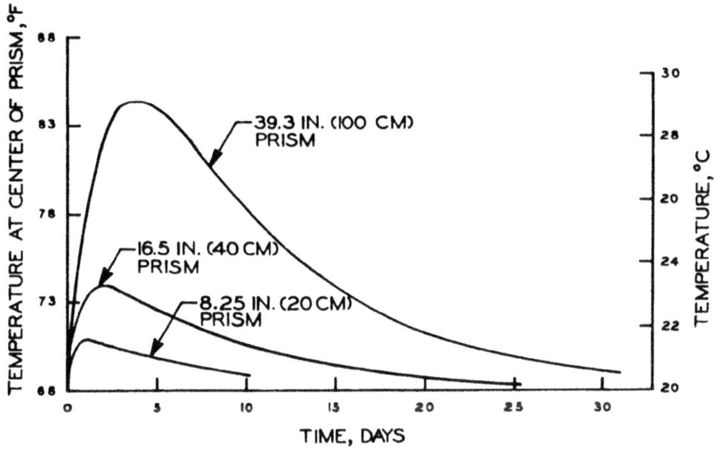

FIGURE 4 TEMPERATURE RISE DUE TO HEAT OF HYDRATION

models size effect well provided that thermal effects are included for larger specimens.

EFFECT OF RELATIVE HUMIDITY ON SHRINKAGE DEVELOPMENT

The method presented in this paper models the effect of ambient relative humidity on shrinkage development solely through the parameter S_∞, the ultimate shrinkage for a particular concrete and a particular atmospheric relative humidity. Keeton (15) measured shrinkage in 6-inch (15.24 cm) concrete cylinders dried under controlled conditions of relative humidity ranging from 100% R.H. to 20% R.H. Typical experimental data are plotted in Figure 5 and compared to analytical results from SHRINC and SASHTEC. The diffusivity function shown in Figure 2 and S_∞ of 1000×10^{-6} for 50% R.H. and 675×10^{-6} for 75% R.H. were used in the analyses. The excellent correlation shown in Figure 5 verifies the validity of predictions of shrinkage development under conditions of varying ambient relative humidity.

EFFECT OF CONCRETE MECHANICAL MATERIAL PROPERTIES ON SHRINKAGE DEVELOPMENT

Mechanical material properties determine stress levels caused by moisture volume changes and hence determine the degree of shrinkage cracking. Nonuniform shrinkage gives rise to a self-equilibrating stress field even in specimens not restrained externally. Typically, tensile stresses build near the surface and

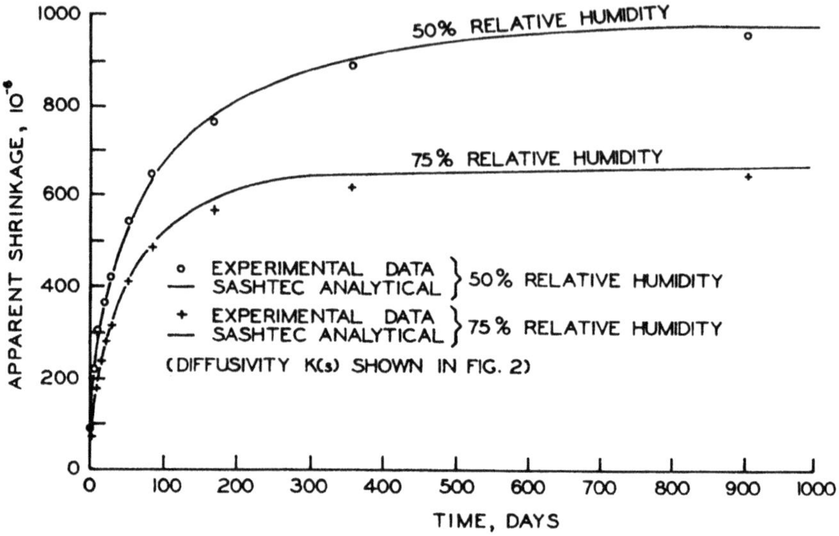

FIGURE 5 SHRINKAGE IN 6 IN. (15.2 CM) DIAMETER COLUMNS AT 50% RELATIVE HUMIDITY AND 75% RELATIVE HUMIDITY

induce cracking in concrete, in turn causing a redistribution of stress and a noticeable change in the apparent shrinkage history of the specimen. The extent of shrinkage cracking significantly affects the subsequent behavior of members when design loads are applied.

Among the structural material properties that affect the extent of cracking, concrete tensile strength f_t, the stress level at which cracking is initiated, is most important. Concrete creep is of next importance since associated relaxation of tensile stresses can be considerable due to the slowness of the shrinkage process. Analyses in which creep behavior is ignored underestimate apparent shrinkage at early stages of drying and overestimate shrinkage cracking. Lastly, the stiffness of concrete, E_c, also affects the stresses that result from shrinkage movement and hence cracking.

The 6-inch (15.24 cm) cylindrical specimens described earlier (Figure 1) were analyzed again using a range of values for f_t in order to determine the sensitivity of shrinkage stresses and strains to variations in concrete tensile strength. Strain and stress results are plotted in Figures 6 and 7.

Cracking slowed the development of apparent shrinkage strain and reduced the buildup of shrinkage stress. The effect of varying the tensile strength is more apparent in the stress than in the

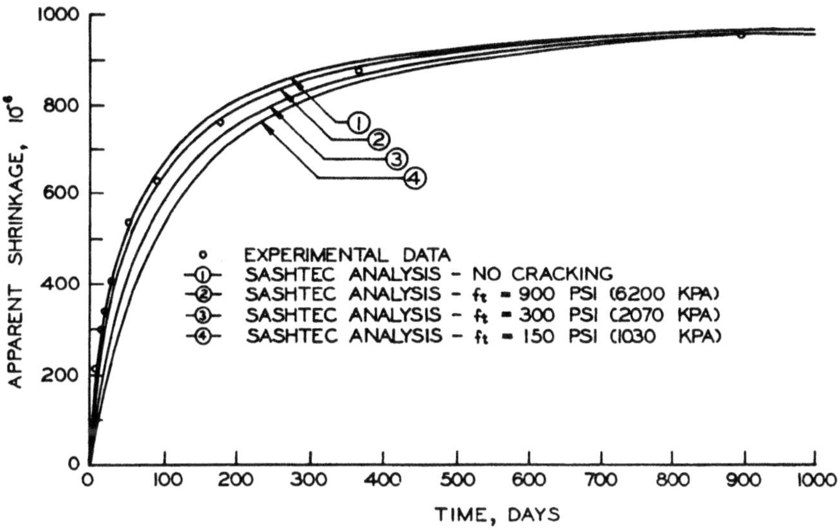

FIGURE 6 EFFECT OF CRACKING ON APPARENT SHRINKAGE — 6 IN. (15.2 CM) DIAMETER CYLINDER

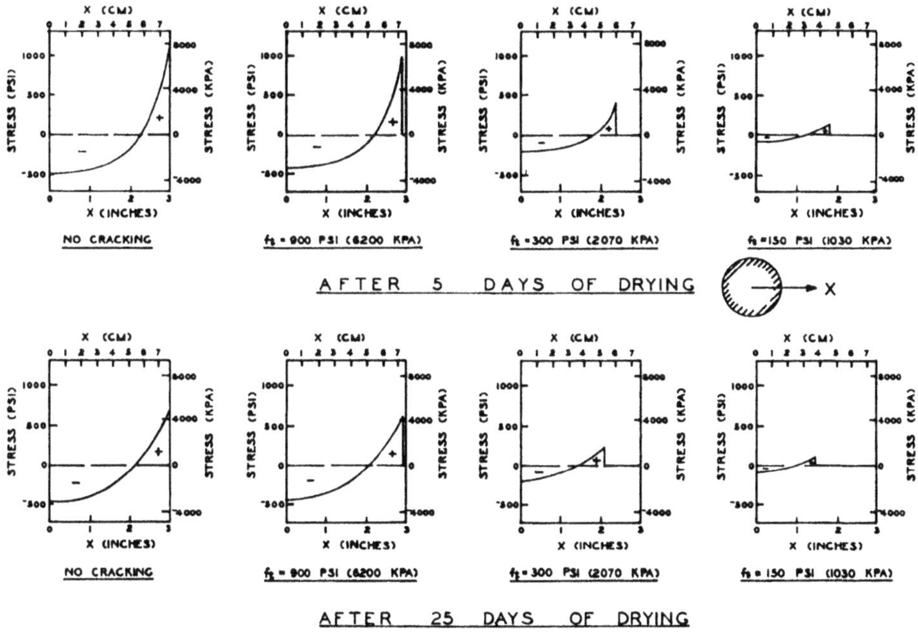

FIGURE 7 EFFECT OF CRACKING ON SHRINKAGE STRESSES – 6 IN. (15.2 CM) DIAMETER CYLINDER

shrinkage strain history. However, since in neither case was shrinkage behavior especially sensitive to f_t , adequate analytical predictions can be obtained using approximate values for tensile strength.

CONCLUSIONS

A practical method has been described in this paper by which the shrinkage stresses and movements that develop during the early life of a structure can be predicted. With approximate identification of moisture diffusion and mechanical properties, satisfactory agreement with experimentally reported data is possible. The method has also been demonstrated to model size effect and the effect of ambient relative humidity on shrinkage development in a consistent and rational way.

ACKNOWLEDGMENT

This research was supported in part by the National Science Foundation, which support is gratefully acknowledged.

REFERENCES

1. R. E. Davis and G. E. Troxell, "Volumetric Changes in Portland Cement Mortars and Concretes," Proceedings, American Concrete Institute, Vol. 25 (1929).

2. W. H. Glanville, "Studies in Reinforced Concrete: II--Shrinkage Stresses," Building Research & Technology Paper No. 11, Department of Scientific & Industrial Research, Great Britain (1930)

3. A. D. Ross, "Shape, Size, and Shrinkage," Concrete & Constructional Engineering, London, Great Britain (1944).

4. G. Pickett, "Shrinkage Stresses in Concrete," Journal of the American Concrete Institute, Vol. 42 (1946).

5. T. C. Hansen and A. H. Mattock, "Influence of Size and Shape of Member on the Shrinkage and Creep of Concrete," Journal of the American Concrete Institute, Vol. 63, No. 2 (1966).

6. Comité Européen du Béton, International Recommendations for the Design and Construction of Concrete Structures, English Edition, Cement & Concrete Institute, London, Great Britain (1970).

7. B. Bresler and R. H. Iding, "Effects of Normal and Extreme Environment on Reinforced Concrete Structures," American Concrete Institute Publication SP-55, Douglas McHenry International Symposium on Concrete and Concrete Structures, Detroit (1978).

8. S. V. Aleksandrovski, Design of Concrete and Reinforced Concrete Structures for Variable Temperature and Humidity Including Effects of Creep, Second Edition (in Russian), Stroiizdat, Moscow, U.S.S.R. (1973).

9. B. Bresler, D. Helmich, and L. V. Ramakrishna, "Nonuniform Drying Shrinkage in Reinforced Concrete," Preliminary Publication, IABSE Symposium: Design of Concrete Structures for Creep, Shrinkage and Temperature Changes, Madrid (1970).

10. R. W. Carlson, "Drying Shrinkage of Large Concrete Members," Journal of the American Concrete Institute, Vol. 8, No. 3 (1937).

11. R. H. Iding and B. Bresler, "SHRINC--A Computer Program for Nonuniform Drying Shrinkage in Concrete Members," Report No. UC-SESM 75-8, Division of Structural Engineering and Structural Mechanics, Department of Civil Engineering, University of California, Berkeley (1975).

12. R. H. Iding and B. Bresler, "SASHTEC--A Computer Program for Structural Analysis of Shrinkage and Temperature Effects in Reinforced Concrete Frames," Division of Structural Engineering and Structural Mechanics, Department of Civil Engineering, University of California, Berkeley (in preparation).

13. R. H. Iding et al., "FIRES-RC II--A Computer Program for the FIre REsponse of Structures - Reinforced Concrete Frames: Second (Revised) Version," Report No. UCB-FRG 77-8, Fire Research Group, Division of Structural Engineering and Structural Mechanics, Department of Civil Engineering, University of California, Berkeley (1977).

14. Z. Nizamuddin, "FIRES-SL--A Computer Program for the FIre REsponse of Structures - SLabs," Ph.D Dissertation, Department of Civil Engineering, University of California, Berkeley (1976).

15. J. R. Keeton, "Study of Creep in Concrete, Phase 1 (I-BEAM)," Technical Report R333-1, U.S. Naval Civil Engineering Laboratory, Port Hueneme, California (1965).

16. M. Mamillan, "Influence de la dimension des eprouvettes sur le retrait," Anales de L'Institute du Batiment et des Travaux Publics, No. 258 (1969).

17. R. H. Iding, B. Bresler, and Z. Nizamuddin, "FIRES-T3--A Computer Program for the FIre REsponse of Structures - Thermal - 3-Dimensional Version," Report No. UCB-FRG 77-15, Fire Research Group, Division of Structural Engineering and Structural Mechanics, Department of Civil Engineering, University of California, Berkeley (1977).

COMPUTERIZED STRUCTURAL ANALYSIS APPLIED TO LARGE SPAN BRIDGES

P.E. Roelfstra[1)]

Rijkswaterstaat Directie Bruggen, The Netherlands

ABSTRACT

Large span bridges constructed according to the cantilever method may show considerable deflections because of creep and shrinkage. In order to achieve correct alignment the shuttering of every corbelling section in the construction of a bridge of this type should be adjusted to a certain excess height. The computation of these excess heights is very complex because of the many stages and load changes during the construction. To simplify these computations a computer program has been developed. With this program the construction of a bridge can be simulated completely. The time-dependent deformations are computed according to the CEB-FIP Model Code for Concrete Structures 1978 (1).

Bruggen met grote overspanningen, die gebouwd zijn volgens de uitbouwmethode, kunnen door krimp en kruip aanzienlijk gaan doorbuigen. Om toch een mooie belijning te verkrijgen wordt bij de bouw de bekisting van iedere uitbouwmoot met een bepaalde overhoogte afgesteld. Het berekenen van deze overhoogten is erg complex vanwege de vele bouwstadia en veranderingen van de belasting. Er is nu een computerprogramma ontwikkeld om deze berekeningen te vereenvoudigen. Met dit computerprogramma kan de bouw van een brug volledig worden gesimuleerd. De tijdsafhandkelijke vervormingen worden hierbij berekend volgens de CEB-FIP Model Code for Concrete Structures 1978 (1).

---

1. Present address : EPFL, Laboratoire des Matériaux de Construction, Chemin de Bellerive 32, 1007 Lausanne (Switzerland).

INTRODUCTION

In Holland most of the large span bridges are constructed according to the cantilever method. This type of bridge may show considerable deflections because of creep and shrinkage. In order to achieve correct alignment the shuttering of every corbelling section in the construction should be adjusted with a certain excess height. Figure 1 shows how these excess heights are determined.

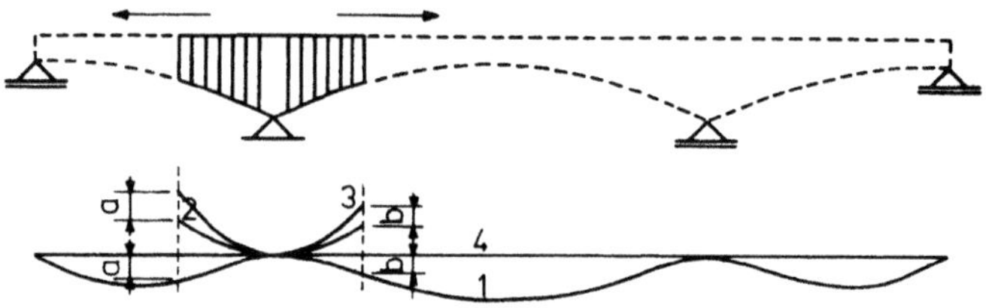

FIG. 1

Determination of the excess heights :

line 1 deflection without excess heights at $t=\infty$

line 2 deflection without excess heights at a certain construction stage

line 3 deflection adjusted with excess heights at a certain construction stage

line 4 achieved alignment at $t=\infty$

a and b excess heights

The computation of these excess heights is very complex because of the many construction stages and load changes. To simplify these computations a computer program based on the finite element analysis has been developed. With this computer program the construction of a bridge can be simulated completely. Prestressing, reinforcement, time-dependent deformations (creep and shrinkage), humidity and temperature changes are taken into account.

The time-dependent deformations are computed according to the CEB-FIP Model Code for Concrete Structures 1978 (1). In this paper some aspects of this computer program will be explained and the executed simulation of the construction of the bridge near Ravenstein (Netherlands) will be discussed.

ELEMENT TYPE

The elements used in this computer program are one dimensional beam elements with three degrees-of-freedom at each of their two end nodes. Figure 2 shows how these beam elements may be built up by bars of concrete, prestressing and reinforcing steel.

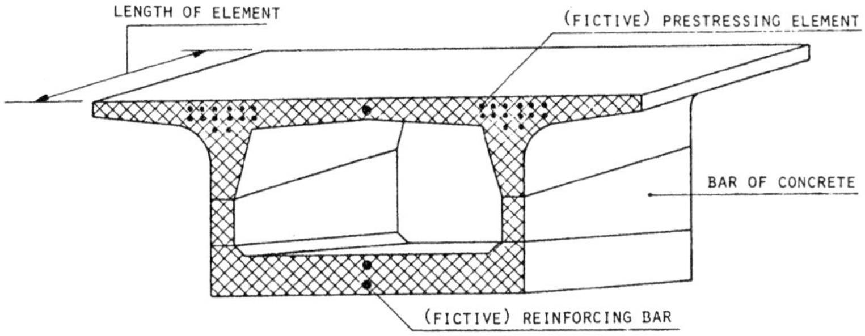

SUBDIVISION INTO BAR ELEMENTS

FIG. 2

One dimensional beam element

This subdivision enables to take in account :

- the influence of the inequality of creep and shrinkage in the cross section
- the influence of reinforcement on creep and shrinkage
- the losses of prestress caused by creep and shrinkage

All the bars in a beam element work together according to the Hypothesis of Bernoulli. Figure 3 shows the assumed displacement of a general reinforcing bar, related to the generalized displacements of the nodes at the ends of the beam.

Figure 4 shows the stiffness matrix of such a bar element. This stiffness matrix is a 6x6 matrix of coefficients, expressing the forces and moments acting on the degrees-of-freedom of the nodes at both ends of a beam.

It needs no explanation that the assumption has been made that the bars of reinforcing and prestressing steel have no bending properties of their own.

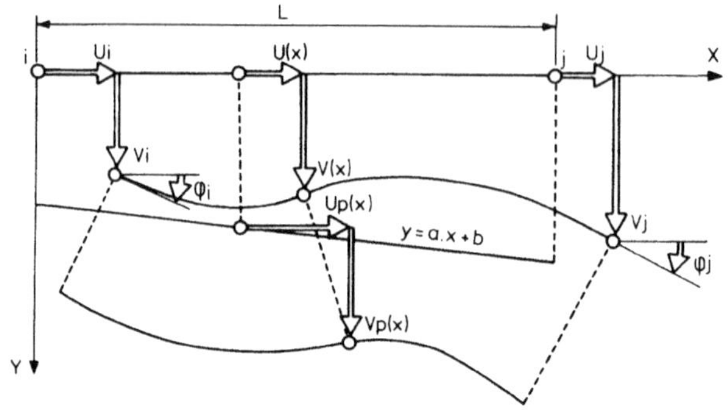

$$\begin{bmatrix} Up(x) \\ Vp(x) \end{bmatrix} = \begin{bmatrix} 1 & -y\frac{d}{dx} \\ 0 & 1 \end{bmatrix} \begin{bmatrix} U(x) \\ V(x) \end{bmatrix}$$

FIG. 3

Assumed displacement of a reinforcing bar

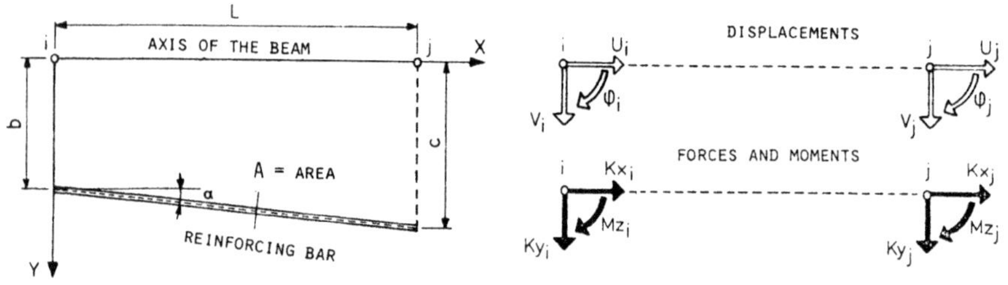

$$\frac{EA\cos^3\alpha}{L}\begin{bmatrix} 1 & \frac{c-b}{L} & -b & -1 & -\frac{c-b}{L} & c \\ \frac{c-b}{L} & \frac{24b^2+12bc+24c^2}{5L^2} & \frac{17b^2+6bc+7c^2}{5L} & -\frac{c-b}{L} & -\frac{24b^2+12bc+24c^2}{5L^2} & \frac{7b^2+6bc+17c^2}{5L} \\ -b & \frac{17b^2+6bc+7c^2}{5L} & \frac{38b^2+14bc+8b^2}{15} & b & -\frac{17b^2+6bc+7c^2}{5L} & \frac{13b^2+4bc+13c^2}{15} \\ -1 & -\frac{c-b}{L} & b & 1 & \frac{c-b}{L} & -c \\ -\frac{c-b}{L} & -\frac{24b^2+12bc+24c^2}{5L^2} & -\frac{17b^2+6bc+7c^2}{5L} & \frac{c-b}{L} & \frac{24b^2+12bc+24c^2}{5L^2} & -\frac{7b^2+6bc+17c^2}{5L} \\ c & \frac{7b^2+6bc+17c^2}{5L} & \frac{13b^2+4bc+13c^2}{15} & -c & -\frac{7b^2+6bc+17c^2}{5L} & \frac{8b^2+14bc+38c^2}{15} \end{bmatrix} \begin{bmatrix} U_i \\ V_i \\ \varphi_i \\ U_j \\ V_j \\ \varphi_j \end{bmatrix} = \begin{bmatrix} Kx_i \\ Ky_i \\ Mz_i \\ Kx_j \\ Ky_j \\ Mz_j \end{bmatrix}$$

FIG. 4

Stiffness matrix of a general reinforcing bar

PRESTRESSING ANALYSIS

It has been assumed that there will be no more than 100 (fictive) tendons in a bridge. The following operations will be made for each of these (fictive) tendons :

Tendon Path

The tendon path in a bridge is computed with the help of some grid points. The rotations in these grid points can be given or not. If the rotation in a grid point is not given, the computer program will compute this rotation according to the method shown in Figure 5.

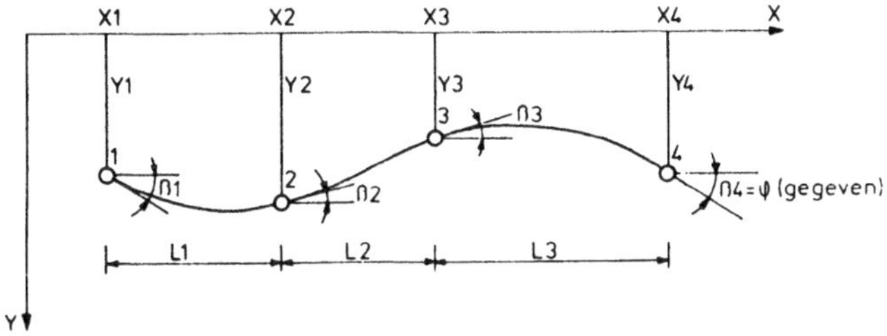

METHOD OF COMPUTING A TENDON PATH

$$\begin{bmatrix} \frac{4}{L1} & \frac{2}{L1} & 0 & 0 \\ \frac{2}{L1} & \frac{4}{L1}+\frac{4}{L2} & \frac{2}{L2} & 0 \\ 0 & \frac{2}{L2} & \frac{4}{L2}+\frac{4}{L3} & 0 \\ 0 & 0 & 0 & 1 \end{bmatrix} \begin{bmatrix} tg\Omega1 \\ tg\Omega2 \\ tg\Omega3 \\ tg\Omega4 \end{bmatrix} = \begin{bmatrix} \frac{6(Y2-Y1)}{L1^2} \\ \frac{6(Y2-Y1)}{L1^2}+\frac{6(Y3-Y2)}{L2^2} \\ \frac{6(Y3-Y2)}{L2^2}+\frac{6(Y4-Y3)}{L3^2}-\frac{2tg\varphi}{L3} \\ tg\varphi \text{ (gegeven)} \end{bmatrix} \qquad (3)$$

FIG. 5

Method of computing a tendon path

A big advantage of this method is that only a few grid points suffices to compute a tendon path and this path will always be continuous.

Tendon Forces

With the help of the determined tendon path the tendon forces are computed. Taken into account are the losses due to friction and anchorage slip. (The losses due to time-dependent deformations will be explained elsewhere). Figure 6 shows the computed tendon forces due to the different ways of tensioning.

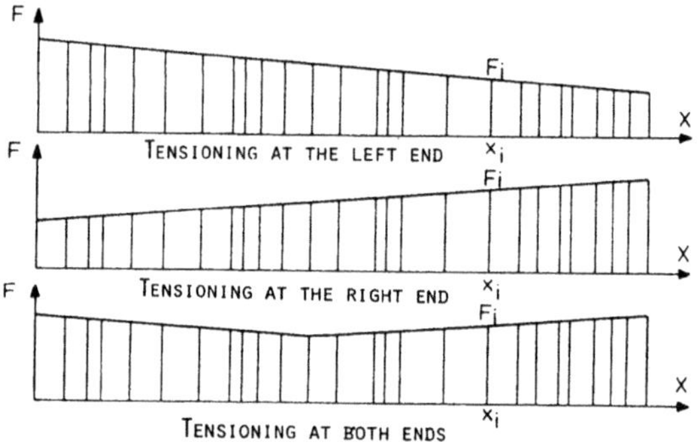

FIG. 6

Tendon forces due to different ways of tensioning

Figure 7 shows the computed tendon forces if anchorage slip is involved.

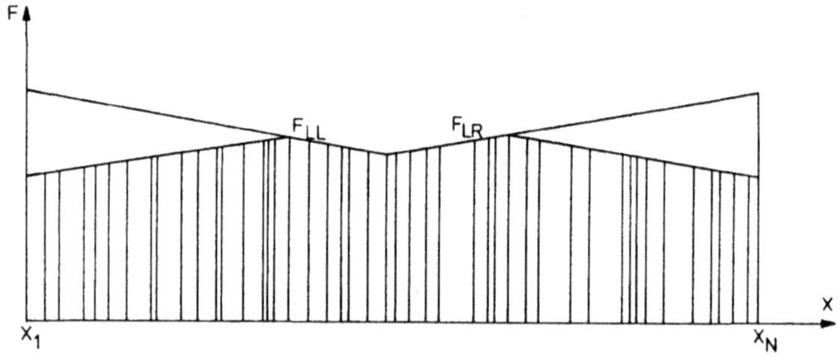

FIG. 7

Tendon forces if anchorage slip is involved

Transposition into Bar Elements

It has been supposed that a tendon may be transposed into straight bar elements (Figure 8). The stiffness matrix and the initial displacement vector of each bar is computed. When the tendon is tensioned the initial displacement vector is corrected with the nodal displacements of the present beam elements.

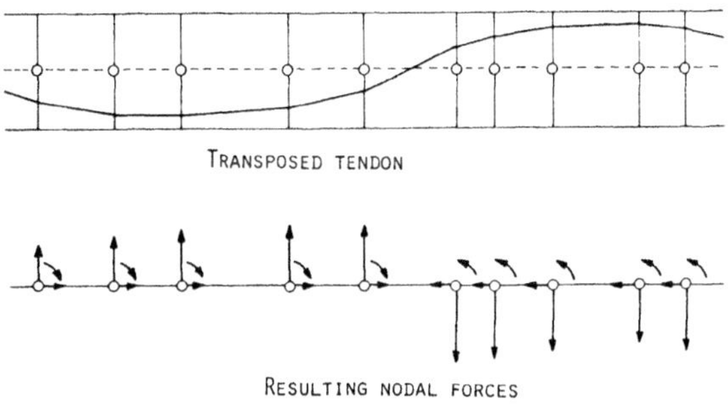

FIG. 8

Transposed tendon and the resulting nodal forces

Losses of Prestress

The losses of prestress due to elastic and time-dependent deformations are determined from the incremental nodal displacements of the beam elements. The assumption has been made that the losses due to the relaxation of the prestressing steel are not time-dependent. So the initial strains are reduced with these losses at once.

SIMULATION OF THE CONSTRUCTION IN A NUTSHELL

The input of the computer program contains the complete data concerning the construction, the material properties and the climatological conditions. Every important time (times of concreting, times of tensioning, times of loading and discharging, times of erecting and removing of supports) is tabulated chronologically. For every mentioned time the complete stiffness matrix is computed, concerning the elements and supports present at that moment. Also is computed the load vector, calculated out of loadings due to :

- initial deformations such as creep, shrinkage and expansion or contraction caused by changes of temperature
- elastic deformations (due to closure in deformed conditions)
- prestressing (decreased by relaxation)
- the dead weight of the bridge construction
- nodal forces (e.g. the weight of the removable cantilever formworks)

With the set of equations, thus compiled, the (unknown) displacements of every nod is computed. And with these nodal displacements the reaction forces, moments, shearing forces and stresses are calculated. At last the increments of shrinkage and creep in every concrete bar is computed up to the next mentioned time. This procedure is repeated until the last mentioned time has been taken into account. The construction of a bridge is thus simulated completely.

SIMULATION OF THE CONSTRUCTION OF A CANTILEVER-BRIDGE NEAR RAVENSTEIN (NETHERLANDS)

Figure 9 shows the side-view and a general cross-section of this bridge. The two spans across the foreland of the river

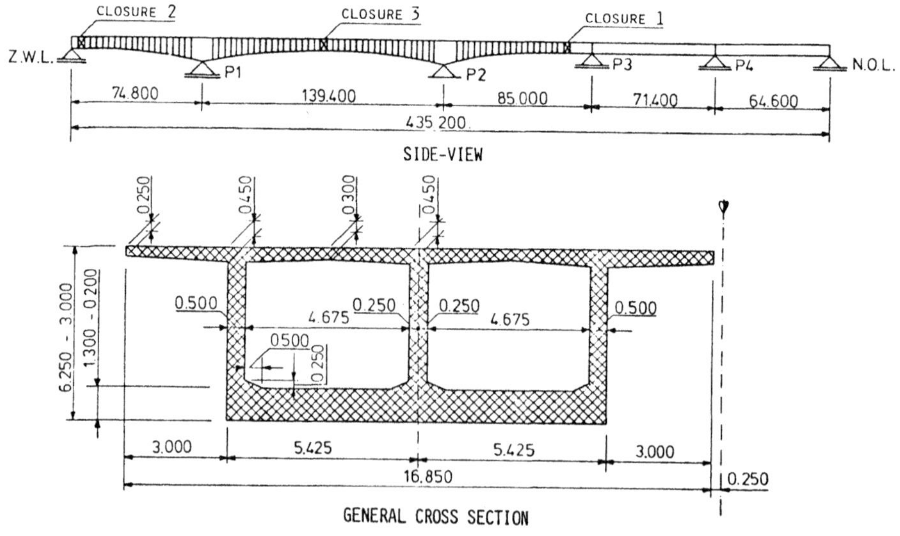

FIG. 9

Maas (P3-P4-NOL) have been concreted in situ in two parts. Next the cantilever sections on pier P2 have been built. Closure 1 connects the corbelling with the spans across the foreland. Afterwards the cantilever sections on pier P1 have been constructed. The second closure was executed at the abutment ZWL. The third and last closure took place above the middle of the river.

Input

To simulate the process of construction the bridge has been divided into 83 beam elements. According to the building schedule all the times of concreting and tensioning have been determined in respect of the start of the construction. Taking advantage of the properties of symmetry only the half of the cross section has been put in. Figure 10 shows in which way this part has been divided into 8 bars of concrete and 5 bars of reinforcement with a fictive diameter. For the rest there are 50 groups of fictive prestressing tendons. The first 38 groups (2x19) form the so-called corbelling tensioning, the remaining 12 groups form the continuity prestressing.

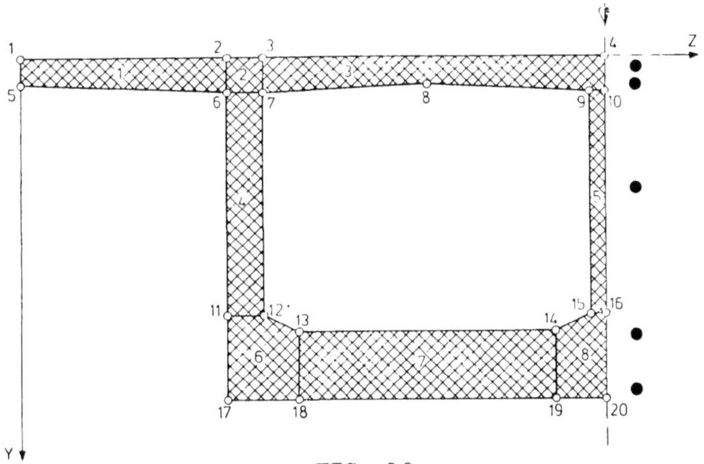

FIG. 10

Subdivision into bar elements

Output

The output consists of tabulations of the deformations, the reaction forces, the shearing forces, the moments and stresses in every part in every stage of construction. To arrange the output in a surveyable way a computer program has been developed by which the results of the calculation are reproduced in graphics.

Graphs

Figure 11 shows some results of the simulation of the process of construction of the Ravenstein bridge. The real sizes of this drawing are 1000x800 mm. The spans across the foreland have been omitted.

Explanation of Figure 11

- To check simply if the input was correct, the geometry, the tendon layout and the supports are drawn.
- The drawn deflections of the bridge in every stage of construction have been corrected with the deflection line for t=10.000 days, as is shown in Figure 1. It is evident that the cantilever section on pier P2 drop as a result of creep and shrinkage during the construction of the cantilever section on pier P1.

  In the deflection figure shows :

  line 1 = the deflections at the time the formwork of element 77 has been removed and the fictive tendon 17 has been tensioned.
  line 2 = the deflections after the concreting of element 78.
  line 3 = the deflections at the time the formwork of element 78 bas been removed and the fictive tendon 18 has been tensioned.
  line 4 = the deflections at the time the third and last closure has been concreted.
  line 5 = the deflections after tensioning of the contimuity prestressing and the removing of the formwork.
  line 6 = the deflections after the concreting of the verges and the paving with asphalt.

  The remaining smooth lines are the deflection lines on different times after the paving with asphalt. The straight line is the deflection line for t=10.000 days.
- The stresses on the upper and lower sides of the bridge are drawn in every stage of construction

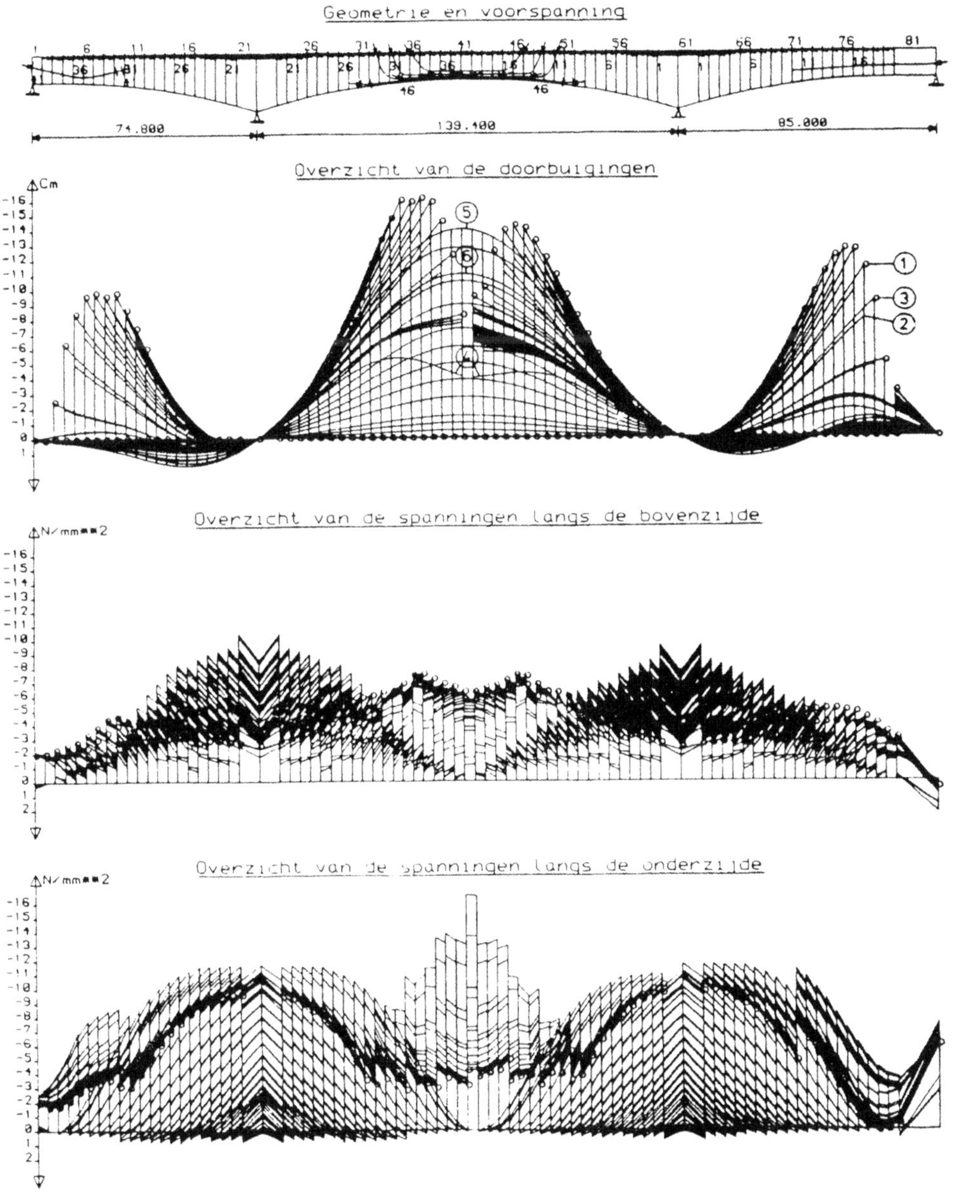

FIG. 11

Geometry, tendon layout, supports, deflections and stresses on the upper and lower sides of the bridge.

in one scheme, in order to see at a glance where these stresses exceed. Detailed drawings can be made of the parts with problems. In these detailed drawings the stresses are figured against an axis of time; the extremes are indicated. The stresses at the time t=10.000 days on the real drawing of Figure 11 are drawn with a red coloured line.

COMPARING THE RESULTS OF THE COMPUTATION WITH TESTS IN SITU

During and after the construction of the bridge near Ravenstein some deformations have been measured. The results of these tests have been used to verify the results of computation.

Strains

Halfway between the first and the last cantilever sections on pier P1 a number of bolts has been placed. relative distance 500 mm. During and after the construction of the bridge the distance between these bolts has been measured accurately. Figure 12 shows where the bolts have been placed.

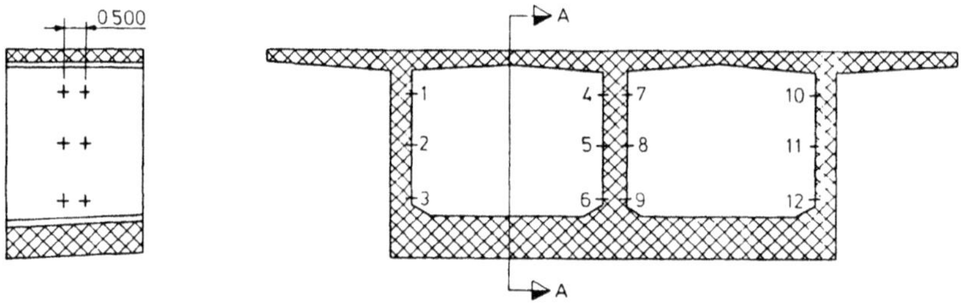

FIG. 12

Situation of the bolts

Figure 13 shows the curves of the measured strains and the computed strains.

Conclusion

Both curves conform qualitatively. Both curves show that during the construction the cantilever section has been bent upwards. This bending initially increased up to the concreting of element 10; it decreased in the following construction stages. This effect was caused by the prestressing moment in the element. which in the earlier part of the construction increased more than the moment by dead weight and the weight of the removable cantilever formwork.

In quantity there are differences, caused mainly by the big spreading of the measured strains at the start of the construction process. Figure 13 shows the tabulated numerical values of the measured strains 7 days after the start.

The increase of the strains after this point of time does not conform to the computed strains. The cause of this difference ought to be found in the method of computing the time-dependent deformations. From this comparison may be concluded prudently, that the time-dependent deformations, computed in accordance with the CEB-FIP Model Code, are too small.

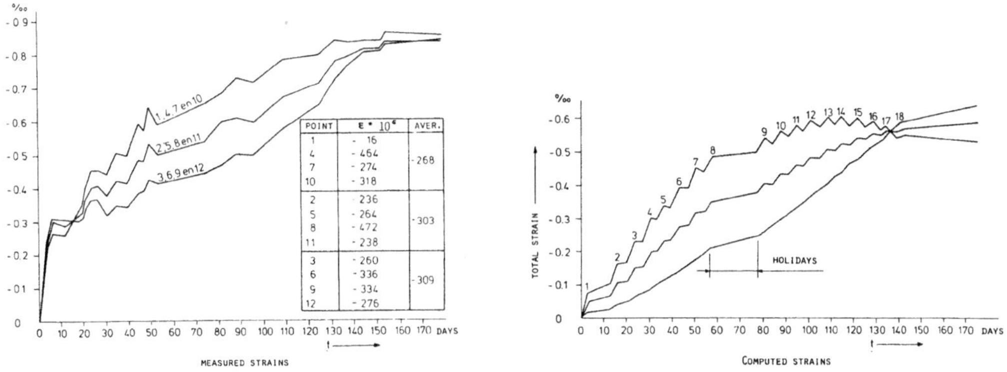

POINT	ε · 10⁶	AVER.
1	- 16	
4	- 464	-268
7	- 274	
10	- 318	
2	236	
5	- 264	-303
8	- 472	
11	- 238	
3	- 260	
6	- 336	-309
9	- 334	
12	- 276	

FIG. 13

Measured strains Computed strains

Deflections in the Middle of the Main Span

Figure 14 shows the computed, the measured and the deflections to be expected in the middle of the main span after the paving with asphalt. In the computation the measured changes in humidity and temperature have been taken into account.

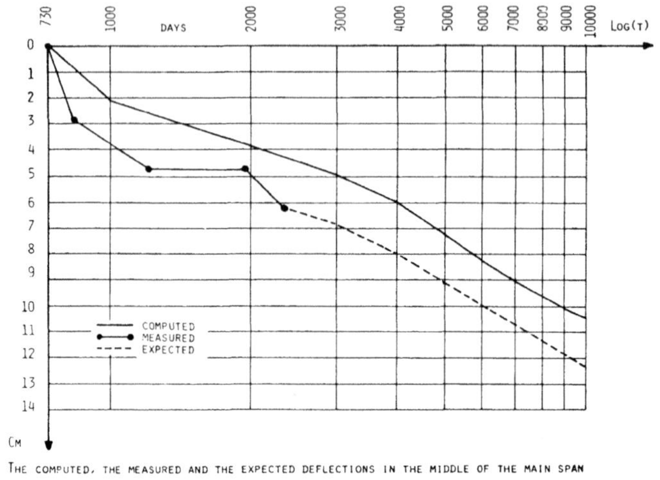

FIG. 14

The computed, the measured and the deflections to be expected

Conclusion

From this comparison, too, may be concluded that the time-dependent deformations, as computed according to the CEB-FIP Model Code, are too small. The difference amounts in broad outline a coefficient = 1,2.

EPILOGUE

Based on the conclusions drawn from the comparison between the computed and measured deformations of the Ravenstein-bridge, the values concerning creep stored in the computer program, have been multiplied by a coefficient = 1,2.

In the mean time the excess heights of the bridges near Heumen, Boxmeer, Well, Neerbosch, Houten and Utrecht (designed by RWS Directie Bruggen, Netherlands, and in construction now or before long) have been computed with the computer program discussed in this paper. The excess heights of the Orwell bridge in England (main span 190 m) have also been computed with this program.

Hitherto it produced good results for the bridges in construction.

All the measured data concerning these bridges are being collected to draw better founded conclusions after some time about the correctness of the values for shrinkage and creep, given in the CEB-FIP Model Code (1978). Future research of time-dependent behaviour on micro level ought to emphasize this conclusion.

REFERENCES

1. CEB-FIP Model Code for Concrete Structures (Comité Euro-International du Béton/Fédération Internationale de la Précontrainte), "International System of Unified Standard Codes of Practice for Structures", Vol. II, (1978).

NUMERICAL ANALYSIS OF CREEP AND SHRINKAGE IN CONCRETE STRUCTURES

W. Haas

Rechen- und Entwicklungsinstitut für EDV im Bauwesen - RIB e. V.,
Stuttgart, German Federal Republic

ABSTRACT

A typical case study is used to demonstrate the basic requirements which must be met by structural programs for creep and shrinkage analysis. Specifically, the condition that the sequence of construction-phases and the corresponding load history are to be modelled properly plays a dominant role in the design of the program architecture. In addition, some results of the numerical analysis are shown compared with those obtained by measurements.

ZUSAMMENFASSUNG

Die Anforderungen an ein Programm zur Berechnung der Auswirkungen des Kriechens und Schwindens werden an einem Beispiel aus dem Brückenbau erörtert. Insbesondere die Forderung, die abschnittsweise Herstellung und Belastung des Bauwerks im Programm berücksichtigen zu können, hat weitreichende Auswirkungen auf das Leistungsspektrum eines derartigen Programms. Berechnungsergebnisse werden erläutert und mit gemessenen Werten verglichen.

1 INTRODUCTION

With regard to the design of structural analysis programs including the effects of the time-dependent behaviour of concrete (as e.g., creep and shrinkage) attention is usually restricted to the mechanical and numerical aspects of the problem, such as suitable stress-strain laws, or stable and accurate time-integration schemes.

In considering these undoubtedly important problems one often neglects the organizational aspects of the task resulting from the fact that concrete structures are usually erected in a sequence of construction phases consisting of different structural systems, or, that cross-sections may be subsequently supplemented by concrete or prestressing cables. Moreover, the dead load and the prestressing are usually set up in stages corresponding to the construction phases. Loading histories of this type may continue over the course of several years. Hence, the first purpose of this contribution consists of reviewing some stress-strain laws for the time-dependent behaviour of concrete from the point of view-mentioned above. Secondly this paper adresses some organizational aspects of programs for the analysis of creep and shrinkage effects in concrete structures. This is done in conjunction with a case history. Finally, some results of the numerical analysis are discussed and compared with those, obtained by in situ measurements.

2 STRESS-STRAIN LAWS FOR THE TIME-DEPENDENT BEHAVIOUR OF CONCRETE

As most design codes (1, 2, 3, 4) consist of statements related to uniaxial stress-strain conditions, the present discussion is restricted to one-dimensional stress-strain laws. In addition since shrinkage is independent of stresses, it will not be taken into account in the following equations,although it is accounted for in the program PRAKSI 4 which produced the results presented below.

In formulating a stress-strain law describing the creep behaviour of concrete, the validity of the principle of superposition (5, 6) or, equivalently linearity (7) can be assumed if the state of stress is restricted by levels up to about 30 % of the ultimate strength. The linearity of the stress-strain law in conjunction with the usual assumption of the Bernoulli-Euler beam theory (i.e., plane section remain plain during deformation) result in a linear stress distribution over the cross-sections.

The linear stress-strain law can be expressed as Stieltjes' integral (8):

$$\varepsilon(t) = \int_0^t \frac{1}{E(\tau)}[1+\varphi(t,\tau)]d\sigma(\tau) \qquad (1)$$

Here ε denotes the strain, σ the stress, E the time-dependent modulus of elasticity, φ the creep function, τ the time at which the stress variation $d\sigma$ occurs and t the time of observation.

Due to the large amounts of data involved the integral representation of equation (1) is unsuitable for implementation into a program (9, 10).

An extremely simple stress-strain law can be obtained if the equation for a constant stress σ_0 applied at time t_0

$$\varepsilon(t) = \sigma_0 \frac{1+\varphi(t,t_0)}{E(t_0)} \qquad (2)$$

is also used in conjunction with time-dependent stress histories $\sigma(t)$.
This leads to the method of the effective modulus E_{eff} which is based on the following equations:

$$\varepsilon(t) = \frac{\sigma(t)}{E_{eff}} \qquad (3)$$

with

$$E_{eff} = \frac{E(t_0)}{1+\varphi(t,t_0)} \qquad (4)$$

This equation allows to account for the effects of creep and shrinkage as in a pseudo-elastic analysis using an effective modulus of elasticity $E = E_{eff}$. The effect of the different times of load applications in the loading history on the effective modulus E_{eff} (s. eq. (4)) is usually neglected and the analysis can then be performed with one effective modulus for a complete loading history.

The following example of a prestressed concrete bridge over the Rhine-river shows the inaccuracies which may occur when the effective modulus method is applied in this way.

The creep analysis of this bridge was done in two different ways. One creep analysis was done "by hand" using an effective modulus method with one effective modulus for one loading history and one time of observation t. A second analysis was done using the computer program PRAKSI 4 which is based on a rate-type creep law. After 200 days since the beginning of the construction of the bridge the first analysis provided a deflection of $w = 5,4$ cm. In contrast the computer analysis predicted for the same time a deflection of $w = 2,05$ cm. The differences in these results can be explained if the history of the deflection is regarded (FIG. 1).

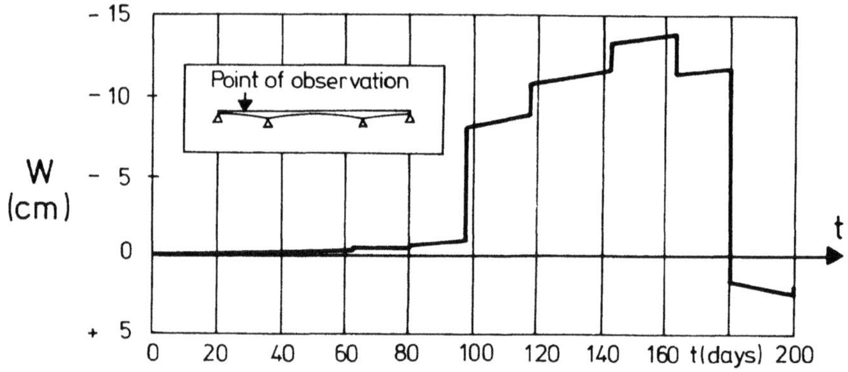

FIG. 1
History of the deflection according to the computer analysis with PRAKSI 4

The steps in the deflection-time plot shown in FIG. 1 are due to the loading history of the bridge which was erected in stages by cantilevering out. For $t = 200$ days the analysis with PRAKSI 4 (rate type creep law) provides an elastic deflection $w_{el} = 4,27$ cm and a creep deflection $w_c = -2,22$ cm. This combination of positive elastic deflection and negative creep deflection is due to the loading history which results in negative elastic and creep deflection until $t = 180$ days and a positive elastic deflection between the times $t = 180$ days and $t = 200$ days. The first analysis based on a constant effective modulus for the complete loading history provides a creep deflection of $w_c = 1,13$ cm for $t = 200$ days. This is due to the fact that the negative increments in creep deflections Δw_c occuring between $t = 0$ days and $t = 180$ days can not be accounted for by this type of analysis.

These inaccuracies can be avoided by analysing each loading case of the loading history separately employing different effective modulus $E_{eff}(t,t_o)$ which depend on the time of observation t and on the time of loading t_o.

However, if the loading history consists of a large number of loading cases with different t_0 as in the example of the case-history to be described later the procedure becomes complex and uneconomical. Apart from this the procedure becomes inaccurate if a considerable change in stress occurs between the loading time t_0 and the time of observation t and if the concrete ages considerably (11).

A similarly simple analysis of the effects of creep and shrinkage for one loading case in one step can be done with Trost's ϱ-formula (6).

$$\varepsilon(t) = \frac{\sigma(t_0)}{E(t_0)}\left[1+\varphi(t,t_0)\right] + \frac{\sigma(t)-\sigma(t_0)}{E(t_0)}\left[1+\varrho\varphi(t,t_0)\right] \tag{5}$$

The relaxation factor ϱ allows the influence of the time dependence of the stress variation and the aging of concrete to be accounted for more accurately than by the effective modulus method. For this reason this procedure is also known in the USA as the "age-adjusted effective modulus method" (12).

The great advantage of equations (3) and (5) is that they allow to compute the stress resultants and displacements at the time of observation t for one loading case in one step (one-step procedures). However, if various construction phases consisting of different structural systems are to be considered the changes in stresses and deformations within each construction phase must be computed using the respective structural system. For this reason, the total changes in the stress resultants and in the deformations can not be calculated in one step (i.e., using one particular structural system). In such cases the above mentioned advantages of the equations (3), (5) may no longer be relevant and thus a rate type stress strain law is preferable. The computer program PRAKSI 4 uses a rate-type creep law based on the differential equation of the aging three-parameter solid (see FIG. 2).

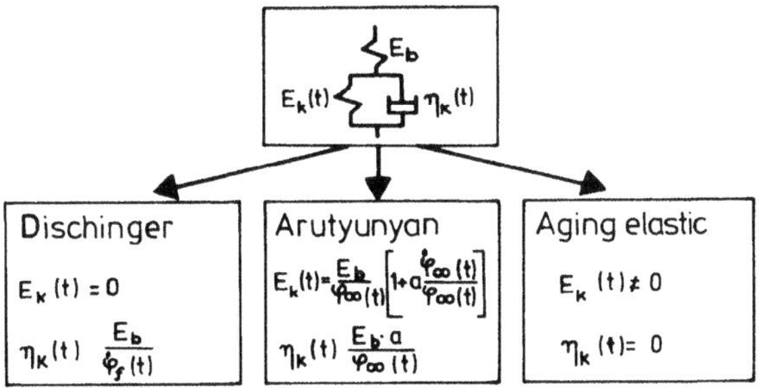

FIG. 2
Some creep laws which can be represented by the aging three-parameter solid

Until now the aging three-parameter solid has been used almost exclusively in the limited form of Arutyunyan's creep law (13,14). However, as shown in (15) the differential equation of the three-parameter solid is able to represent a great variety of creep laws including Dischinger's creep law (rate of creep method), the German guidelines for prestressed concrete structures (3) and the CEB recommendations (1) with a completely sufficient accuracy provided that $E_K(t)$ and $\eta_K(t)$ are choosen appropriately.

3 CREEP AND SHRINKAGE ANALYSIS OF THE KOCHER VALLEY BRIDGE

The Kocher Valley Bridge is a prestressed concrete motorway bridge. Its maximum height above the valley floor is 185 m. The structural system of the bridge consists of a frame structure, the four central columns of which are rigidly connected with the superstructure. The regular width of the spans of the superstructure is 138 m (FIG. 3).

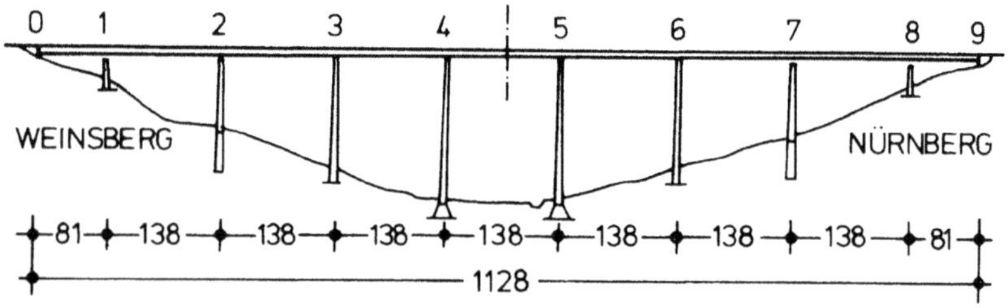

FIG. 3
Side view of the Kocher Valley Bridge

The six-lane highway is supported by a box girder 6,5 m high with widely overhanging plates (FIG. 4).

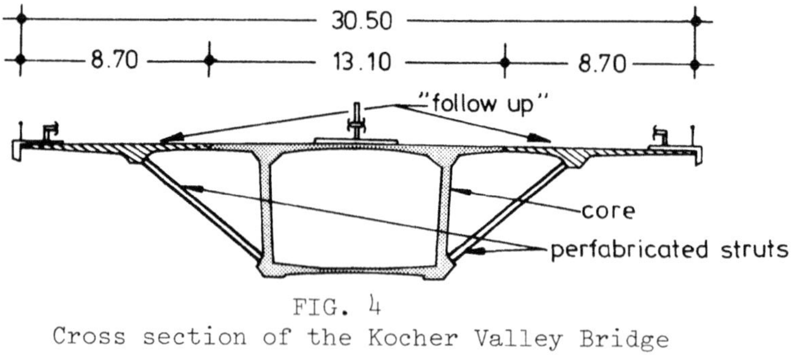

FIG. 4
Cross section of the Kocher Valley Bridge

The superstructure is constructed in stages by cantilevering out using an advancing girder. In this procedure the box girder section without the lateral plates (i.e., the core cross-section) is first constructed by cantilevering out in both directions from the tops of the piers. As soon as one section extending from one centre of a span to the centre of the next span has been completed, one begins to concrete the lateral plates on both sides of the core structure, the so-called "follow-up" structure, thus completing the final cross-section of the bridge. These concrete plates are braced against the core by means of prefabricated struts. Simultaneously, the next section in the cantilever construction of the core proceeds as before. Due to the short period of time available, the construction of the superstructure is advanced, starting from both abutments simultaneously. The delayed addition of the "follow-up" cross-section to the core cross-section causes stress redistributions in the cross-section ("cross-sectional creep") and deformations which,in turn, produce stress redistributions in the entire structural system ("system creep"). Both types of creep-effects can, in this case, no longer be considered independently (14).

The construction of the bridge in 14 construction phases with different structural systems also yields stress redistributions in the total structure. In order to keep these effects to a minimum, so-called joint-expansion moments and forces are applied by means of presses to the coupling joints before these are closed. These joint-expansion moments and forces provide the internal stresses which would be present if all loads had been applied right from the start to the final structural system. But as the bridge contains concrete of varying ages this measure cannot completely suppress system creep. In addition, the temporal progression of the deflections has to be calculated in order to determine the required camber of the bridge.

In calculating the effects of creep and shrinkage it suffices to analyse only half of the bridge due to the symmetry of the structural problem. It is idealized as a plane frame with 13 beams (FIG. 5). Each of these beams may have varying cross-sectional properties along its axis and each cross-section may be a composite cross-section. With this extraordinarily efficient beam we are able to account for the gradual increase in cross-sectional values due to the concreting-on of the "follow-up" cross-section and the successive prestressing and injecting of the tendons.

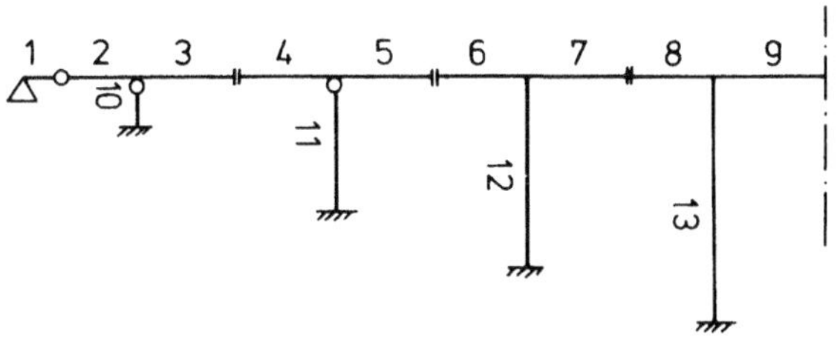

FIG. 5
Structural system for the computer analysis

The following loading cases must be considered:

- 32 loading cases of dead load due to the weight of the core cross-section
- 32 loading cases of dead load due to the weight of the "follow-up" cross-section
- 39 loading cases for various positions of the cantilevering out scaffold and the advancing girder
- 9 loading cases due to joint expansion
- 9 loading cases related to pavement, railing etc.
- 42 loading cases due to prestressing of 175 tendons in the core cross-section and 90 tendons in the "follow-up" cross-section

These loading cases were applied at 42 different instants of time.

The analysis was based on the improved three-parameter solid. The integration in time was carried out in 90 time steps. It is interesting to note that 83 steps were required to describe accurately the different construction phases and the loading history and only 7 steps to determine the changes in the internal forces and displacements during the period beginning with the opening of the bridge to traffic (t = 1120 days) up to the assumed date of termination (t = 12.000 days).

FIG. 6 shows the changes of the normal forces in the core cross-section and the "follow-up" cross-section versus time. It indicates that the "follow-up" cross-section gradually carries a considerable proportion of the load. At $t \rightarrow \infty$ 68 % of the normal force N = - 31 MN in the "follow-up" structure are caused by the effects of creep and shrinkage. Another result of creep and shrinkage is a reduction of the normal force in the core of 50 MN. Of this reduction of the normal force in the core, 29 MN are due to prestress losses and 21 MN have been taken over by the "follow-up" cross-section.

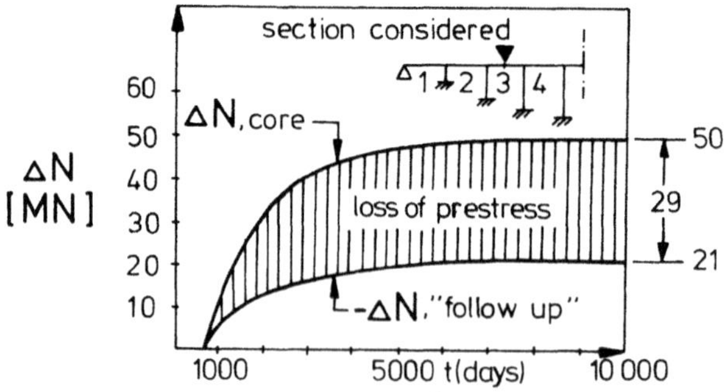

FIG. 6
Changes of the normal forces in the core and the "follow-up"

FIG. 7 shows the changes of the bending moment in the middle of the third span. The oscillations at the beginning of the curve are due to the different construction phases during this period. The total change in the bending moment ΔM = 12.1 MN at the time $t \rightarrow \infty$ is about 5 % of the bending moment due to the total dead weight of the superstructure on the final system. Thus, the expansion of the joints has greatly reduced system creep.

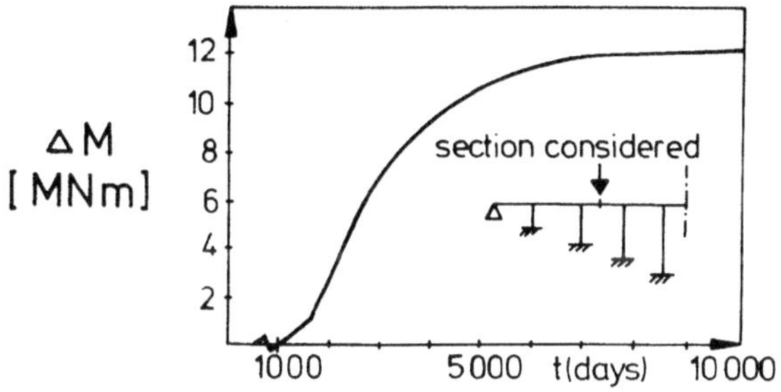

FIG. 7
Changes of the bending moment in the middle of the third span of the bridge

FIG. 8 shows the calculated and measured camber of the bridge for three different points in time. The figure contains the cantilevering out section which belongs to pier no. 7 (see FIG. 3).

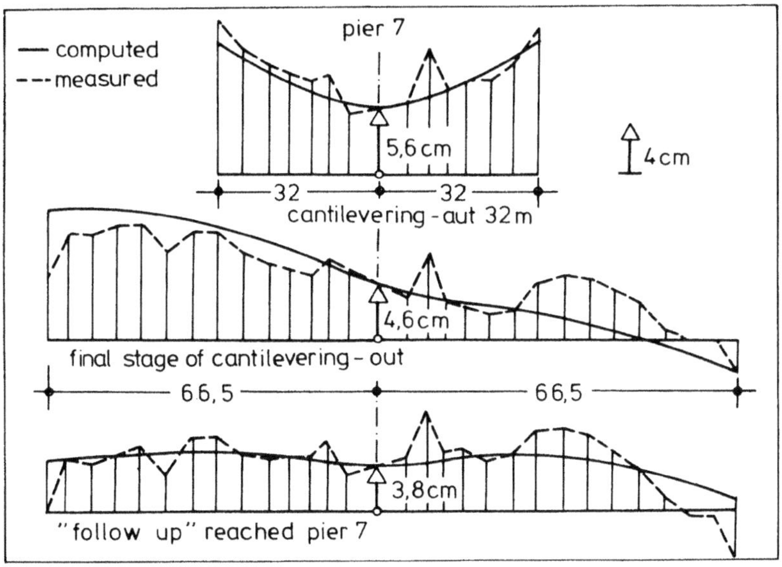

FIG. 8
Camber of the cantilevering out section of pier no. 7

In order to eliminate the inaccuracies in the calculation of the settlements of the soil and the thermal strains of the pier the curves of the calculated camber were shifted in such a way that on top of the pier no. 7 the values of the measured and calculated camber were identical. The maximum shift was 4 cm.

In FIG. 8 the maximum difference between the calculated and the measured values is 5,5 cm. In comparison to the inaccuracies which always occur during the construction, of the assumed creep and shrinkage curves the correspondence of measured and calculated values is good for the relatively large span ℓ = 138 m between two piers.

4 REFERENCES

1. Comité Européen du Beton - CEB, "Time Dependent Behaviour of Concrete. State of Art Report", 1973.

2. ACI Commitée 209/II, "Prediction of Creep, Shrinkage and Temperature Effects in Concrete Structures", ACI - SP27.Designing for Effects of Creep, Shrinkage and Temperature. Detroit, 1971.

3. "Richtlinien für Bemessung und Ausführung von Spannbetonbauteilen", Fassung Juni 1973, printed in Betonkalender 1975.

4. DIN 1045 - "Beton- und Stahlbetonbau, Bemessung und Ausführung", Ausgabe Januar 1972.

5. Mc Henry, D., "A New Aspect of Creep in Concrete and its Application to Design", Proc. ASTM 43 (1943).

6. Trost, H. "Auswirkungen des Superpositionsprinzips auf Kriech- und Relaxationsprobleme bei Beton und Stahlbeton", Beton- und Stahlbetonbau 61 (1967), H. 10, 11.

7. Onat, E. T., "Description of Mechanical Behaviour of Inelastic Solids", Proc. of the 5th U.S. National Congress of Applied Mechanics, p. 421.

8. Bažant, Z. P., "Theory of Creep and Shrinkage in Concrete Structures, A Precis of Recent Developments", Mechanics Today, Vol. 2. New York, Toronto, Oxford, Sydney, Braunschweig, Pergamon Press Inc. 1975.

9. Zienkiewicz, O. C., Watson, M., "Some Creep Effects in Stress Analysis with Particular Reference to Concrete Pressure Vessels", Nuclear Engineering and Design 4 (1966) p. 406.

10. Haas, W., "Über ein für die EDV geeignetes Verfahren zur Erfassung des Kriechens und Schwindens von Beton", Dissertation, Universität Stuttgart, 1974.

11. Bažant, Z. P., Najjar, L. J., "Comparison of Approximate Linear Methods for Concrete Creep", Proc. ASCE, J. Struct. Div. 99 (1973) no. St9, p. 1851.

12. Bažant, Z. P., "Prediction of Concrete Creep, Effects using Age-Adjusted Effective Modulus Method", ACI Journal 69 (1972), p. 212.

13. Arutyunyan, N. Kh., "Some Problems in the Theory of Creep in Concrete Structures", Pergamon Press, 1966.

14. Schade, D., Haas, W., "Elektronische Berechnung der Auswirkungen von Kriechen und Schwinden bei abschnittsweise hergestellten Verbundstabwerken", Heft 244 des DAfStb, Berlin 1975.

15. Haas, W., "Normengerechte Berücksichtigung des zeitabhängigen Verformungsverhaltens von Beton, dargestellt an zwei Beispielen aus dem Hoch- und Brückenbau", Finite Elemente in der Baupraxis, 290 - 299, Verlag Wilhelm Ernst und Sohn, 1978.

THE INFLUENCE OF THERMAL CREEP ON THE OPERATIONAL BEHAVIOUR OF COMPLEX STRUCTURES

Y Bangash, Thames Polytechnic, London, U.K.
G L England, King's College, London U.K.

ABSTRACT

A three-dimensional finite element creep analysis has been developed in which a special provision has been made for creep and creep recovery. Computer programmes have been developed to assess the influence of thermal creep on the operational behaviour of complex structures. The work has application to prestressed concrete nuclear reactor pressure vessels and to concrete gravity platforms for the storage of oil at sea. The results obtained illustrate the significance of creep and temperature on the stress behaviour of nuclear and oil containment structures.

On a effectué une analyse du fluage par élément fini tri-dimensionnel, analyse dans laquelle une mention spéciale a été accordée du fluage et a la récupération du fluage. On a mis sur ordinateur des programmes ayant pour but d'évaluer l'influence du fluage thermal sur le comportement opérationnel des structures complexes. Ce travail s'applique aux "containments" sous pression de réacteur nucléaire en béton précontraint et aux plates-formes gravité en béton destinées à emmagasiner le pétrole en mer. Les résultats obtenus illustrent l'importance du fluage et de la température sur le comportement sous pression des structures de "containments" nucléaires et de petrole.

KEY WORDS

Creep, creep recovery, Finite Element, temp, stress, prestress, concrete, vessels, platforms, oil containment, computer application.

NOTATION

Bi	the matrix of derivatives of shape functions of node i
B	strain-displacement matrix of the element
D	material matrix
e	elastic parameter
E	Young's modulus
G	force/volume
K	stiffness matrix
N	shape function of the element
Ni	shape function of ith node in the element
I	identity matrix

where n = total number of mesh nodes around the element

P = nodal forces

$\{\sigma_0\},\{\epsilon_0\}$ = initial stresses and strains

ϵ^e_f = flow strain

ϵ^e_d = delayed elastic strain

τ = age of loading

t = age of concrete

J_f, J_d = creep functions

t' = pseudo time

ν = Poissons ratio

T = temperatures

INTRODUCTION

In prestressed concrete reactor vessels and gravity platforms, thermal gradients exist during normal operation and accident conditions. In order to estimate the stress and deformational behaviour of these structures with a reasonable degree of precision, a three-dimensional finite element analysis has been developed in which a special provision has been made to incorporate the influence of time-dependent thermal creep strains. Comparison of results with and without creep for the case of concrete reactor vessels shows that creep does cause concern in some local areas. Other results obtained illustrate the significance of creep recovery component of strain.

ANALYSIS USING DISPLACEMENT FINITE ELEMENT

The displacement finite element model (1-5, 9-11) has been adopted on account of the efficient solution procedures available for the solution of large problems. It is used here in conjunction with an incremental time step approach to the transient creep problem, to evaluate limiting strain rates in the steady-state stress problem for sustained temperatures, and average strain rates in the cyclic temperature creep shakedown problem.

For the transient creep solution the unknown variables are the displacements at the nodes of the finite element mesh and are expressed in the form of an element displacement vector

$$\{u\}^e = \begin{Bmatrix} \{u_1\}_1 \\ \{u_2\}_1 \\ \vdots \\ \{u_n\}_n \end{Bmatrix} \tag{1}$$

The displacement field within each element can be expressed as

$$\{u\} = [N]\{u\}^e = \sum_{i=1}^{n} (N_i [I] \{u\}_i) \tag{2}$$

With the displacements known at all points within each element, the strains can be expressed in the following form:

$$\{\epsilon\} = \sum_{i=1}^{n} ([B_i]\{u\}_i^e = [B]\{u\}^e \tag{3}$$

The stresses can be determined within each element from the strains as

$$\{\sigma\} = [D](\{\epsilon\} - \{\epsilon_0\}) + \{\sigma_0\} \tag{4}$$

Appendix I.1 gives the three-dimensional anisotropic material matrix [D] for concrete.

Now virtual displacement $\{du\}^e$ are applied on nodes, the sum of work done, dW, by the stresses, distributed body and surface forces over the element volume, Vol, and surface, S, is given by

$$dW = (\{du\}^e)^T \left(\int_{vol} [B]^T \{\sigma\} dV - \int_S [N]^T \{p\} ds - \int_{vol} [N]^T \{G\} dV \right) \tag{5}$$

In order to maintain equilibrium within the element, a system of external nodal forces $\{P\}^e$ must be applied which will reduce virtual work (dW) to zero. Equation (5) will take the form

$$(\{du\}^e)^T \{P\}^e = (\{du\}^e)^T \left(\int_{vol} [B]^T \{\sigma\} dv - \int_S [N]^T \{p\} ds - \int_{vol} [N]^T \{G\} dv \right) \tag{6}$$

For anisotropic stress-strain relationship, Appendix I gives the matrix [D] of the form

$$\{\epsilon\} = [D]\{\sigma\} \tag{7}$$

Equation (6) is valid for any virtual displacement $\{du\}^e$, thus it can be eliminated from both sides of Equation (6). Substituting Equation (3) and (4) into Equation (6)

$$\{P\}^e = (\int_{vol} [B]^T [D][B] dv)\{u\}^e - \int_{vol} [B]^T [D]\{\epsilon_0\} dv + \int_{vol} [B]^T \{\sigma_0\} dv - \int_s [N]^T \{p\} ds - \int_{vol} [N]^T \{G\} dv \quad (7)$$

Equation (7) is the force-displacement relation with stiffness transformation. The following terms can be explained in the above equation:

The element stiffness matrix $[K]^e = \int_{vol} [B]^T [D][B] dv$. (7a)

Nodal force due to body force, $\{P_b\} = - \int [N]^T \{G\} dv$ (7b)

Nodal force due to surface force, $\{P_s\}^e = - \int_s [N]^T \{p\} ds$ (7c)

Nodal force due to initial stress, $\{P_{\sigma_0}\}^e = - \int_{vol} [B]^T \{\sigma_0\} dv$ (7d)

Nodal force due to initial strain, $\{P_{\epsilon_0}\}^e = - \int_{vol} [B]^T [D]\{\epsilon_0\} dv$ (7e)

Equation (7) can be rewritten as

$$\{P\}^e = [K]^e \{u\}^e + \{P_b\}^e + \{P_s\}^e + \{P_{\sigma_0}\}^e + \{P_{\epsilon_0}\}^e \quad (8)$$

Equation (8) is the force-displacement relation for each element, for whole structure the stiffness matrix and load vectors are assembled according to the nodal incidences, overall equilibrium equations can be written in the following form:

$$[K]\{u\} = \{P\} \quad (9)$$

Equation (9) is solved for unknown displacements $\{u\}$. The element strains can be obtained from nodal displacements. For linear material stresses are obtained from Equation (4). For anisotropic stress-strain behaviour, the strain values are computed from Equation (7).

SIMULATION OF THREE-DIMENSIONAL CREEP

(i) Transient creep behaviour

The influence of creep on the time-dependent behaviour of concrete has been studied using various theories (6 - 9). In the time-dependent stress and strain analyses described here, a creep formulation has been used which allows independent variation of irreversible and delayed recoverable strains, with respect to temperature and time. Simultaneous adoption of a pseudo-time (12) concept simplifies the mathematical form of the modelling and allows the four element Burgers representation to be employed in pseudo-time. This approach has been used numerically to predict strain from stress (13); using experimental data to predict stresses in prestressed concrete reactor pressure vessels using a semi-analytical approach with a differential creep formulation (14); and in an integral formulation in other non-uniform temperature problems (15). The integral formulation is described here and used later in the analysis of a nuclear reactor containment vessel. The reasons governing its suitability to this type of problem have been given elsewhere (16).

When loads are long sustained, the solution is eventually governed by the Maxwell elements of the model. A Maxwell representation for the transient creep effects is also possible at elevated temperatures when creep recovery is small compared to the irrecoverable creep.

Figures 1 to 6 show the components of creep which form part of the four-element Burgers model representation of creep. Their equivalent mathematical description is:

$$\epsilon_f^c(t) = \sum_0^t J_f(t-\tau)\overset{*}{\Delta\sigma} \tag{10}$$

$$\epsilon_d(t) = \sum_0^t J_d(t-\tau)\overset{*}{\Delta\sigma} \tag{11}$$

Where '*' represents the incremental value of   at

The above equations can be written in pseudo-time, t'.

$$\epsilon_f^c(t') = \sum_0^{t'} J_f(t'-\tau')\overset{*}{\Delta\sigma} \tag{12}$$

$$\epsilon_d^c(t') = \sum_0^{t} J_d(t'-\tau')\overset{*}{\Delta\sigma} \tag{13}$$

where $J_f(t'-\tau') = t'-\tau'$

Where '*' is defined as before but in pseudo time. The components of creep strains in pseudo-time having special provisions for reversible and irreversible creep effects can now be written in the form

$$J_d(t') = \sum_{i=1}^{n} J_{d\alpha_i}(1-e^{-J_f/Q_i}) = Q[1-e^{(-\beta(t'-\tau'))}]\ mm^2/N \tag{14}$$

Total strain is then the summation of elastic and creep strain and is given by

$$\epsilon(t') = \epsilon_e(t') + \epsilon_c(t') \tag{15}$$

Using creep Burger model in pseudo-time, the incremental flow component is written as

$$\Delta J_f(t') = \Delta t' \, \rho(T) = [(t'+\Delta t') - t'] \, \rho(T) \tag{16}$$

where $\Delta t'$ = incremental pseudo-time = 1/50
$\rho(T)$ = temperature function = $\frac{1}{50}(T+30)\rho(T_\phi)$

where $\rho(T_\phi)$ = temperature at ambient level (in the vessel normally 20°C)

The incremental delayed elastic strain component is given by

$$\begin{aligned}\Delta J_d(t') &= [J_d(t'+\Delta t') - J_d(t')] \\ &= (1-e^{-\beta\Delta t'})[Q - J_d(t')]\end{aligned} \tag{17}$$

Equation (15) is now written as

$$\Delta J(t') = \Delta J_f(t') + \Delta J_d(t') \tag{18}$$

Pseudo-time is converted to a real time in Equation (19) using data (20) obtained for a limestone concrete

$$t' = [-15.37346 + 7.88977 \log_e(t)] \times 10^{-6} \tag{19}$$

For cases of anisotrophy and cracking in addition to creep, Appendix I gives the material matrix, $[D]$ of Equation (20).

$$\{\epsilon\} = [D]\{\sigma\} \tag{20}$$

Strains and stresses are computed both for linear and non-linear cases using initially $\Delta J(t') = 1/E_1$ and

$$\{\epsilon\} = 1/E_1 \, [D^*]\{\epsilon\} \tag{21}$$

or in the incremental form

$$\{\epsilon_c\} = \Delta J(t')[D^*]\{\epsilon\} \tag{21a}$$

The initial value of E is E_{20} which is the Young's modulus at ambient temperature and which is taken from the following equation:

$$\begin{aligned}E(T) &= E_{20}\left[1 - \frac{T-20^\circ}{137}\right] \quad \text{for } 20^\circ \le T \le 50^\circ \\ E(T) &= E_{20}(0.78)\left[1 - \frac{T-50^\circ}{341}\right] \quad \text{for } 50^\circ \le T \le 85^\circ\end{aligned} \tag{22}$$

In general three Poisson's ratios are specified and the remaining three, ν_{21}, ν_{32} & ν_{31} are calculated from the symmetry of compliances given in Appendix I.

(ii) Steady-state creep solutions

Temperature-dependent creep causes stress redistribution to occur with time under load. Stresses eventually stabilise to steady-state values and may be evaluated by direct calculation using elastic solution techniques in which the elastic modulus is simply replaced by the reciprocal of the normalising creep-temperature function (17), $\rho(T)$.

In situations where temperatures repeat cyclically in time between two or more states it is possible to identify repeating states of stress in the long term which, because of creep, are dissimilar to the initial elastic stresses. These creep shakedown stresses may be evaluated using elastic procedures after suitable substitution of the normal elastic parameters representing elastic modulus and initial strain. This formulation has been described in detail elsewhere (18), and is described briefly below. The theory has application to the calculation of stresses in concrete oil storage structures at sea which are subjected to periodic filling and emptying of hot crude oil, during the storage and pump-out cycles of operation.

Adopting a Maxwell formulation for creep and elastic strains in the shakedown state, and assuming that the stresses during any constant temperature portion of the temperature cycle may be regarded as time-invariant over the interval, the stress states are linked by the thermal stress change between any two states. Considering two repeating temperature states, T_1 and T_2, with corresponding stress states, $\tilde{\sigma}_1$ and $\tilde{\sigma}_2$, leads to the statement that,

$$\tilde{\sigma}_2 - \tilde{\sigma}_1 = \tilde{\sigma}_\alpha \qquad (23)$$

where   is the thermal stress change corresponding to the temperature change from T_1 to T_2.

When the ratio of the hold times at the different temperatures is $k_2{:}k_1$, it has been shown that the shakedown stresses, $\tilde{\sigma}_2$, may be evaluated from an analogous elastic calculation of the form

$$\{\sigma\} = E\,[D^*]^{-1}\{\epsilon - \epsilon_0\} \qquad (24)$$

The shakedown stresses $\tilde{\sigma}_2$ can be derived from Equation (24) using a stiffness formulation in conjunction with a finite element analysis, by replacing $[D^*]$ by its creep counterpart, E by $1/\rho(T^*)$ where T^* is defined as the average temperature over the cycle interval, and ϵ_0 by $-(k_1 T_1 \tilde{\sigma}_\alpha)/(k_1 + k_2)$.

A finite element analysis evaluates the average displacement rates at the node points and these are used in turn to evaluate the initial strain rates and associated shakedown stresses in the temperature state T_2.

APPLICATION TO COMPLEX ENGINEERING PROBLEMS

The above analysis is applied to two different problems, namely a prestressed concrete reactor vessel and a cell of a concrete gravity platform. The numerical data for these problems is given in Table I.

A computer program is written using three-dimensional finite element in Fortran IV. A combination of solid, panel and line elements representing concrete, steel liner and prestressing reinforcement steel is used. Both elastic and plastic/cracking conditions can be investigated. This program is also interfaced with a specially developed finite element mesh generation.

Figure 7 shows typical combinations of these elements used in the current analysis.

(a) Prestressed concrete pressure vessel

Figure (8) shows a three-dimensional finite element mesh generation of a prestressed concrete vessel designed for high temperature gas-cooled reactor. Both internal and external loads have been computed in accordance with Equations (7) and (8). Program CREEP, which is a part of Program ISOPAR, gives results when the vessel considers the effect of creep. First a normal operation condition (prestress + pressure + temperature) is considered. This vessel is analysed with and without creep. The axial, radial and circumferential stresses have been computed. The influence of creep when the Young's modulus is constant or variable is considerable. In some local areas due to creep, the stresses change significantly with passage of time. Figure (9) shows stresses at the top cap with and without creep for up to 20 years. Figure (10) shows the behaviour of the vessel under operating and thermal shutdown conditions with and without the influence of creep.

The vessel is then taken to elastoplastic and cracking conditions. Figure (11) shows the relationship between gas load, prestressing strains, and deflections and cracks with and without the influence of creep. These results show that where the influence of creep is not considered, the load carrying capacity is over-estimated by as much as 25%. With creep considerable changes were discovered in the cracking pattern of the vessel and the overstress conditions in local areas (stand pipes and boiler/ circulator penetrations).

(b) Concrete gravity platforms

Figure (12) represents a single cell oil containment structure of cylindrical shape. Its behaviour has been examined when subjected to filling and emptying, with hot oil, over the central section. The stress results obtained for the limiting shakedown stress state are typical of those which can be created in a multi-cellular containment, with the exception that here, temperature crossfall reversals have not been considered.

In this problem the cyclic nature of the temperature changes varies with cylinder depth, and in consequence the equivalent initial strains of Equation (24) are radius- and depth-dependent quantities in the creep shakedown stress analysis.

The longitudinal stresses in the containment wall during the hot condition - state T_2 of Figure 12 - are shown in Figure 13 for various radial and depth locations. In the cold condition it is necessary to subtract the thermal stresses. For this problem the thermal stresses are differentiated into two components, those due to the intrinsic cylindrical geometric constraints of the containment and a temperature change from T_1 to T_2; and those due to the thermal stress perturbation associated with the moving oil/water interface at the division between the hot and cold vertical zones. These latter stresses are treated in detail in reference (18).

The combined influence of creep and temperature on the behaviour of oil-containing structures of single and multi-cellular types is described in detail, with suggested design procedures, in a recently published report by the Underwater Engineering Group of the Construction Industry Research and Information Association (19).

REFERENCES

1. Zienkiewicz, O.C., The Finite Element Method, McCraw Hill Book Co. (UK), 1977.
2. Bangash, Y., "A basis for the design of bonded reinforcement in prestressed concrete pressure vessels". Paper 7978S Suppl. (viii) 1972. The Institution of Civil Engineers, London. Also Discussion Suppl. (xviii) 1972, pp 157-163.
3. Bangash, Y., "A circumferential prestressing load analysis due to wire/strand winding systems on prestressed concrete reactor vessels". Proc. Institution of Civil Engineers, London, Vol.57, Part 2, Sept. 1974, pp 437-450. Also Discussion, Part 2, Vol.59, March 1975, pp 195-200.
4. Bangash, Y., "Movements in prestressed concrete reactor vessels", Int. Journal Nuclear Engineering & Design, Nov. 1978.
5. Bangash, Y., "Safety and reliability of prestressed concrete reactor vessels". Int. Journal of Nuclear Engineering & Design, Jan. 1980.
6. Ross, A.A., "Creep of concrete under variable stress". Journal of the American Concrete Institute, Vol. 54, Mar. 1958, pp 739-758.
7. England, G.L., "Numerical creep analyses applied to concrete structures". Proceedings of American Concrete Institute, Journal Vol. 64, No. 6, June 1967, pp 301-311.
8. Bangash, Y., "Prestressed concrete reactor vessel time-saving ultimate load analysis". J. Inst. Nuclear Engineering, Vol. 13, No. 4, July/Aug. 1972.
9. Bangash, Y., "Safe analysis of cooling pipes for prestressed concrete reactor vessels". Nuclear Engineering Design, Vol 55, p 305-313, 1979.
11. Bangash, Y., Discussion to paper on "Finite element non-linear analysis of concrete structures" (D.V. Phillips & O.C.Zienkiewicz), Institution of Civil Engineers, London, Vol. 61, Part 2, Dec. 1976, pp 829-890.
12. England, G.L. and Jordaan, I.J., "Time-dependent and steady-state stresses in concrete structures with steel reinforcement, at normal and raised temperatures". Mag. of Conc. Res., Vol 27, No. 92, Sept. 1975, pp 131-142.
13. England, G.L. and Illston, J.M., "Methods of computing stress in concrete from a history of measured strain". Civil Eng. and Public Works Review, London, England, Vol. 60, 1965, pp 513-517, 692-694, 846-847.
14. England, G.L. and Allen, S.J., "The calculation of time-dependent stresses in PCPVs by an approximate method". Proc. of 1st Conference on Structural Mechanics in Reactor Technology, Berlin, Germany, 1971. Vol. 4, Part H, pp 289-313.
15. Jordaan, I.J., "Analysis of creep in concrete structures under general states of stress". Seminar on Concrete Structures Subjected to Triaxial Stresses, Bergamo, Italy, May 1974. Papers 111-8, pp 1-18.

REFERENCES (continued)

16. Jordaan, I.J., England, G.L., and Khalifa, M.A., "Creep of concrete: A consistent engineering approach". Proceedings of American Society of Civil Engineers, Structural Division, Vol. 103, No. ST3, Mar. 1977, pp 475-491.
17. England, G.L., "Steady-state stresses in concrete structures subjected to sustained temperatures and loads". Nuclear Engineering and Design, Vol. 3, Parts I and II, 1966, pp 54-65, 246-255.
18. England, G.L., Moharram, A., and Andrews, K.R.F., "Time-dependent stress redistribution in concrete oil storate structures". Proceedings of International Conference on Offshore Structures, Rio de Janeiro, Brazil, October 1979.
19. Richmond, B., England, G.L., and Bell, T., "Designing for temperature effects in concrete oil-containing structures". Report of the Underwater Engineering Group, CIRIA, July 1980. No. UR17, p 116.
20. Browne, R.D., "Properties of concrete in reactor vessels". Proc. of International Conference on PCPV. The Institution of Civil Engineers, London, England. 1967, pp 131-151.

Anisotropic stress-strain relations can be written in the form:

$$\epsilon_x = \frac{\sigma_x}{E_1} - \frac{\nu_{21}}{E_2}\sigma_y - \frac{\nu_{31}}{E_3}\cdot\sigma_z$$

$$\epsilon_y = \frac{\sigma_y}{E_2} - \frac{\nu_{12}}{E_1}\sigma_x - \frac{\nu_{32}}{E_3}\sigma_z$$

$$\sigma_z = \frac{\sigma_z}{E_3} - \frac{\nu_{13}}{E_1}\sigma_x - \frac{\nu_{23}}{E_2}\sigma_y$$

$$\gamma_{xy} = \tau_{xy}/G_{12}$$

$$\gamma_{yz} = \tau_{yz}/G_{23}$$

$$\gamma_{zx} = \tau_{zx}/G_{13} \qquad \text{(AI.1)}$$

Due to symmetry of compliances, relations can be included

$$\{\epsilon\} = [D]\{\sigma\} \qquad \text{(AI.2)}$$

In matrix form,

APPENDIX I

A1 Anisotropic material matrix [D]

$$\begin{bmatrix} D_{11} & D_{12} & D_{13} & 0 & 0 & 0 \\ D_{21} & D_{22} & D_{23} & 0 & 0 & 0 \\ D_{31} & D_{32} & D_{33} & 0 & 0 & 0 \\ 0 & 0 & 0 & D_{44} & 0 & 0 \\ 0 & 0 & 0 & 0 & D_{55} & 0 \\ 0 & 0 & 0 & 0 & 0 & D_{66} \end{bmatrix} \qquad \text{(AI.1)}$$

Where

$$D_{11} = (1-\nu_{23}\nu_{32})E_1/\gamma_K \; ; \; D_{22} = (1-\nu_{13}\nu_{31})E_2/\gamma_K$$

$$D_{33} = (1-\nu_{12}\nu_{21})E_1/\gamma_K \; ; \; D_{12} = (\nu_{12}+\nu_{13}\nu_{32})E_2/\gamma_K$$

$$D_{21} = (\nu_{21}-\nu_{23}\nu_{31})E_1/\gamma_K \; ; \; D_{13} = (\nu_{13}+\nu_{12}\nu_{23})E_3/\gamma_K$$

$$D_{31} = (\nu_{31}+\nu_{21}\nu_{32})E_1/\gamma_K \; ; \; D_{23} = (\nu_{23}+\nu_{13}\nu_{21})E_3/\gamma_K$$

$$D_{32} = (\nu_{32}+\nu_{31}\nu_{12})E_2/\gamma_K \; ; \; D_{44} = G_{12}\cdot a'$$

$$D_{55} = G_{23}\cdot a'$$

$$D_{66} = G_{13}\, a'$$

a' – for interlocking effects

$$\gamma_K = 1-\nu_{12}\nu_{21}-\nu_{13}\nu_{31}-\nu_{23}\nu_{32}-\nu_{12}\nu_{23}\nu_{31}-\nu_{21}\nu_{13}\nu_{32}$$

Due to symmetry of compliances, relations can be written.

$$\begin{aligned} E_1\nu_{21} &= E_2\nu_{12} \\ E_2\nu_{32} &= E_3\nu_{23} \\ E_3\nu_{13} &= E_1\nu_{31} \end{aligned} \qquad \text{(AI1.1)}$$

The values of G12, G23 and G13 are calculated in terms of modulus of elasticity and poisson's ratios as follows:

$$G_{12} = \tfrac{1}{2}\left[\frac{E_1}{2(1+\nu_{12})} + \frac{E_1}{2(E_1/E_2+\nu_{12})}\right]$$

$$G_{23} = \tfrac{1}{2}\left[\frac{E_2}{2(1+\nu_{23})} + \frac{E_2}{2(E_2/E_3+\nu_{23})}\right] \qquad \text{(AI1.2)}$$

$$G_{13} = \tfrac{1}{2}\left[\frac{E_3}{2(1+\nu_{31})} + \frac{E_3}{2(E_3/E_1+\nu_{31})}\right.$$

TABLE I

INPUT DATA FOR VESSEL AND GRAVITY PLATFORM CELLS

Internal diameter	11.125 m (36' 6")
Internal height	21.946 m (72' 0")
Refuelling hole	1.371 m (4' 6")
Control rod zone	7.010 m (23' 0") O/D
E(Steel)	193×10^3 MN/m^2
E(Concrete) Short term	38×10^3 MN/m^2; 34×10^3MN/n^2(*)
Long term	20.7×10^3 MN/m^2
E(Control Rod Zone)	0.74E
Poissons Ratio	0.15
Coefficient of Thermal Expansion of Concrete	$10 \times 10^{-6}/^{o}C$; $12 \times 10^{-6}/^{o}C$(*)
Short Term Specific Creep	$1830 \times 10^{-9}/MN/m^2/^{o}C$
Long Term Specific Creep	$2407 \times 10^{-9}/MN/m^2/^{o}C$
Minimum Crushing Strength at 28 days	41.7 MN/m^2
Creep parameters	$\beta = 0.2814 \times 10^6$ N/mm^2
	$Q = 14.21457 \times 10^{-6}$ mm^2/N
Prestressing Tendons and Strand-Winding	
Longitudinal prestressing:-	
Ultimate Load/Tendon	9.75 MN; 7 MN/m^2(*)
Tendon Load when locked off at 75% U.T.S.	7.31 MN
Coefficient of friction	0.25
Wobble factor	3.278×10^{-4}/m
Circumferential prestressing:-	
Ultimate force in a single strand	0.128 MN
Force in strain at 75% U.T.S.	0.096 MN

(*) Values used for oil containment analysis.

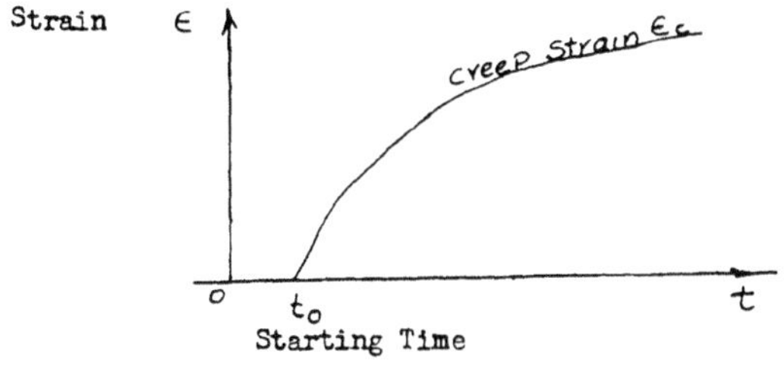

FIG I Total Time-Dependent Strain

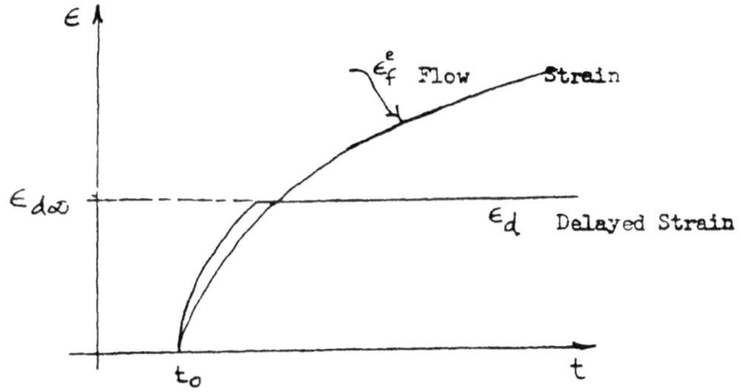

FIG 2 Flow(Irrecoverable) And Delayed Elastic(Recoverable) Strain

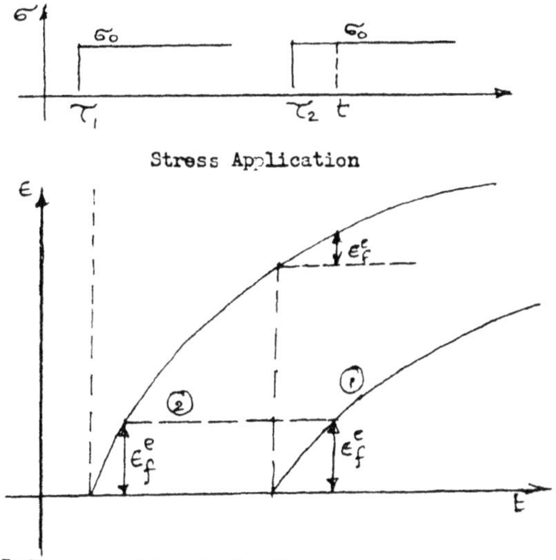

FIG 3 Influence of Age At Loading On Flow Component Of Strain.

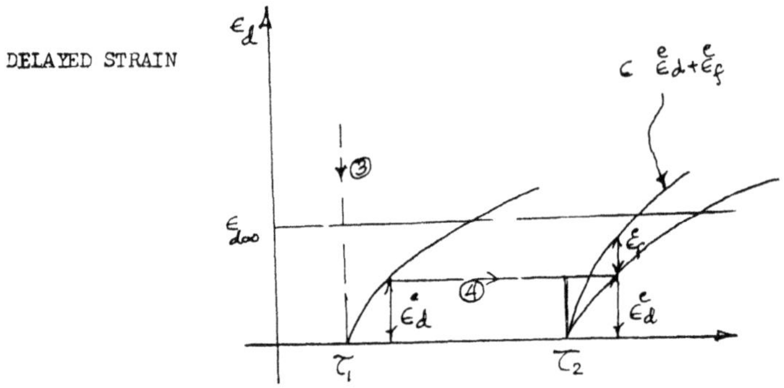

FIG 4. Creep Strain At Different Age Of Loading

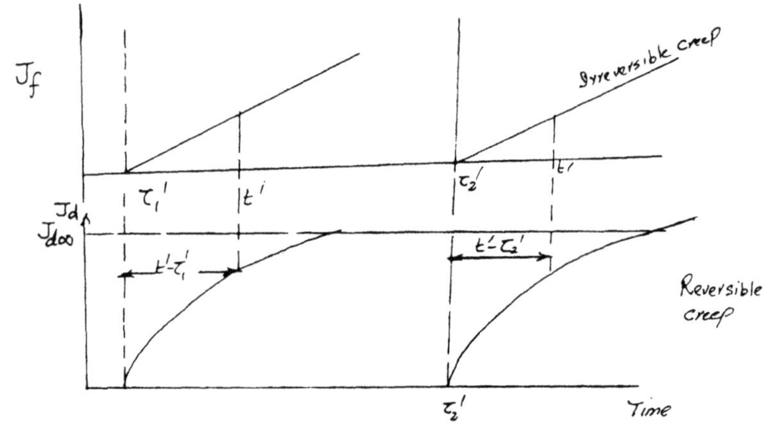

FIG 5 Component Of Creep Strain In Psuedo Time

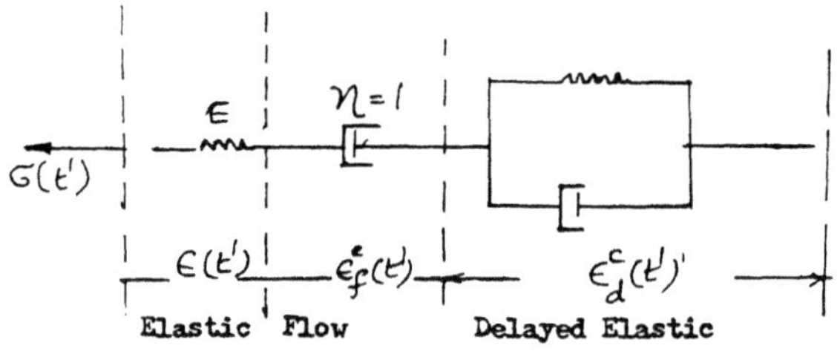

FIG 6. Burgers Model (Psuedo Time)

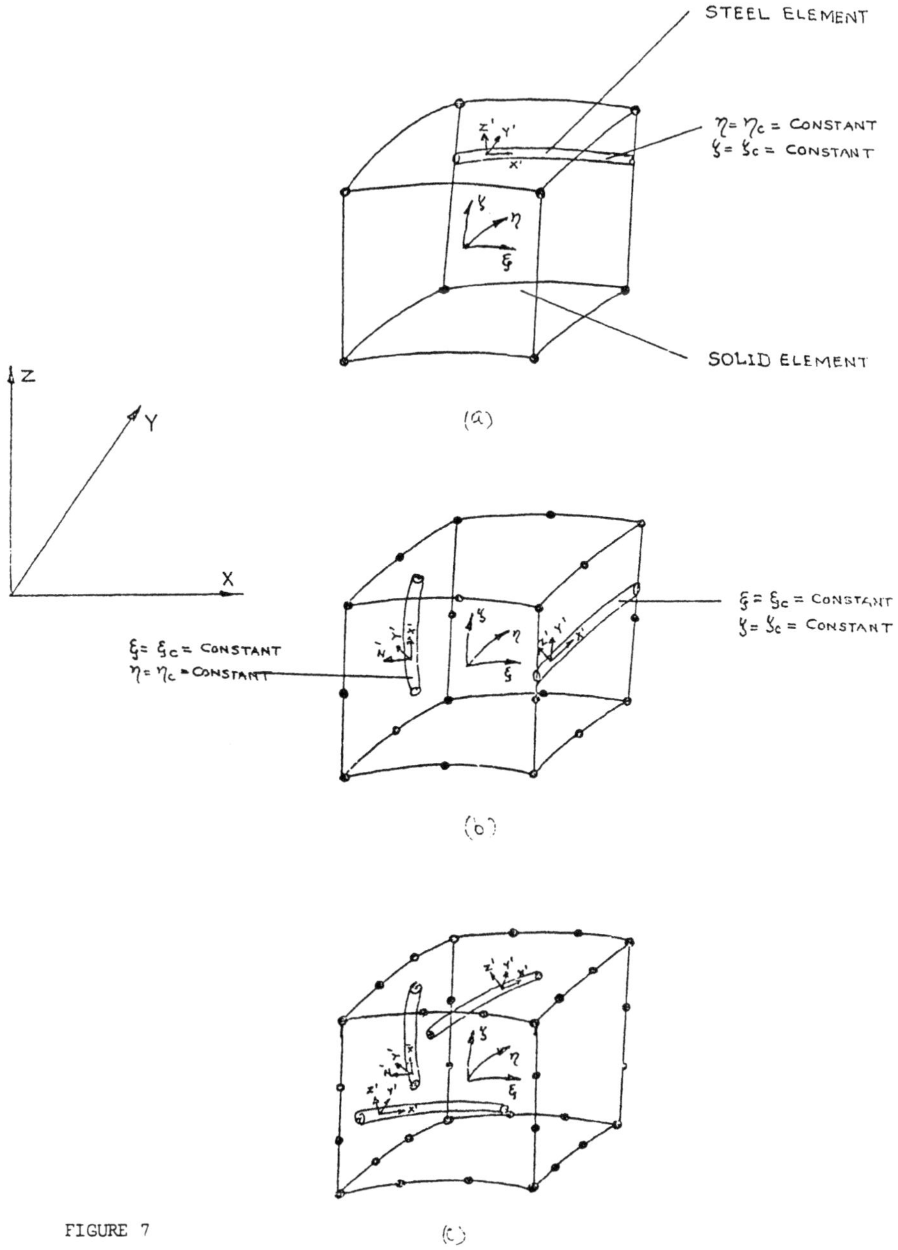

FIGURE 7

Combination of finite elements

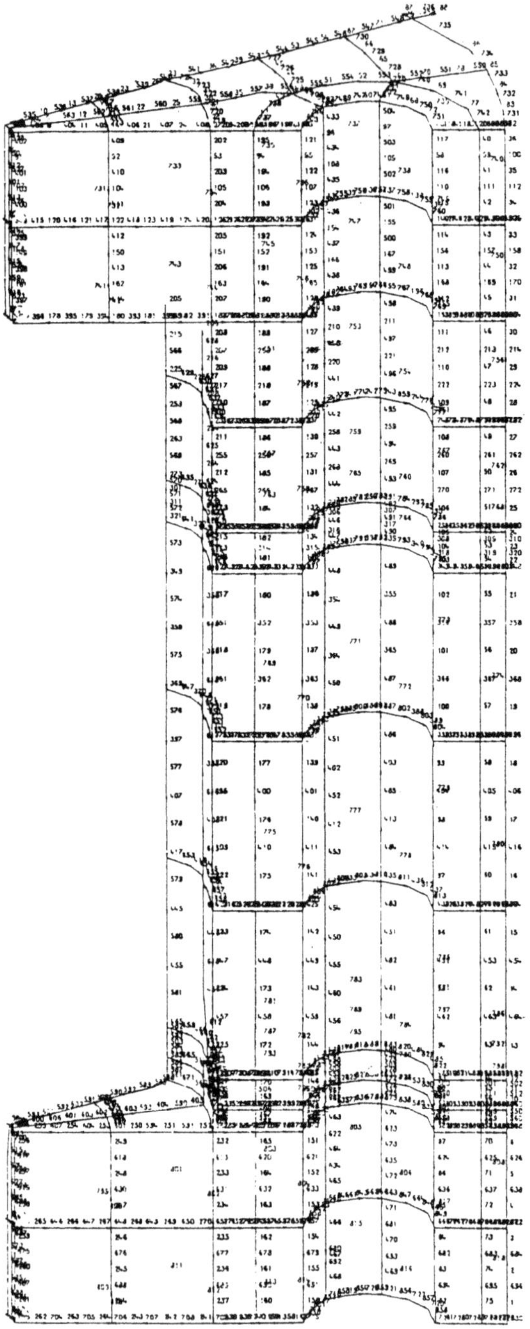

FIGURE 8

PRESTRESSED CONCRETE CYLINDRICAL MULTI-CAVITY REACTOR VESSEL (H.T.R)
GENERAL VIEW

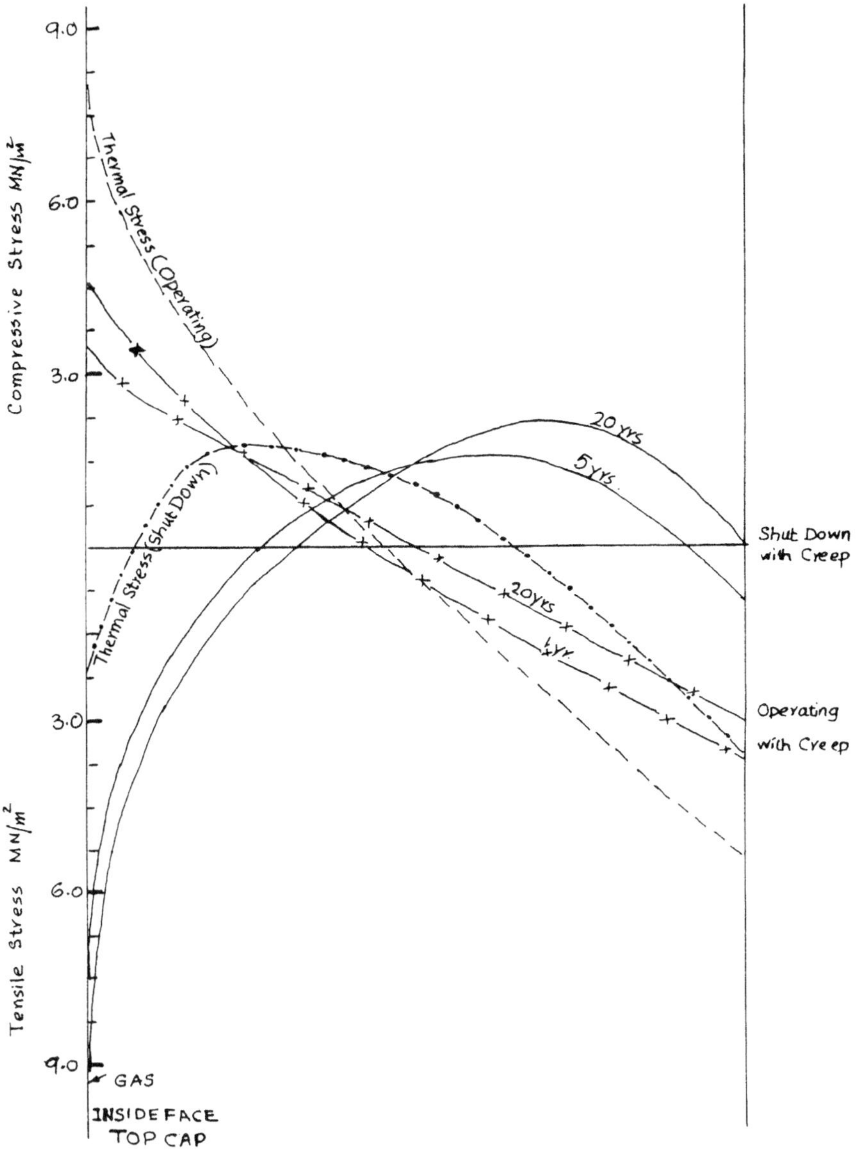

FIGURE 9

Stresses through top cap: operating and temperature shut-down.

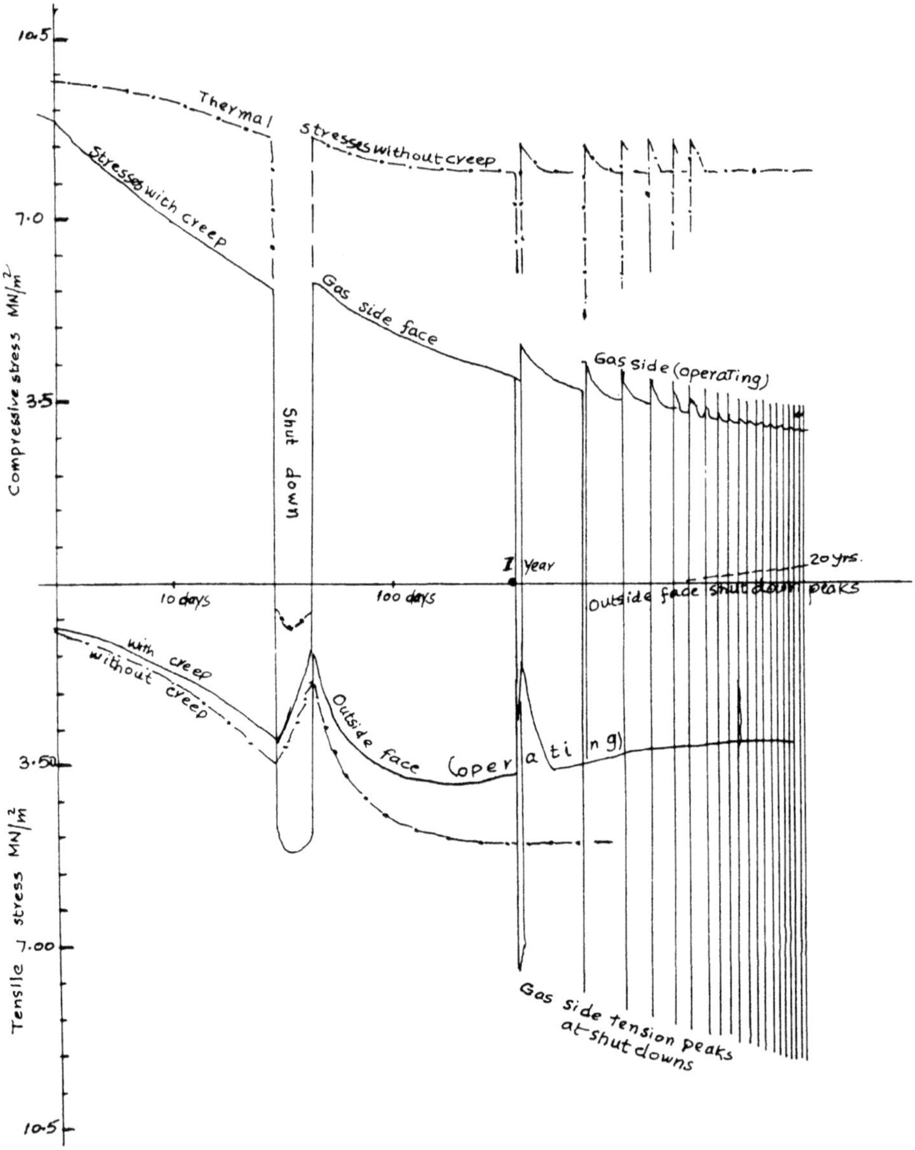

FIGURE 10

Stresses in top cap, showing influence of shut-down

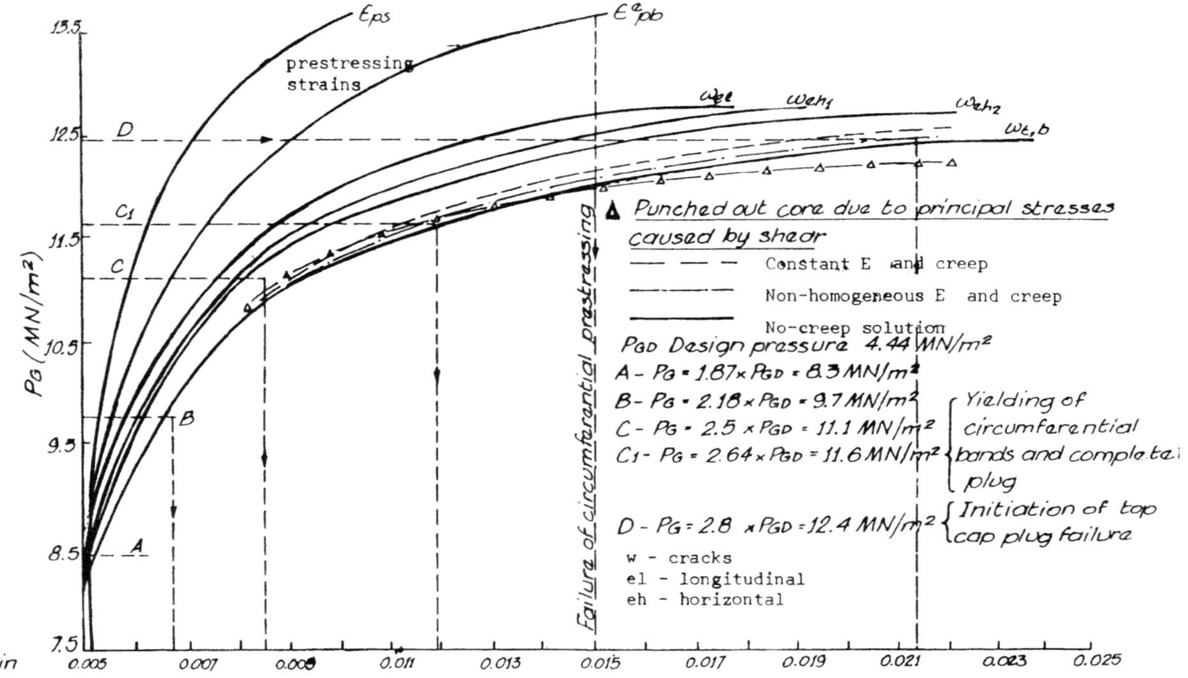

Crack width top cap & wall (mm × ...)	5	10	15	20	25	30	35	40	45	50	55	60	65	70	75	80	85	90	95	100
Crack width bottom cap (mm × ...)		8		16		24		32		40		48		56		64		70		78
Top cap central defl. (mm)		0.9		1.2		1.5		1.8		2.1		2.4		2.7		3.0		3.2		3.6
Bottom cap central defl. (mm)		0.75		1.0		1.25		1.50		1.75		2.0		2.25		2.50		2.75		3.0
Wall horizontal defl. (mm)		0.75		1.0		1.25		1.50		1.75		2.0		2.25		2.50		2.75		3.0
$\hat{x}$ (m)	0.40					0.30						0.20						0.10		Constant →

FIGURE 11

Vessel deflection and cracking with and without creep plotted against gas pressure.

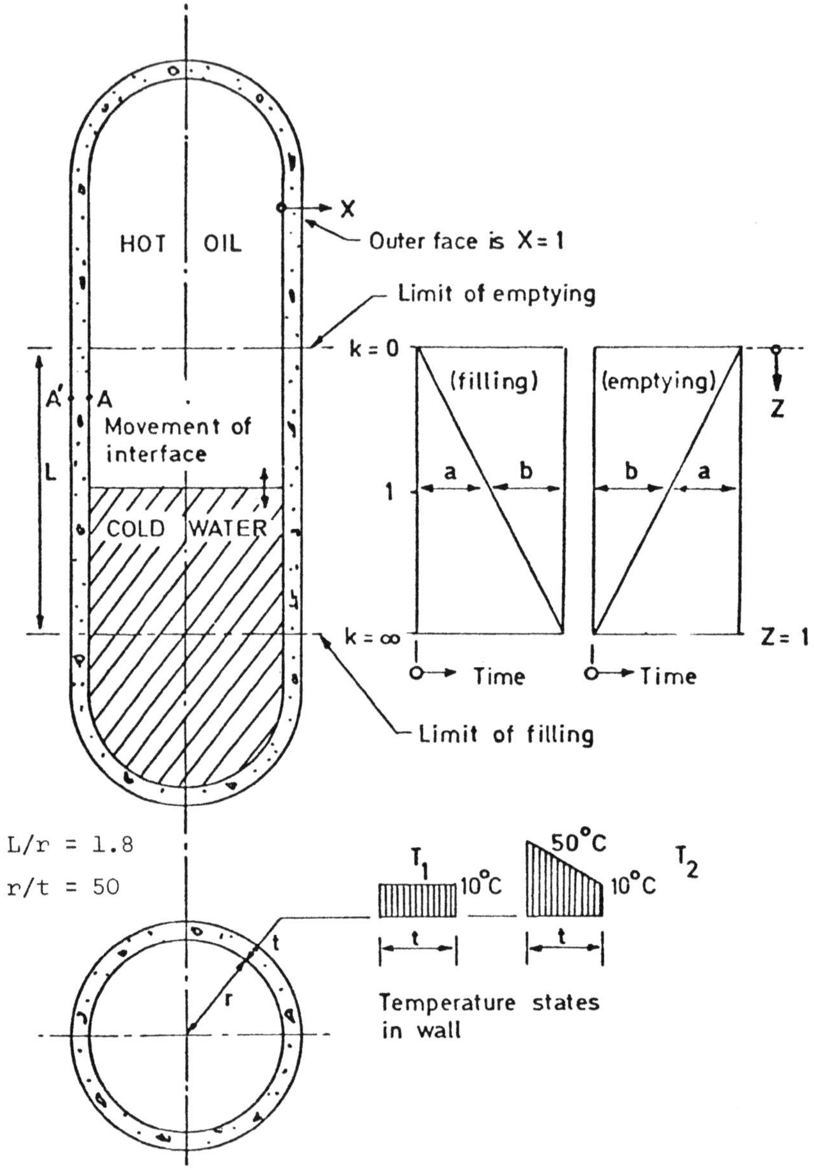

FIGURE 12

Single cell oil containment structure subjected to regular filling and emptying cycles, with hot oil over central portion. There are no 'hold' times in either the full or empty condition. Ordinates 'a' represent time in cold condition. $a{:}b=k_1{:}k_2$.

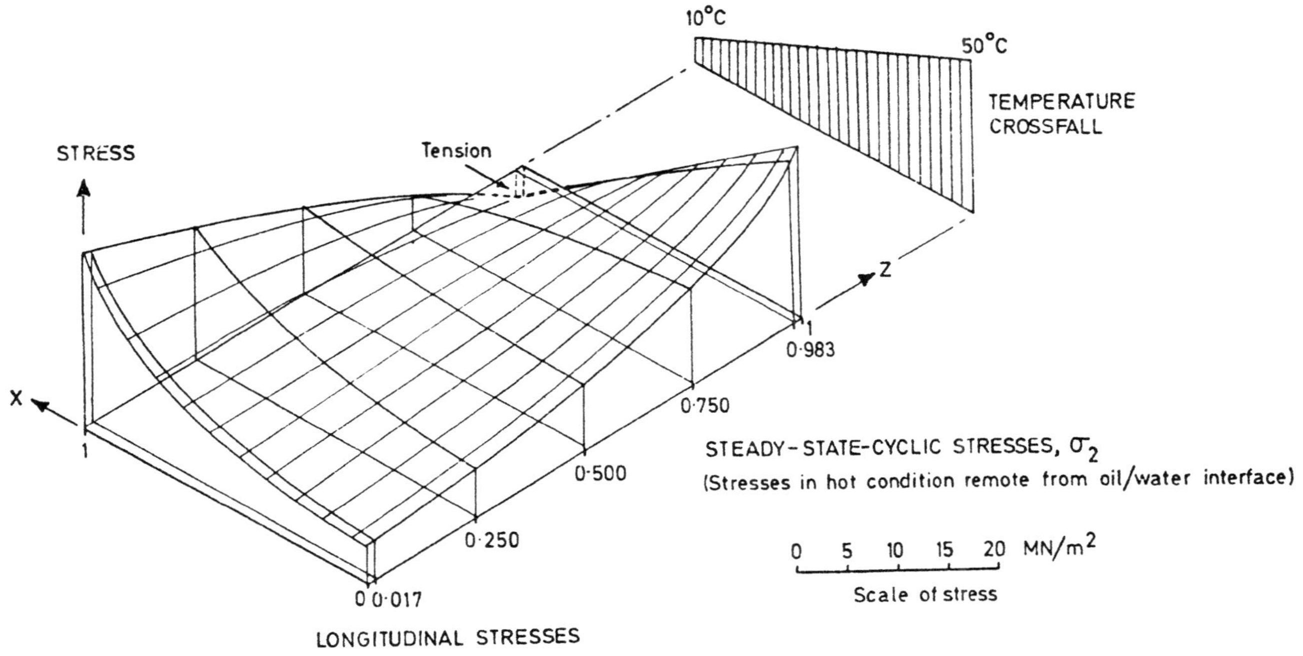

FIGURE 13

Creep shakedown stresses for cylindrical containment of Fig.14, showing variation with depth and radius. Stresses in cold state, T_1, obtained by subtraction of thermal stresses ℄.

VISCOELASTIC STABILITY ANALYSIS OF SHELLS

Silvio Merazzi,
Ecole Polytechnique Fédérale de Lausanne,
CH-1015 Lausanne, Switzerland

Peter Stehlin,
The Aeronautical Research Institute of Sweden,
S-16111 Bromma, Sweden

Frank A. Brogan,
Lockheed MSC,
Palo Alto, California 94304, USA

SUMMARY

Geometrically nonlinear viscoelastic deformations of shell structures are analyzed. On the basis of linear viscoelastic materials the stress-strain relations of orthotropic layers are developed. The inelastic effects of viscoelasticity are inserted into a computer code by means of the principle of virtual work, and the resulting integration technique leads to nonlinear equations which are solved by the modified Newton-Raphson method. Examples of the viscoelastic collapse of shell structures are given.

1. INTRODUCTION

Viscoelastic deformations of solids lead to complicated differential equations even for very simple structural models. Therefore numerical methods have to be provided to solve these types of equations. The method proposed in this report is based on a numerical finite difference solution of the nonlinear elastic problem [1] as well as the viscoelastic integral equations.

The material is assumed to be linear viscoelastic with memory [2]. This type of material is suitable for the description of the viscoelastic behaviour of epoxi resins and concrete. Materials of this type may be modeled by combining springs and dashpots of Maxwell and Kelvin type [1]. This method has serious disadvantages for the numerical treatment because of the resulting differential equations. Instead of representing the material as a series of simple viscoelastic models it can be approximated directly using mathematical functions. In fact, exponential series are very well suited to approximate creep and relaxation functions of general shape. Furthermore, their properties contribute to transform the convolution integrals of the viscoelastic stress-strain relations.

The theory may be applied to arbitrary solids. In this context it has been implemented in the STAGS shell analysis code [1] in order to analyse the viscoelastic stability of shells. The following assumptions are made :

- The shell is composed of thin orthotropic sheets.
- Strain-displacement relations are nonlinear.
- External loads do not change their direction during deformation except for pressure loads.
- The proportionality and superposition laws hold for the material.

The first step in nonlinear analysis is generally a bifurcation buckling analysis. Then the effects of different initial deformations are studied on the elastic model by incrementing the load as a fraction of the critical load, retaining the solution vectors as starting values for the viscoelastic stability analysis. The examples presented in this report have all been computed using this procedure.

2. VISCOELASTIC CONSTITUTIVE RELATIONS

The stress-strain relations of linear viscoelastic (thermorheologically simple) materials for an orthotropic sheet are of the form

$$\sigma_i(t) = \int_0^t C_{ij}(t-\tau) \cdot d\varepsilon_j(\tau) \qquad i,j = 1,3 \tag{2.1}$$

where C_{ij} are the relaxation functions of the material considered. Eq.(2.1) may be integrated by parts, provided that the C_{ij} and their derivatives are continuous functions of t

$$\sigma_i(t) = C_{ij}(0)\cdot\varepsilon_j(t) - \int_0^t \frac{\partial C_{ij}(t-\tau)}{\partial\tau} \cdot \varepsilon_j(\tau)\cdot d\tau \qquad (2.2)$$

Eq.(2.2) contains two parts, namely the elastic contributions $C_{ij}(0)\cdot\varepsilon_j(t)$ and the inelastic parts contained in the integral. The elastic contributions represent the elasticity constants of plane strain.

In the case of isotropy

$$\left.\begin{aligned} C_{11}(t) &= C_{22}(t) \\ C_{12}(t) &= C_{21}(t) \\ C_{13}(t) &= C_{23}(t) = 0 \end{aligned}\right\} \qquad (2.3)$$

It is obvious that Eq.(2.2) if applied without transformations lead to very cumbersome numerical procedures, because the convolution integrals require re-evaluation of the whole strain history for each new time step. Therefore the relaxation functions are approximated by exponential series [3]

$$C_{ij}(t) = -\sum_{k=1}^{m} C_{ij}^{k}\cdot(1-e^{-\kappa_k\cdot t}) \qquad i,j = 1,3 \qquad (2.4)$$

The constants C_{ij}^{k} may be obtained by collocation [3],[4].

Substituting Eq.(2.4) into Eq.(2.2)

$$\sigma_i(t) = C_{ij}(0)\cdot\varepsilon_j(t) - \sum_{k=1}^{m}\left\{\kappa_k\cdot e^{-\kappa_k\cdot t}\cdot C_{ij}^{k}\cdot\int_0^t e^{\kappa_k\cdot\tau}\cdot\varepsilon_j(\tau)\cdot d\tau\right\} \qquad (2.5)$$

The convolution integrals of Eq.(2.2) are now transformed into simple integrals. Their evaluation is described in the next chapter.

3. SOLUTION PROCEDURE

The numerical procedure used in the STAGS code [1] is based on the direct minimization of the potential energy of the structure with respect to the unknown displacement variables d_i obtained by finite difference mesh discretization

$$\frac{\partial}{\partial d^i} U(d_j) = \frac{\partial V(d_i)}{\partial d^i} = P_i \qquad i,j = 1,n \qquad (3.1)$$

$U(d_j)$ denotes the elastic energy of the structure and $V(d_i)$ the energy of the external loads.

In the presence of viscoelastic and plastic strains, Eq.(3.1) cannot be applied, but the principle of virtual work leads to a similar formulation

$$\frac{\partial}{\partial d^i}\,\overline{U}(d_j) = P_i + Q_i \qquad i,j = 1,n \qquad (3.2)$$

Now $\overline{U}(d_j)$ denotes the elastic energy computed with elastic stresses and the actual total strains. Q_i contains the inelastic contribution of Eq.(2.2). The elastic stresses are the stresses obtained from the total strains assuming elastic material behaviour.

A brief description of the numerical procedure used in presence of inelastic time-dependent strains follows. Consider the time steps t_{n-2}, t_{n-1}, t_n. Eq.(2.5) may be written at time t_n, where ℓ denotes the iteration index

$$\sigma_i^{\ell}(t_n) = C_{ij}(0)\cdot\varepsilon_j^{\ell-1}(t_n) - \sum_{k=1}^{m}\left\{\kappa_k\cdot e^{-\kappa_k\cdot t_n}\cdot C_{ij}^{k}\cdot\int_0^{t_n} e^{\kappa_k\cdot\tau}\cdot\varepsilon_j^{\ell-1}(\tau)\cdot d\tau\right\} \qquad i,j = 1,3 \qquad (3.3)$$

To avoid the re-evaluation of the integrals in Eq.(3.3) the strains $\varepsilon_j^{\ell-1}(\tau)$ are approximated during each iteration by a parabola using the preceding values $\varepsilon_j^{\ell-1}(t_n)$, $\varepsilon_j(t_{n-1})$, $\varepsilon_j(t_{n-2})$ [5]. All three stress components are thus evaluated at each finite difference meshpoint. Three points through the shell thickness have to be evaluated to obtain the inelastic contributions Q_i. Therefore, during the analysis, 27 strain components have to be retained at each meshpoint.

The strategy used during nonlinear time-dependent creep analysis is essentially the same as for nonlinear elastic collapse analysis [1].

4. EXAMPLES

All examples have been computed with the STAGS-A code [1]. One-term exponential series are used to characterize the material. In a first step the linear elastic buckling load p_{cr} is determined for the present loads. This gives an idea about the load intensity to be applied during nonlinear analysis. The second step consists of computing the geometrically nonlinear elastic deformations. Then, for selected values of p , a restart is made, keeping the load intensity p constant and integrating Eq.(3.3) in time.

4.1. Example 1 - Cylindrical arch

The linear buckling load $p_{cr} = 0.00216\ \mathrm{N/mm^2}$ for this example (Fig.1) has been computed using elastic constants

$$E = 2000\ \mathrm{N/mm^2}\ ; \quad \nu = 0$$

Relaxation functions were of the form

$$C_{11}(t) = E - 1000 \cdot (1 - e^{-0.7 \cdot t})$$
$$C_{22}(t) = C_{11}(t)\ ,$$

all other functions $C_{ij}(t)$ of Eq.(2.2) being zero. Time-response of the displacements for different load intensities is displayed in Fig.2. If p is kept sufficiently small the displacements $w_1(t)$ reach stable values due to the asymptotic form of $C_{11}(t)$. If p is increased the displacements $w_2(t)$ reflect the primary creep, secondary creep where the creep rate is approximately constant, and the collapse. For values of p near p_{cr} the collapse occurs immediately.

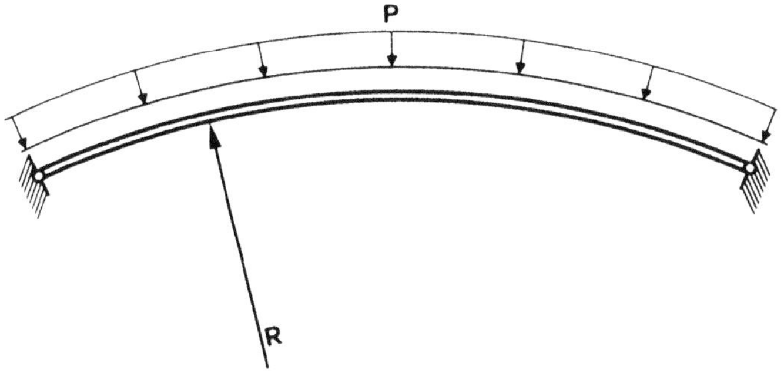

Fig.1. - Cylindrical arch; R = 2000 mm, angle of sector = 11.6°, thickness = 3,5 mm, width = 20 mm.

4.2. Example 2 - Paraboloid shell

The shell (Fig.3) is subjected to external live pressure loads. Elastic material constants are

$$E = 2600\ \mathrm{N/mm^2}\ ; \quad \nu = 0.36$$

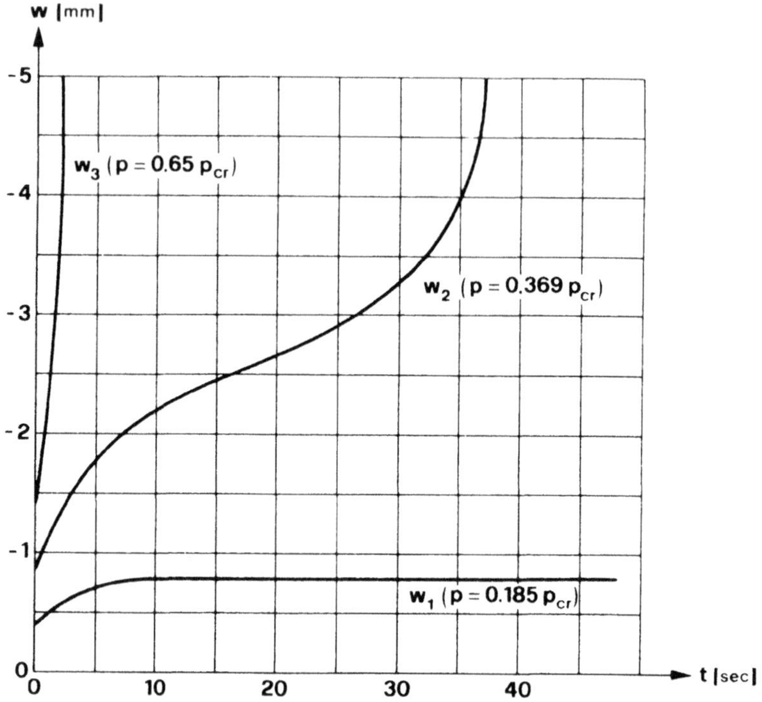

Fig.2. - Displacements w at the center of the arch.

and the relaxation functions

$$\left.\begin{aligned} C_{11}(t) &= 2990 - 1350 \cdot f(t) \\ C_{22}(t) &= C_{11}(t) \\ C_{33}(t) &= 1490 - 496 \cdot f(t) \\ C_{12}(t) &= 1075 - 486 \cdot f(t) \\ C_{21}(t) &= C_{12}(t) \end{aligned}\right\} \quad [N/mm^2]$$

$$f(t) = 1 - e^{-0.1 \cdot t} \qquad [sec]$$

The bifurcation buckling load has been determined to be $p_{cr} = 0.06\ N/mm^2$

The buckling pattern suggests that a quarter of the shell may be analyzed. Furthermore, imperfections of the form

$$w_0(x,y) = -\,0.5 \cdot \sin\frac{\pi \cdot x}{100} \cdot \cos\frac{\pi \cdot y}{22.5} \qquad [mm]$$

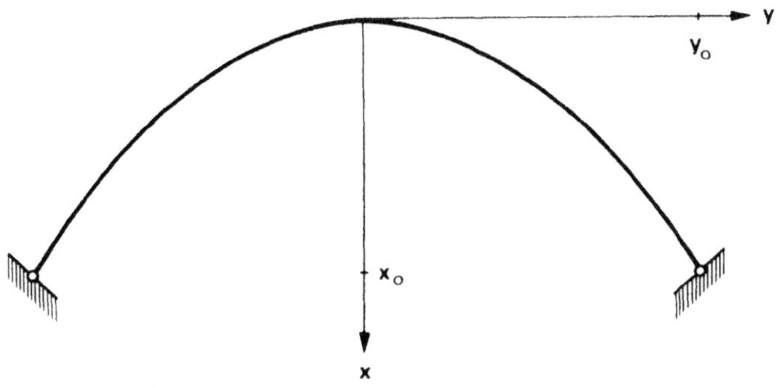

Fig.3. - Paraboloid shell; $x_0 = 100$ mm, $y_0 = 130$ mm, thickness = 1 mm.

have been introduced. The final model for nonlinear analysis has resulted in 468 equations. Displacements w perpendicular to the shell for different values of p are displayed in Fig.4.

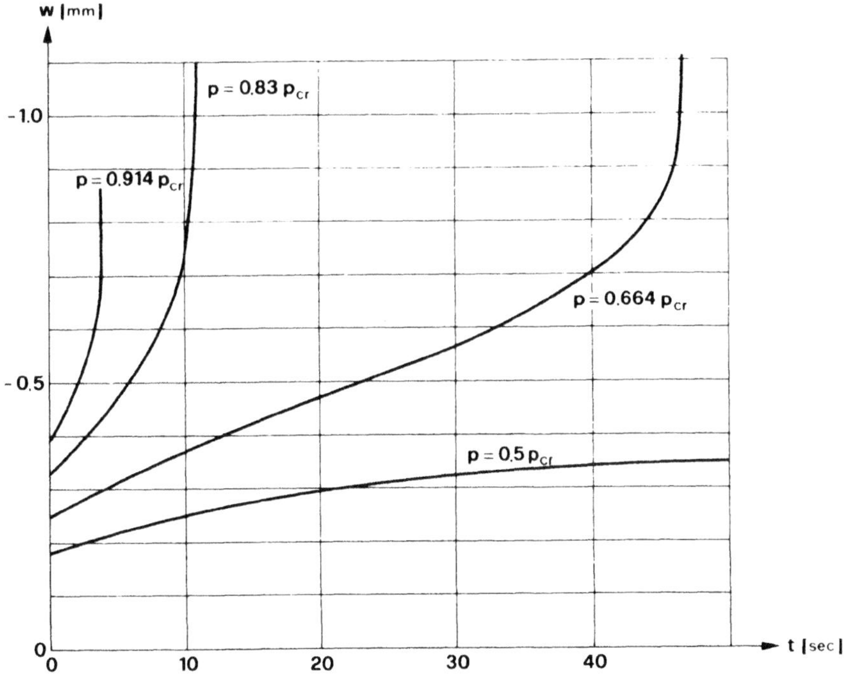

Fig.4. - Displacements w at meshpoint x = 85.7, y = 0.

REFERENCES

1. Almroth, B.O., Brogan, F.A. et al. "Collapse Analysis for Shells of General Shape". Lockheed, Palo Alto Research Laboratory TR AFFDL TR-71-8 (1973).

2. Flügge, W. Viscoelasticity. Springer-Verlag, Berlin (1975).

3. Schapery, R.A. "A Simple Collocation Method for Fitting Viscoelastic Models to Experimental Data". California Institute of Technology (1961).

4. Gradowczyk, M.H., Moavenzadeh, F. "Characterization of Linear Viscoelastic Materials". Trans. of the Soc. of Rheology 13:2 (1969).

5. Merazzi, S., Stehlin, P. "Viscoelastic Stability of Shells - Numerical Aspects". ISNM 37, Birkhäuser-Verlag, Basel (1977) 167-193.

SYMBOLS

$C_{ij}(t)$	relaxation functions
C_{ij}^k	k-th relaxation modulus in exponential series
d_i	i-th displacement variable
$\varepsilon_i(t)$	time-dependent strain component of orthotropic sheet
κ_k	k-th retardation constant in exponential series
n	number of displacement variables
P_i	i-th external load
Q_i	i-th component of inelastic contributions to the load vector
$\sigma_i(t)$	time-dependent stress components of orthotropic sheet
t,τ	time
U	elastic energy of the structure
$\bar{U}$	modified elastic energy
V	energy of the external loads
$w(x,y)$	shell displacements normal to the shell surface
x,y	shell surface coordinates.

STRESSES AND DISPLACEMENTS OF COMPOSITE AND PRESTRESSED STRUCTURES

V.Lazić and J. Lazić

Faculty of Transport and Traffic Engineering and
Faculty of Civil Engineering, Belgrade, Yugoslavia

ABSTRACT

If the axial force and the bending moment are linearly dependent on the concrete relaxation function, the accurate values of stresses and displacements can be obtained through algebraic expressions. This procedure is a generalization of the well-known Bažant method{1}, which yields the accurate algebraic stress-strain relation for concrete, provided the strain is a linear function of the creep coefficient.

RESUME

Si l' effort normal et le moment de flexion dépendent linéairement de la fonction de relaxation du béton, les valeurs exactes de la contrainte et des déplacements peuvent être obtenues par des expressions algébriques. Cette procédure représente la généralisation de la méthode connue de Bažant{1}, selon laquelle la relation algébrique décrit exactement le comportement rhéologique du béton, si la déformation dépend linéairement du coefficient du fluage.

1. RHEOLOGICAL PROPERTIES OF MATERIALS

The following form of the concrete creep function is adopted:

$$\Phi(t,\tau)= \frac{1}{E_c^o}\left| \frac{1}{e(t)} +f^*(t,\tau) \right|= \frac{1}{E_c^o}F^*(t,\tau) \qquad (1.1)$$

where

Wittmann, F.H. (ed.) Fundamental Research on Creep and Shrinkage of Concrete. ISBN 90-247-2549-6.

$$e(t)=\frac{E_c(t)}{E_c^o}\ ,\quad E_c^o=E_c(t_o),\ e(t_o)=1, \tag{1.2}$$

$$f^*(t,\tau)=\frac{E_c^o}{E_c(\tau)}-\frac{E_c^o}{E_c(t)}+\frac{E_c^o}{E_{c28}}\varphi(t,\tau),$$

$\varphi=\varphi(t,\tau)$ is the creep coefficient {2}.

When the stress is a continuous time function, then, on the basis of the superposition principle, the normal strain is given by the expression:

$$\varepsilon(t,t_o)+\varepsilon_S(t,t_o)=\frac{1}{E_c^o}\left|\frac{1}{e(t)}\sigma_c(t,t_o)+\int_{t_o}^{t}\frac{\partial f(t,\tau)}{\partial\tau}\sigma_c(\tau,t_o)d\tau\right| . \tag{1.3}$$

Introducing the linear integral operator:

$$\hat{F}=\frac{1}{e}\hat{1}+\hat{f}\ ,\qquad \dot{f}=-\dot{f}^*, \tag{1.4}$$

this integral relation is written in the symbolic form:

$$\varepsilon+\varepsilon_S=\frac{1}{E_c^o}\hat{F}\sigma_c. \tag{1.5}$$

The comparison of expressions (1.5) and (1.3) indicates the significance of operator $\hat{f}$, associated to function $\dot{f}$, and of the identity operator $\hat{1}$, associated to the Dirac function. Apart from the already introduced notation $\dot{g}=\partial g(t,\tau)/\partial\tau$, the integral of function $\dot{g}$ will be defined in the following manner:

$$g^*=g^*(t,t_o)=\hat{g}\ 1^*=\int_{t_o}^{t}\frac{\partial}{\partial\tau}g(t,\tau)H(\tau-t_o)d\tau,$$

$1^*=H(t-t_o)$ is the Heaviside step function.

The solution of equation (1.5) is:

$$\sigma_c = E_c^o \hat{R}\varepsilon + \sigma_S, \qquad \sigma_S = E_c^o \hat{R}\varepsilon_S, \tag{1.6}$$

where

$$\hat{R} = e\,\hat{1} - \hat{\Psi}, \tag{1.7}$$

represents the operator inverse to operator $\hat{F}$:

$$\hat{R}\,\hat{F} = \hat{1}, \qquad \hat{R}\,\hat{F} = \hat{F}\,\hat{R}. \tag{1.8}$$

Upon integration the following is obtained:

$$\hat{R}\,F^* = 1^*, \quad \hat{R}\,F^* = \hat{F}\,R^*. \tag{1.9}$$

$F^* = F^*(t,\tau)$ is a nondimensional creep function and $R^* = R^*(t,\tau)$ is a nondimensional concrete relaxation function.

The relaxation property of prestressing steel was taken into account in the way presented in {3}:

$$\sigma_p = E_p \hat{R}_p \varepsilon, \qquad \hat{R}_p = \hat{1} - \hat{\Psi}_p. \tag{1.1o}$$

The steel relaxation function:

$$R_p^* = \hat{R}_p\, 1^* = 1^* - \Psi_p^*, \tag{1.11}$$

is presented in the form which sufficiently accurately approximates the function flow R_p^* from t_o to the observed time t {3}:

$$R_p^* = \rho'\,1^* + \rho R^*, \qquad \rho' = 1 - \rho. \tag{1.12}$$

For all observed times t and t_o the value of parameter ρ is different. In the special case when $\rho = 0$, eq.(1.1o) determines the elastic material.

Reinforcing steel (m) and steel member (n) obey Hooke´ s law:

$$\sigma_j = E_j \varepsilon, \qquad j = m,n. \tag{1.13}$$

In a special case the girder does not contain the prestressing steel or the steel member (Fig. 1).

2. THE BASIC EQUATIONS OF THE PROBLEM

The initial system of equations is: the stress-strain relations for concrete (1.6) and for all kinds of steel (1.1o),(1.13), Navier´ s assumption, equilibrium conditions of external and internal forces and the principle of virtual forces. By elimination of stresses and common strain ε, the basic equations relevant to an arbitrary cross section {3}, are obtained:

$$\begin{aligned} E_u A_i \hat{R}_{11} \eta + E_u S_i \hat{R}_{12} \kappa &= N, \\ E_u S_i \hat{R}_{21} \eta + E_u J_i \hat{R}_{22} \kappa &= M. \end{aligned} \tag{2.1}$$

The notations are: E_u is the relative modulus of elasticity; $A_i = A_i(x)$, $J_i = J_i(x)$ are transformed cross-section area and moment of inertia of this area about the z-axis, respectively (Fig.1); x is a coordinate of the observed cross-section at the bar axis; $S_i = \sqrt{A_i J_i}$; $\eta = \eta(x,t,t_o)$ is a normal strain of the bar axis; $\kappa = \kappa(x,t,t_o)$ is the change in the curvature of the bar axis; $N = N(x,t,t_o)$ is the axial force; $M = M(x,t,t_o)$ is the bending moment. Operators $\hat{R}_{hl}$ are:

$$\begin{aligned} &\hat{R}_{hl} = \hat{R}_{hl}(x,t,\tau) = e_{hl}\hat{1} - \gamma_{hl}\hat{\psi}, \qquad h,l = 1,2, \\ &e_{hl} = e_{hl}(x,t) = \delta_{hl} + \gamma_{hl}(e-1), \quad \delta_{hl} = \begin{cases} 1 & \text{for } h = l \\ 0 & \text{for } h \neq l \end{cases}, \end{aligned} \tag{2.2}$$

where nondimensional quantities $\gamma_{hl} = \gamma_{hl}(x)$ are determined through the reduced geometrical properties of the cross-section and parameter ρ{3}. The principal values of matrix $\|\gamma_{hl}\|$ are: γ_1 and γ_2 $(0 < \gamma_2 < \gamma_1 < 1)$. The following notations will be also used:

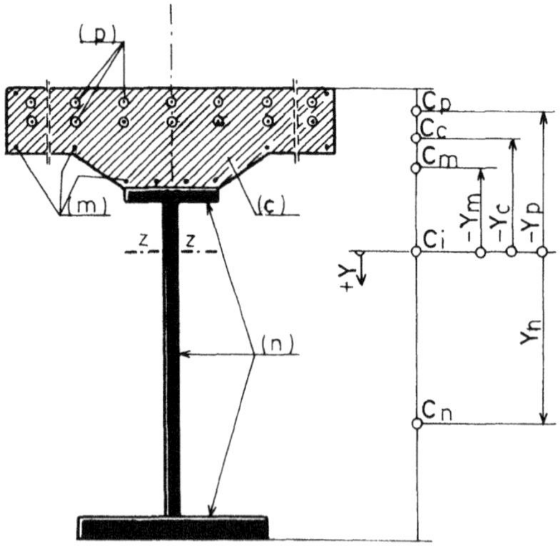

Fig.1. Cross-section of a girder

$$\gamma_h'=1-\gamma_h,\delta\gamma_1=\gamma_1-\gamma_{11},\quad \delta\gamma_2=\gamma_{11}-\gamma_2,\ \Delta\gamma=\gamma_1-\gamma_2. \tag{2.3}$$

In the general case, the laws of algebra for ordinary numbers, except the comutativity law, are valid for the linear integral operators. Especially, the operators used here have a comutativity property, so that the system of equations (2.1) is formally resolved as a system of algebraic equations, by determinants.

The principal values of the operator matrix of system (2.1) are:

$$\overset{\frown}{R}_h=\overset{\frown}{R}_h(\gamma_h,t,\tau) = e_h\overset{\frown}{1} -\gamma_h\overset{\frown}{\psi}, \qquad h = 1,2, \tag{2.4}$$

$$e_h=e_h(\gamma_h,t) = 1+\gamma_h(e-1).$$

Their inverse operators are:

$$\overset{\frown}{F}_h=\overset{\frown}{F}_h(\gamma_h,t,\tau) = \frac{1}{e_h}\overset{\frown}{1} +\gamma_h\overset{\frown}{f}_h, \qquad h = 1,2. \tag{2.5}$$

The following holds for them:

$$\overset{\frown}{R}_h\overset{\frown}{F}_h= \overset{\frown}{1}, \quad \overset{\frown}{R}_h\overset{\frown}{F}_h=\overset{\frown}{F}_h\overset{\frown}{R}_h, \qquad h= 1,2. \tag{2.6}$$

The solution of system (2.1) is expressed through inverse operators $\overset{\frown}{F}_h$ {3}:

$$E_u\eta= \frac{1}{\Delta\gamma}\left|\overset{\frown}{F}_1(\delta\gamma_2 \frac{N}{A_i} + \gamma_{12}\frac{S_i}{A_i}\frac{M}{J_i})+\overset{\frown}{F}_2(\delta\gamma_1\frac{N}{A_i} - \gamma_{12}\frac{S_i}{A_i}\frac{M}{J_i})\right|$$

$$E_u\kappa= \frac{1}{\Delta\gamma}\left|\overset{\frown}{F}_1(\gamma_{12}\frac{S_i}{J_i}\frac{N}{A_i} + \delta\gamma_1\frac{M}{J_i})+\overset{\frown}{F}_2(-\gamma_{12}\frac{S_i}{J_i}\frac{N}{A_i} + \delta\gamma_2\frac{M}{J_i})\right| \tag{2.7}$$

Functions $B_h^* = B_h^*(\gamma_h,t,t_o)$ expressing the estresses and displacements will be defined:

$$\overset{\frown}{B}_h= \overset{\frown}{R}\,\overset{\frown}{F}_h, \qquad B_h^*=\overset{\frown}{B}_h1^* = \overset{\frown}{R}\,F_h^*, \qquad h = 1,2. \tag{2.8}$$

It can be shown that they are the solutions of the integral equations:

$$K_h^*(\gamma_h,t,t_o)B_h^*(\gamma_h,t_o,t_o) + \\ + \int_{t_o}^{t} K_h^*(\gamma_h,t,\tau)\frac{\partial}{\partial\tau} B_h^*(\gamma_h,\tau,t_o)d\tau= 1, \qquad h=1,2. \tag{2.9}$$

Functions K_h^* are known because they depend on the nondimensional concrete creep function and on the geometrical cross section properties:

$$K_h^* = \gamma_h 1^* + \gamma_h' F^*. \qquad h=1,2. \qquad (2.1o)$$

Especially for $\gamma_2=0$, $K_2^* \equiv F^*$ and $B_2^* \equiv R^*$. The nondimensional concrete relaxation function R^* is therefore the solution of the same integral equation (2.9) for $\gamma_2=0$.

3. STRESSES AND DISPLACEMENTS

If the axial force and the bending moment change according to the law:

$$N = N_o+N_1 R^*, \qquad M = M_o+M_1 R^*, \qquad (3.1)$$

where N_o, N_1, M_o and M_1 are arbitrary time independent quantities, then stresses linearly depend on functions R^* and B_h^* relevant to the observed cross-section; displacements depend on functions B_h^* relevant to the finite number of girder cross sections.

3.1. Stresses

In the statically determinate structures N and M follow the law (3.1) due to the following permanent external influences H:

a) the self-weight (H=G):

$$N_G = \mathcal{N}_G 1^*, \qquad M_G = \mathcal{M}_G 1^*, \qquad (3.2)$$

where $\mathcal{N}_G$ and $\mathcal{M}_G$ are time independent quantities;

b) prestressing by forces (H=P):

$$\begin{aligned} N_p &= n_p^o R_p^* = n_p^o \rho' 1^* + n_p^o \rho R^*, \\ M_p &= m_p^o R_p^* = m_p^o \rho' 1^* + m_p^o \rho R^*, \end{aligned} \qquad (3.3)$$

where the values besides functions 1^* and R^* are time independent {3};

c) the shrinkage (H=S) if it is assumed that:

$$\varepsilon_S = r|F^* - 1^*| , \qquad (3.4)$$

i.e. that function ε_S has the same flow as $E^o \varphi/E_{c28}$ in the same interval from t_o to the observed time t. It is obvious that for all observed times t and t_o the value of parameter r is different. Such an approximation is customary. Then:

$$N_S = \overline{n}_S \,|1^*-R^*| , \qquad M_S=\overline{m}_S |1^*-R^*| , \tag{3.5}$$

where $\overline{n}_S$ and $\overline{m}_S$ are time independent quantities.

Due to the same influences, in a statically indeterminate structure to the n-th degree, N and M change according to the law (3.1) if the redundant forces change according to the law:

$$X_{\lambda H}=X^{o}_{\lambda H} + \Delta X_{\lambda H}|1^*-R^*|, \quad \lambda=1,2,\ldots,n \tag{3.6}$$

where $X^{o}_{\lambda H} = X_{\lambda H}(t_o,t_o)$ and $\Delta X_{\lambda H}$ are time independent quantities (H = G,P,S).

Navier' s assumption determines the common strain ε of the cross-section points with the ordinate y(Fig.1):

$$\varepsilon =\eta+\kappa y. \tag{3.7}$$

Therefrom by means of relations (2.7), (1.6), (1.1o), (1.13), (3.2),(3.3) and (3.5) the expressions of stresses in all parts of the cross-section (n,m,p,c), due to the influence H, are obtained in the form:

$$\sigma_{kH}=\nu_k \sum_{h=1}^{2} \overset{\frown}{\Lambda}_{hk}L_{hH}+\delta_{kH}\sigma_H, \tag{3.8}$$

k=n,m,p,c; H = G,P,S, where also an additional condition for σ_{pP} {3}was used. In expression (3.8) we encounter:

$$\nu_k= \frac{E_k}{E_u} , \qquad k = n,m,p, \qquad \nu_c= \frac{E^{o}_c}{E_u} , \tag{3.9}$$

$$L_H = \frac{N_H}{A_i} + \frac{M_H}{J_i} y, \qquad L_{1H}=L_H-L_{2H},$$

$$L_{2H}= \frac{1}{\Delta\gamma}\left| (\delta\gamma_1 \frac{N_H}{A_i} -\gamma_{12} \frac{S_i}{A_i} \frac{M_H}{J_i})+(-\gamma_{12}\frac{S_i}{J_i} \frac{N_H}{A_i} +\delta\gamma_2 \frac{M_H}{J_i})y\right| , \tag{3.1o}$$

$$\overset{\frown}{\Lambda}_{hj}= \overset{\frown}{F}_h, \; j = n,m; \quad \overset{\frown}{\Lambda}_{hp}=\overset{\frown}{R}_p \overset{\frown}{F}_h; \quad \overset{\frown}{\Lambda}_{hc}=\overset{\frown}{R} \overset{\frown}{F}_h = \overset{\frown}{B}_h; \tag{3.11}$$

symbol δ_{kH} denotes zero in all cases except in δ_{cS} and δ_{pP} when $\delta_{cS}=\delta_{pP}=1$;

$$\sigma_{cS}=\nu_c E_c r|1^*-R^*| \; ; \quad \sigma_{pP}=\nu_p C R^*_p, \tag{3.12}$$

where C is a time-independent quantity.

When the axial force and the bending moment follow the laws (3.2), (3.3), (3.5), i.e. (3.1), in expressions (3.8) the following integrations appear at the determination of stresses:

$$\hat{F}_h 1^*, \hat{R}_p \hat{F}_h 1^*, \hat{B}_h 1^*, \hat{F}_h R^*_p, \; \hat{R}_p \hat{F}_h R^*_p, \; \hat{B}_h R^*_p, \hat{F}_h R^*, \hat{R}_p \hat{F}_h R^* \text{ and } \hat{B}_h R^*.$$

By applying the algebraic operations with operators, the quoted functions and products are transformed in such a way that they reduce to the linear combinations of functions 1^*, R^* and B^*_h.

The stress expressions may then be presented in the unique form:

$$\sigma_{kh}=\nu_k \left|U_{kH}1^*+V_{kH}R^* + \sum_{h=1}^{2} W_{hkH}B^*_h\right| , \; k = n,m,p,c, \tag{3.13}$$

$H = G,P,S$, Table 1 contains the values of coefficients U_{kH}, V_{kH} and W_{hkH} for parts k (k=n,m,p,c), of the cross-section and for the influences H (H=G,P,S). These coefficients are not time functions. In Table 1 we have:

$$U_{pP}=\rho'\left(\frac{P}{A_{pr}} + \sum_{h=1}^{2} \frac{\gamma_h - \rho}{\gamma'_h} L_{hP}\right);$$

$$V_{pP}=\rho\left(\frac{P}{A_{pr}} - \sum_{h=1}^{2} \frac{\gamma_h-\rho}{\gamma_h} L_{hP}\right); \tag{3.14}$$

$$V_{cS}= - E_u r - \sum_{h=1}^{2} \frac{1}{\gamma_h} L_{hS}.$$

The general form (3.13) for stresses holds for statically indeterminate strcutures under conditions (3.6) but coefficients U_{kH}, V_{kH} and W_{hkH} have somewhat different forms.

3.2. Displacements

The generalized displacement $\xi=\xi(x,t,t_o)$ is determined from the principle of virtual forces. The reduced generalized displacement is:

$$\Delta^*=E_u J_u \xi=E_u J_u \int_L \left| \hat{M}(u,x)\kappa(u,t,t_o)+\hat{N}(u,x)\eta(u,t,t_o)\right| du; \tag{3.15}$$

TABLE 1

Coefficients U_{kH}, V_{kH} and W_{hkH}; $k=j,c,p$; $j=n,m$; $H = G,P,S$.

	U_{jH}	V_{jH}	W_{hjH}	U_{cH}	V_{cH}	W_{hcH}	U_{pH}	V_{pH}	W_{hpH}
H=G	$\sum_{h=1}^{2} \frac{1}{\gamma_h'} L_{hG}$	0	$-\frac{\gamma_h}{\gamma_h'} L_{hG}$	0	0	L_{hG}	$\sum_{h=1}^{2} \frac{\rho'}{\gamma_h'} L_{hG}$	0	$-\frac{\gamma_h-\rho}{\gamma_h'} L_{hG}$
H=P	$\sum_{h=1}^{2} \frac{\rho'}{\gamma_h'} L_{hP}$	0	$-\frac{\gamma_h-\rho}{\gamma_h'} L_{hP}$	0	$\sum_{h=1}^{2} \frac{\rho}{\gamma_h} L_{hP}$	$\frac{\gamma_h-\rho}{\gamma_h} L_{hP}$	U_{pP}	V_{pP}	$-\frac{(\gamma_h-\rho)^2}{\gamma_h\gamma_h'} L_{hP}$
H=S	$\sum_{h=1}^{2} \frac{1}{\gamma_h'} L_{hS}$	0	$-\frac{1}{\gamma_h'} L_{hS}$	$E_u r$	V_{cS}	$\frac{1}{\gamma_h} L_{hS}$	$\sum_{h=1}^{2} \frac{\rho'}{\gamma_h'} L_{hS}$	$-\sum_{h=1}^{2} \frac{\rho}{\gamma_h} L_{hS}$	$-\frac{\gamma_h-\rho}{\gamma_h\gamma_h'} L_{hS}$

TABLE 2

Coefficients $P_H^{(a)}$ and $Q_{hH}^{(a)}$; $H = G,P,S$.

$P_G^{(a)}$	$Q_{hG}^{(a)}$	$P_p^{(a)}$	$Q_{hP}^{(a)}$	$P_S^{(a)}$	$Q_{hS}^{(a)}$
$\sum_{h=1}^{2} \frac{1}{\gamma_h'^{(a)}} D_{hG}^{(a)}$	$-\frac{\gamma_h^{(a)}}{\gamma_h'^{(a)}} D_{hG}^{(a)}$	$\sum_{h=1}^{2} \frac{\rho'}{\gamma_h'^{(a)}} D_{hP}^{(a)}$	$-\frac{\gamma_h^{(a)}-\rho}{\gamma_h'^{(a)}} D_{hP}^{(a)}$	$\sum_{h=1}^{2} \frac{1}{\gamma_h'^{(a)}} D_{hS}^{(a)}$	$-\frac{1}{\gamma_h'^{(a)}} D_{hS}^{(a)}$

$\hat{N}(u,x)$ and $\hat{M}(u,x)$ are the axial force and the bending moment respectively at point u, due to the unit generalized force $\hat{P}=1^*$ at point x, corresponding to the generalized displacement ξ; J_u is a relative moment of inertia.

In the general case, the geometrical properties of the girder cross-section change along its axis. In order to determine the displacements, the girder model with constant cross-sections in a finite number of intervals is adopted. To the observed interval (a) correspond length $L^{(a)}$, values $\gamma_h^{(a)}$, functions $F_h^{(a)}$ and so on.

The application of expression (2.7) yields the reduced generalized displacement Δ_H^* from influence H:

$$\Delta_H^* = \sum_{(a)} \sum_{h=1}^{2} \frac{1}{\gamma_h'^{(a)}} \left| \hat{1} - \gamma_h^{(a)} \hat{B}_h^{(a)} \right| D_{hH}^{(a)}, \qquad H = G,P,S. \qquad (3.16)$$

Quantities $D_{hH}^{(a)}$ depend on time through the axial force N_H and the bending moment M_H. For any interval (a) they are determined by well-known methods of Static of Elastic Structures.

When N and M change according to the laws (3.2), (3.3), (3.5), i.e. (3.1) the reduced generalized displacement has the form:

$$\Delta_H^* = \sum_{(a)} \left| P_H^{(a)} 1^* + \sum_{h=1}^{2} Q_{hH}^{(a)} B^{*(a)}_h \right| , \quad H = G,P,S. \qquad (3.17)$$

Table 2 contains the values of coefficient $P_h^{(a)}$ and $Q_{hH}^{(a)}$ which are not time functions. The general form (3.17) for Δ_H^* holds even for statically indeterminate girders under conditions (3.6), but this coefficients have somewhat different forms.

The equations for determining redundant forces $X_{\lambda H}$, which otherwise represent the system of nonhomogeneous integral equations, reduce under assumption (3.6) to the system of algebraic equations.

4. CONCLUSION

Chapter 3 displays the following: If the axial force and the bending moment change according to law (3.1) then stresses (3.13) in a general case linearly depend on functions R* and B_h^*, while displacements (3.17) linearly depend on functions B_h^*; functions R* and B_h^* are solutions of integral equations (2.9). When by the known numerical methods once for ever values of functions R* and B_h^* are calculated for a determinate t_o, $t-t_o$, and γ_h, the same expressions (3.13) and (3.17) represent the accurate algebraic expressions for stresses and displacements in composite and pre -

stressed structures.

For the creep function given by the CEB-FIP Model Code {2}, Table 3 displays the values of functions B_h^* and R^* for given $\varphi_f, h_o, t_o, \Delta t = t - t_o$ and γ_h.

TABLE 3

Values of Functions B_h^* and R^* for $\varphi_f = 4$ and $h_o = 20$ cm.

t_o days	γ_h	$\Delta t = t - t_o$ days: 3	10	30	91	183	365	10^4
3	0.9	0.9332	0.9110	0.8737	0.8326	0.8048	0.7813	0.7206
	0.7	0.8213	0.7664	0.6771	0.5859	0.5292	0.4842	0.3800
	0.5	0.7315	0.6549	0.5343	0.4197	0.3543	0.3059	0.2046
	0.0	0.5707	0.4666	0.3153	0.1946	0.1399	0.1058	0.0482
7	0.9	0.9430	0.9208	0.8830	0.8347	0.8063	0.7793	0.7099
	0.7	0.8457	0.7908	0.7014	0.6009	0.5385	0.4880	0.3726
	0.5	0.7659	0.6895	0.5693	0.4433	0.3716	0.3174	0.2059
	0.0	0.6183	0.5150	0.3642	0.2290	0.1664	0.1258	0.0574
28	0.9	0.9588	0.9462	0.9232	0.8770	0.8465	0.8161	0.7366
	0.7	0.8856	0.8531	0.7955	0.6876	0.6222	0.5615	0.4212
	0.5	0.8225	0.7755	0.6946	0.5515	0.4714	0.4015	0.2564
	0.0	0.6975	0.6293	0.5180	0.3443	0.2626	0.1999	0.0928
100	0.9	0.9706	0.9633	0.9483	0.9217	0.8977	0.8693	0.7877
	0.7	0.9167	0.8791	0.8578	0.7913	0.7344	0.6713	0.5109
	0.5	0.8683	0.8391	0.7815	0.6878	0.6117	0.5319	0.3487
	0.0	0.7667	0.7213	0.6354	0.5069	0.4130	0.3257	0.1600

An approximate procedure for determining functions B_h^* may be established, similar to the Bažant procedure for determining relaxation function R^* {4}.

REFERENCES

1. Z.Bažant, "Prediction of Concrete Effects Using Age-Adjusted Effective Modulus Method", J. ACI, Vol. 69 (1972) 212-217.

2. CEB-FIP Model Code for Concrete Structures, Bulletin n° 124/125-E (1978).

3. V.Lazić and J.Lazić, "Les contraintes et les pertes de précontrainte dans les structures précontraintes", Ann. I.T.B.T.P. TMC No 211-(1977) 33-56.

4. Z.P.Bažant, S.-S. Kim, "Approximate Relaxation Function for Concrete", J.of the Structural Div. , Proc. Am.Soc. of Civil Engrs. , Vol.1o5 (1979) 2695-2705.

SIMPLYFIED CALCULATION OF THE RELAXATION OF STRESS RESPECTING THE DELAYED ELASTICITY

Vladimír Křístek, DSc, C.E.; Zdeněk Šmerda, PhD, C.E.

Senior Research Fellow, Assoc. Professor of Civil Engineering, Czech Technical University, Prague, Czechoslovakia
Senior Research Fellow, Professor of Civil Engineering, Civ. Eng. Research Institute, Bratislava, Czechoslovakia

The solution of the relaxation problems that occurs very often in engineering practice when calculating the concrete structures is very complicated unless the calculations are based in the elementary methods as on the rate-of-creep method or perhaps on the principle of delayed elasticity. The more exact calculation of the stresses relaxation regarding the more complex expressions for the creep is rather tedious because the use of an iterative method is unavoidable. So it is when the creep and shrinkage predictions prescribed in the Recommendations of CEB-FIP (1978) are used, and thus the calculation of the stresses in the case of sustained deformations should be done by some iterative way which of course is not too favourite among the engineers.

Because the relaxation problems of this type are very frequent and important (especially in the bridge structures) it seems to be useful to find the simple analytical way of solution; one of them is presented in this Paper.

1 ANALYTICAL EXPRESSIONS OF THE CREEP AND THE SHRINKAGE

The above mentioned Recommendations of CEB-FIP give the following expressions for calculation of creep and shrinkage:

for the creep (as the sum of contributions of instantaneous nonreversible deformation, delayed elasticity and the rate-of-creep approach)

$$\varphi(t,t_o) = \beta_a(t_o) + \varphi_d \beta_d(t-t_o) + \varphi_f \left[\beta_f(t) - \beta_f(t_o)\right] \qquad (1)$$

with

$$\beta_a(t_o) = 0.8\ (1 - \frac{f_c(t_o)}{f_{c\infty}}) \tag{2a}$$

$$\varphi_d\,\beta_d(t-t_o) = 0.4\,\beta_d(t-t_o) \tag{2b}$$

and $\varphi_f = \varphi_{f1}\varphi_{f2}$ (2c)

for the shrinkage:

$$\varepsilon_s(t,t_o) = \varepsilon_{so}\left[\beta_s(t) - \beta_s(t_o)\right] \tag{3}$$

ε_{so} being expressed as a product $\varepsilon_{s1}\ \varepsilon_{s2}$.

All the necessary quantities, namely

$\frac{f_c\ (t_o)}{f_{c\infty}}$, $\beta_d(t-t_o)$, φ_{f1} (respecting the ambient influences), φ_{f2} , $\beta_f(t)$ and ε_{s1}, ε_{s2} and $\beta_s(t)$, are given in graphs or numerically.

As there is no doubt that the results of calculations using the CEB-FIP procedure of the solution together with the given quantities correspond well with reality, the following way of solution was based upon them. First of all, for the analytical solutions it is necessary to express the numerical quantities by the functional ones; there was made an attempt of their substitution with the following proposals.

a) The expression of $f_c(t_o)/f_c = f$

It is proposed

$$f_p = 1.1226 - 1.1854\ t_o^{-0.3} \tag{4}$$

(t_o is in days).

The agreement between the numerical values recommended by CEB-FIP and the function f_p given by Eq.(4) can be seen in the Tab. 1:

t_o (days)	3	5	10	20	50	100	200	500	1000
CEB-FIP	0.27	0.38	0.52	0.64	0.77	0.83	0.90	0.95	0.97
Eq. (4)	0.27	0.39	0.53	0.64	0.76	0.82	0.88	0.94	0.97

b) The expression of $\beta_d(t-t_o)$

It is proposed:

$$\beta_{dP}(t-t_o) = 0.28 + 0.5 \arctan\left[0.011(t-t_o)\right]^{\frac{2}{3}} \qquad (5)$$

The values of the Recommendation CEB-FIP and those given by the function (5) are shown in the Tab. 2:

$t-t_o$ (days)	1	3	5	10	50	100	500	1000	)1500
CEB-FIP	0.28	0.32	0.35	0.40	0.58	0.69	0.93	0.99	1.00
Eq.(5)	0.31	0.33	0.35	0.39	0.58	0.69	0.91	0.97	0.99

c) The expression of φ_{f2}

It is proposed :

$$\varphi_{f2P} = 1.1089 + 0.8469\, e^{-0.0027h_o} \qquad (6)$$

where $h_o = \frac{2\,A_c}{U}$ (in mm) is the fictive thickness calculated according to the CEB-FIP.

The comparison of functions is given in Tab.3:

h_o (mm)	50	100	200	400	600	800	1600
CEB-FIP	1.85	1.70	1.55	1.40	1.30	1.25	1.12
Eq.(6)	1.85	1.73	1.55	1.33	1.22	1.17	1.12

d) The expression of ε_{s2}

It is proposed :

$$\varepsilon_{s2P} = 0.7001 + 0.6292\, e^{-0.0046\, h_o} \qquad (7)$$

The comparison of function is given in Tab. 4:

h_o (mm)	50	100	200	400	600	800	1600
CEB-FIP	1.20	1.05	0.90	0.80	0.78	0.75	0.70
Eq.(7)	1.20	1.10	0.95	0.80	0.74	0.72	0.70

e) The expression of $\beta_f(t)$ and $\beta_s(t)$

It is proposed: for creep function

$$\beta_f(t) = 1 - e^{-\left(A_f \frac{t}{365}\right)^{B_f}} \qquad (8a)$$

for shrinkage

$$\beta_s(t) = 1 - e^{-\left(A_s \frac{t}{365}\right)^{B_s}} \qquad (8b)$$

in both cases the time t is in days.

The values of constants A_f, B_f and A_s, B_s are (respective to the fictive thickness h_o) given in Tab. 5:

h_o (mm)	A_f	B_f	A_s	B_s
50	3.5	0.43	7.1	0.42
100	2.5	0.41	3.0	0.49
200	1.8	0.39	1.2	0.57
400	1,2	0.38	0.5	0.66
800	0.7	0.36	0.2	0.77
1600	0.3	0.33	0.1	0.74

The comparison of the curves according to CEB-FIP (full lines) and the proposed expressions (8) (dashed lines) is shown in Fig. 1 and Fig. 2 for creep and shrinkage, respectively.

In practical cases it is usually necessary to solve the effect of the creep together with the effect of the shrinkage and then it is convenient to have the similar expression for both phenomena. For this case the following function seems to be the most suitable:

$$\beta_f(t) = \beta_s(t) = 1 - e^{-\left(A \frac{t}{365}\right)^{0.4}} \qquad (9)$$

where $A = e^{5.145 - 0.918 \ln h_o}$ (10)

h_o beeing again the fictive thickness.

The differences from the previous curves (CEB-FIP) and Eqs. (8a) and (8b) are shown in Figs. 1 and 2 (dotted lines).

It can be seen that the approximation of functions $\beta_f(t)$ and $\beta s(t)$ using Eqs. (8) is fairly good and is also satisfactory for the creep function as well as for the shrinkage one with respect to Eq.(9), excepting the last two cases where $h_o \geqslant 800$ mm.

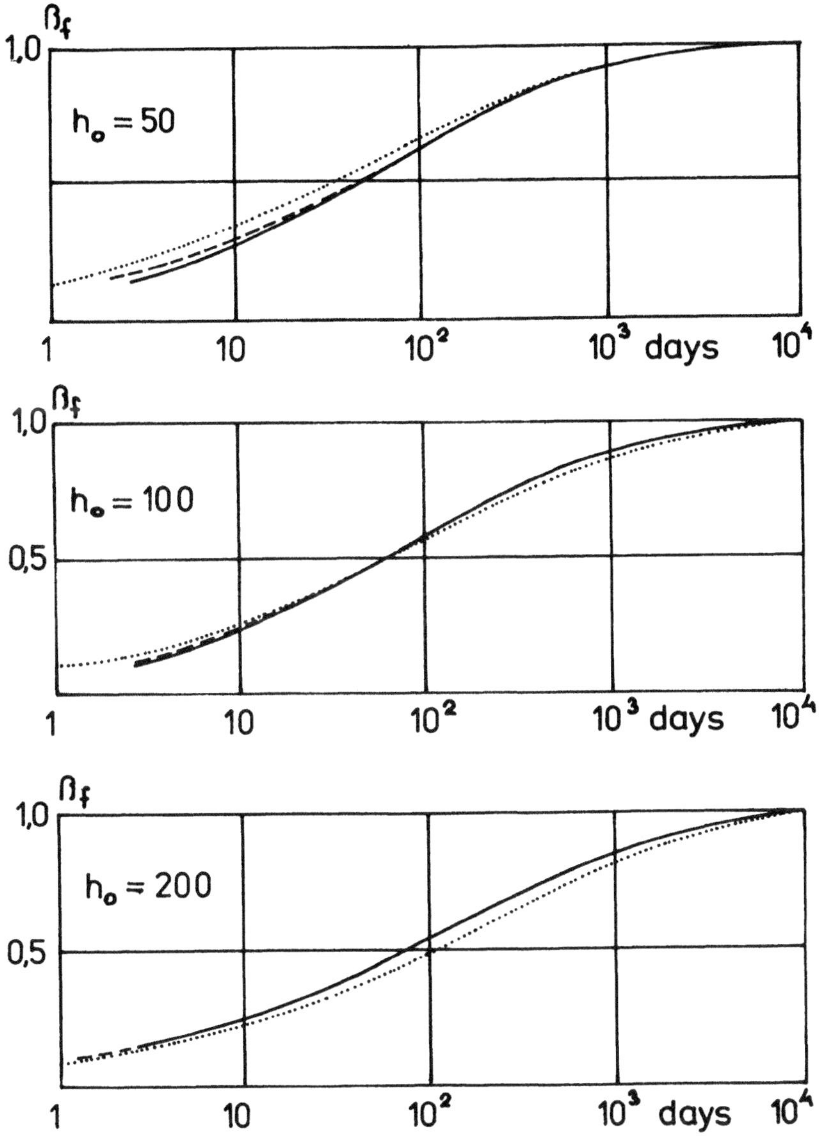

Fig. 1a. Creep functions β_f

CEB-FIP full lines
Eq. (8a) dashed lines
Eq. (9) dotted lines

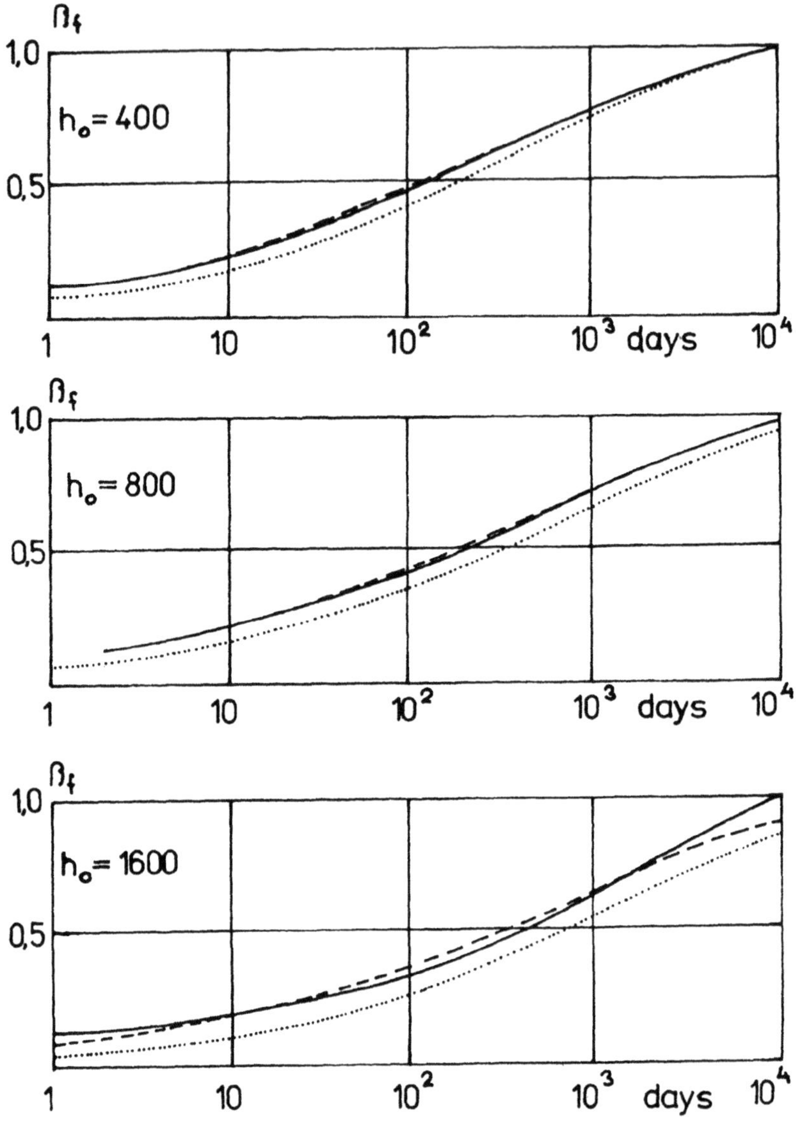

Fig 1b. Creep functions β_f
CEB-FIP full lines
Eq. (8a) dashed lines
Eq. (9) dotted lines

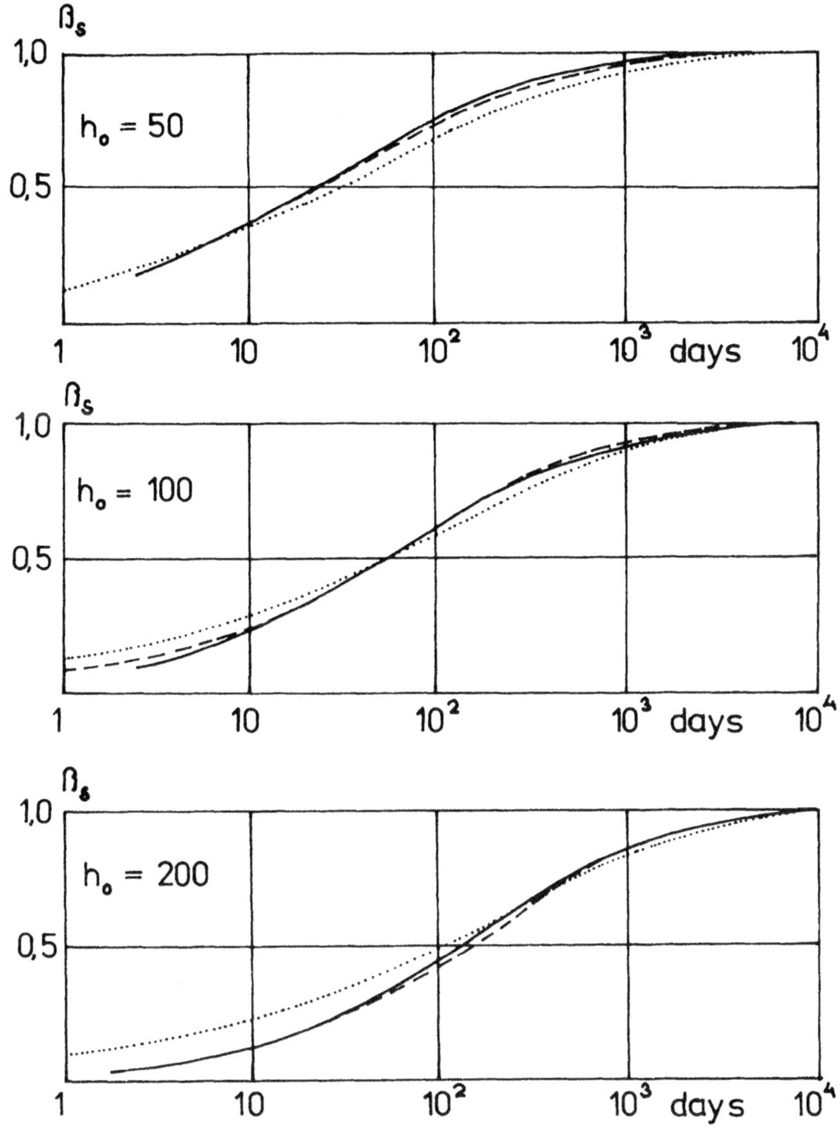

Fig. 2a. Shrinkage functions β_s
CEB-FIP full lines
Eq. (8a) dashed lines
Eq. (9) dotted lines

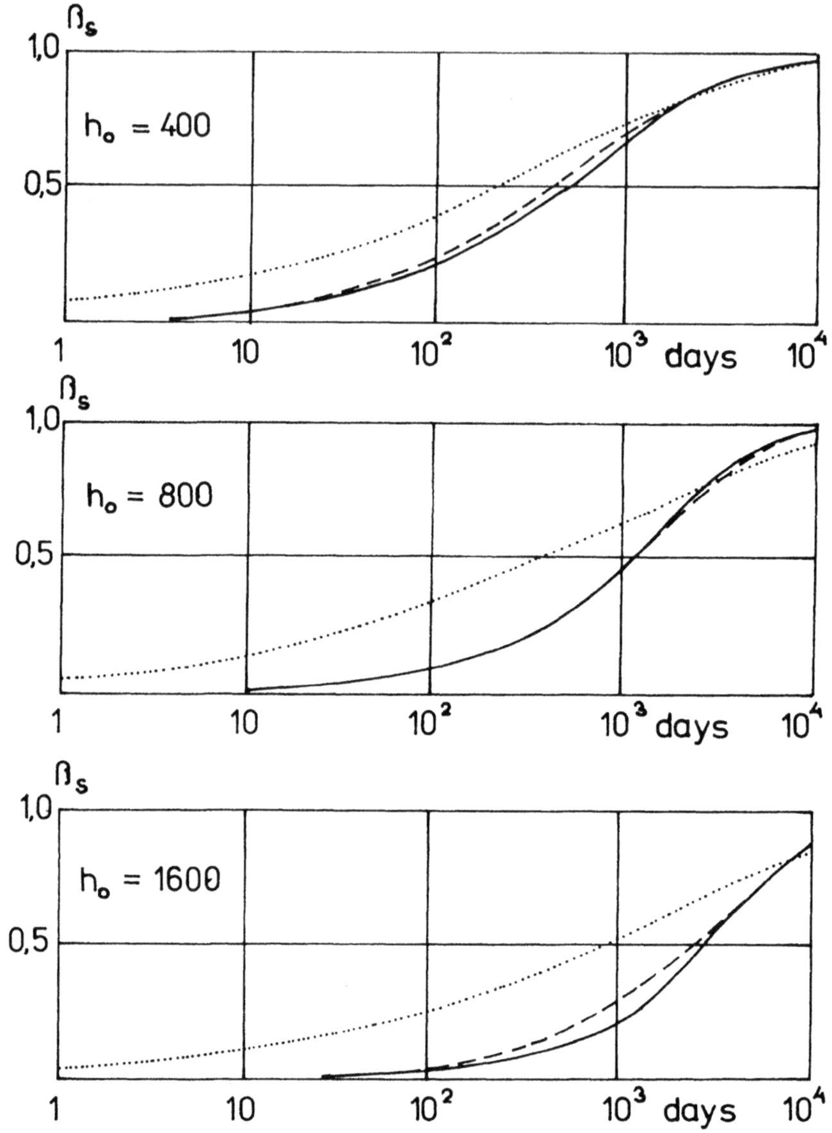

Fig. 2b. Shrinkage functions β_s

CEB-FIP full lines

Eq.(8a) dashed lines

Eq.(9) dotted lines

2 APPROXIMATE CALCULATION OF THE RELAXATION

As it has been noted, a calculation of the relaxation represents one of the most important but simultaneously very laborious task. In order to simplify the calculation, an approximate procedure is proposed.

The strain at time t due to stress σ (generally variable in time), acting from time τ_1, is given by the well known relation

$$\varepsilon(t) = \frac{\sigma(t)}{E} - \frac{1}{E}\int_{\tau_1}^{t} \sigma(\tau)\,\frac{\partial}{\partial \tau}\left[1 + \varphi(t,\tau)\right] d\tau \qquad (11)$$

in which E denotes Young's modulus (approximately considered to be time independent).

Introducing Eq. (1), the relation (11) can be written in the form

$$\varepsilon(t) = \frac{\sigma(t)}{E} - \frac{1}{E}\int_{\tau_1}^{t} \sigma(\tau)\,\frac{\partial}{\partial \tau}\left[\varphi_d\,\beta_d(t-\tau) - \varphi_f\beta_f(\tau)\right] d\tau =$$

$$= \frac{\sigma(t)}{E} - \frac{1}{E}\int_{\tau_1}^{t} \sigma(\tau)\left[\varphi_d\,\frac{d\beta_d(z)}{dz}\,\frac{dz}{d\tau} - \varphi_f\,\frac{d\beta_f(\tau)}{d\tau}\right] d\tau \qquad (12)$$

in which $z = t - \tau$, i.e. $\frac{dz}{d\tau} = -1$.

The stress σ , variable in time, can be considered as consisting of two components as follows

$$\sigma(\tau) = \sigma_o - \sigma_2(\tau) \qquad (13)$$

The component σ_o is the stress value at the time τ_o, when the concrete was loaded; the stress σ_o remains of constant value.

The component σ_2 , on the other hand, is a stress growing during the relaxation period. The course of the stress σ_2 is governed by the initial condition:

$$\sigma_2(\tau_r) = 0 \qquad (14)$$

in which τ_r denotes the beginning of the relaxation.

Substituting Eq. (13) for Eq. (12) it can be written

$$\varepsilon(t) = \frac{1}{E}\left\{\sigma_o - \sigma_2(t) + \int_{\tau_1}^{t}\left[\sigma_o - \sigma_2(\tau)\right]\left[\varphi_d\,\frac{d\beta_d(z)}{dz} + \right.\right.$$

$$+ \varphi_f \frac{d\beta_f(\tau)}{d\tau} \Big] d\tau \Big\} \qquad (15)$$

The term $\beta_d(z)$, expressing the delayed elasticity, is a function of the stress action period $z = (t-\tau)$, i.e. it is not directly a function of the time τ. In order to avoid difficulties consequent from this fact, the course of the delayed elasticity is approximately assumed to be of a linear variation

$$\varphi_d \beta_d (z) = a + k z \qquad (16)$$

The gradient k may be evaluated in dependence on the length of the period of the stress action, related to the duration of the relaxation.

The stress σ_o acts from time τ_o causing a corresponding delayed elasticity strain whose part belonging into the relaxation period $\tau_r \leq \tau \leq t$ may be (with regard to Eq. (1)) rather fairly characterized by the gradient k_o the value of which is

$$k_o = \varphi_d \frac{\beta_d(t-\tau_o) - \beta_d(\tau_r - \tau_o)}{t - \tau_r} \qquad (17)$$

The stress component σ_2 does not manifest itself before the beginning of the relaxation. The gradient k in Eq. (16), corresponding to the stress σ_2, therefore, may be approximately expressed by the relation

$$k_2 = \frac{\beta_d(t - \tau_r) - \beta_d(0)}{t - \tau_r} \varphi_d \qquad (18)$$

the diagram of which in terms of the relaxation duration, considering Eqs. (2b) and (5) is shown in Fig. 3.

Eq. (15) thus can be written in the form

$$\varepsilon(t) = \frac{1}{E}\Big(\sigma_o - \sigma_2(t) + \int_{\tau_1}^{t} \Big\{ \sigma_o \Big[k_o + \varphi_f \frac{d\beta_f(\tau)}{d\tau} \Big] -$$

$$- \sigma_2(\tau) \Big[k_2 + \varphi_f \frac{d\beta_f(\tau)}{d\tau} \Big] \Big\} d\tau \Big) \qquad (19)$$

During the relaxation process the strain ε is unchanged; the stress variation is governed by the following differential equation

$$\frac{d\sigma_2}{d\tau} + \sigma_2 \left[k_2 + \varphi_f \frac{d\beta_f(\tau)}{d\tau}\right] =$$

$$= \sigma_o \left[k_o + \varphi_f \frac{d\beta_f(\tau)}{d\tau}\right] \tag{20}$$

The general solution of the homogeneous equation corresponding to Eq. (20) is

$$\sigma_2 = C\, e^{-[k_2\tau + \varphi_f \beta_f(\tau)]} \tag{21}$$

and, in order to obtain a solution of the non-homogeneous Eq.(20), it holds for the constant C that

$$C = \sigma_o \int \left(k_o + \varphi_f \frac{d\beta_f}{d\tau} \right) e^{(k_2\tau + \varphi_f \beta_f)}\, d\tau \tag{22}$$

As examples of the approach presented two important limit cases are further analyzed.

1. if the stress σ_o is applied at time $\tau_o = \tau_r$ and further changes of strain are immediately restrained, it follows from Eqs. (17) and (18) that $k_o = k_2$. Substituting for Eq. (22) it is simply obtained

$$C = \sigma_o\, e^{(k_2\tau + \varphi_f \beta_f)} + K \tag{23}$$

in which K is a constant.

Substituting Eq. (23) for Eq. (21) and then for Eq. (13) (with regard to the initial condition (14)) the final form of the stress course during the relaxation process is

$$\sigma = \sigma_o\, e^{-\left\{k(\tau - \tau_o) + \varphi_f [\beta_f(\tau) - \beta_f(\tau_o)]\right\}} \tag{24}$$

It should be noted that in this case the relaxation phenomenon manifests itself more pronouncely than if only the elementary rate-of-creep method would be applied.

2. quite different situation arises when the time τ_o of application of the stress σ_o and the beginning of restraining the strain changes (i. e. the starting time of the relaxation process) τ_r are not identical. This case is very frequent in practice, namely in bridges, analyzing the influence of changes of structural system during the construction process

by the initial stress method resp. by the polygonal relaxation method.

If the time difference $\tau_r - \tau_o$ is sufficient to assume that the major part of creep deformations due to delayed elasticity caused by the stress component σ_o are finished, the coefficient k_o may be neglected. The formula for the constant C given by Eq. (22) becomes

$$C = \sigma_o \varphi_f \int \frac{d\beta_f}{d\tau} \; e^{(k_2\tau + \varphi_f\beta_f)} \, d\tau \tag{25}$$

Using the proposed expressions for the creep prediction (Eqs. (6) and (8a), resp. (9), (10)), Eq. (25) may be approximately substituted by the following relation

$$C = \sigma_o \varphi_f \int \left[P(\tau) + e^{\varphi_f\beta_f(\tau)} \frac{d\beta_f(\tau)}{d\tau} \right] d\tau \tag{26}$$

It was found that for the expressions proposed, the function $P(\tau)$ is approximately proportional to $\sqrt{\tau}$. After inegration it may be written

$$C = \sigma_o \left[\varphi_f \; Q(\tau) + e^{\varphi_f\beta_f(\tau)} \right] + K \tag{27}$$

in which $Q(\tau) = \int P(\tau) \; d\tau$, and K is a constant.

Thus, the general solution of Eq. (20) is given by

$$\sigma_2 = \left\{ \sigma_o \left[\varphi_f \; Q(\tau) + e^{\varphi_f\beta_f(\tau)} \right] + K \right\} e^{-[k_2\tau + \varphi_f\beta_f(\tau)]} \tag{28}$$

The constant K can be evaluated using the initial condition (14). Considering Eq. (13), the resulting function describing the stress in the course of relaxation is expressed as

$$\sigma(\tau) = \sigma_o \left(1 - \left\{ \varphi_f \left[Q(\tau) - Q(\tau_r) \right] + e^{\varphi_f\beta_f(\tau)} - e^{\varphi_f\beta_f(\tau_r)} \right\} e^{-[k_2\tau + \varphi_f\beta_f(\tau)]} \right) \tag{29}$$

The accuracy of the calculation of relaxation using Eq. (29) was proved for the case if the original stress σ_o was applied at the time τ_o = 28 days and the relaxation period was from τ_r = 180 days to τ = 1180 days. Using the coefficient k = 0.000274 day^{-1} (see Fig. 3) and values of φ_f and function

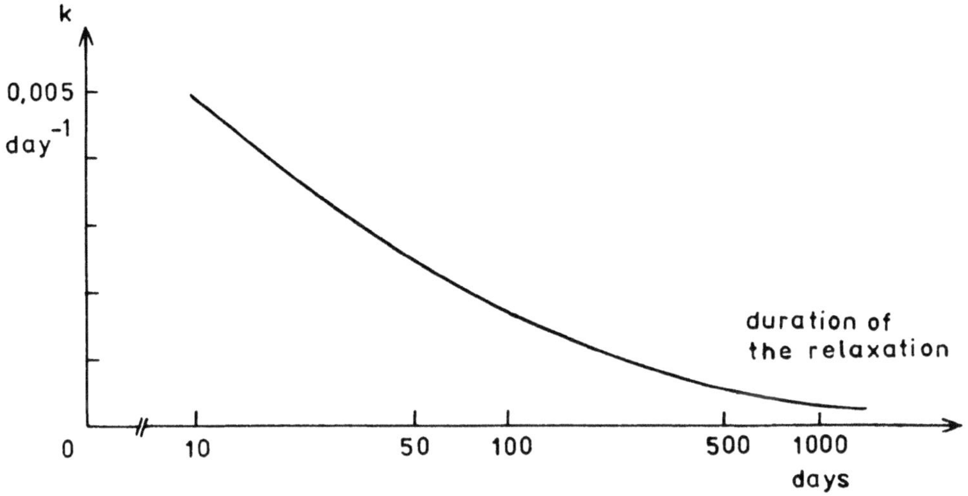

Fig. 3. Values of the coefficient k_2 in dependence on duration of the relaxation

β_f in accordance with Eq. (8a) and Table 4, an excellent agreement with the result on the numerical procedure based on the Recommendations of CEB-FIP was reached.

In contrary to the preceding case 1, the relaxation were less pronounced here at case 2 than if the elementary rate-of-creep theory would be used.

CONTENIDO

El cálculo de los problemas de la relaxacíon de las tensiones es muy complicado en los casos en los cuales se usan las expresiones de la fluencia mas complejas - como se recomienda por ejemplo en las Recomendaciones del Comitet CEB-FIP 1978 - ya que se debe de aprovechar de los métodos iterativos. Para simplificar el trabajo con el cálculo se presenta aquí un método analítico que parece bastante sencillo para el uso práctico. En la parte primera se presentan las sustituciones analíticas de las curvas dadas por el CEB-FIP, en la parte segunda está el método de cálculo en el cual está envuelto como el efecto de la fluencia tanto el efecto de las deformaciones elásticas retardadas.

SHRINKAGE AND LOAD INDUCED CURVATURE OF REINFORCED CONCRETE BEAMS

D.W. Hobbs

Materials Research Department, Cement and Concrete Association, U.K.

ABSTRACT

Using the equivalent tensile force method predictions are made of the shrinkage induced curvature of reinforced concrete beams. It is shown that the predicted curvatures for uncracked concrete beams are in close agreement with published experimental data on singly and doubly reinforced beams. It is also shown that the total predicted curvatures, due to a combination of elastic movement, creep and shrinkage, of partially cracked beams are in approximate agreement with published data obtained by Stevens (6).

En employant la méthode de la force de traction équivalente, on prédit la courbure, induite par le retrait, des poutres en béton armé. On montre que les courbures prédites pour les poutres non fissurées correspond de très pres aux données experimentales publiées sur les poutres uniquement et doublement armées. On montre aussi que les courbures totales prédites pour les poutres partiellement fissurées, attribuables à une combination du mouvement élastique, du fluage et du retrait, s'accordent approximativement avec les données publiées et obtenues par M Stevens (6).

INTRODUCTION

When reinforcement is placed non-symmetrically in an uncracked beam and the concrete shrinks, a bending moment is induced due to non-uniform restraint of the shrinkage by the reinforcement and, as a result, the curvature and deflection of the concrete beam is increased. In the case of a partially cracked beam (i.e. a beam in which the cracked section supports some tensile load) a bending moment will be induced by shrinkage even when the reinforcement is symmetrically placed. The contribution of shrinkage to the curvature and deflection of a concrete beam is believed to be somewhere between 5 and 30% (1-5) but higher contributions can occur if the applied bending moments are low (5).

In the present paper the equivalent tensile force method is used to estimate the shrinkage induced curvature of reinforced concrete beams. It is shown that this method gives close agreement with published data on uncracked concrete beams. Unfortunately it is not possible to check the predictions against data obtained on partially cracked beams because such data are not currently available. However it is shown that the predicted total curvatures produced by a combination of elastic movement, creep and shrinkage are in approximate agreement with the experimental curvatures obtained by Stevens (6) on a range of loaded concrete beams stored internally and externally.

THE EQUIVALENT TENSILE FORCE METHOD

The equivalent tensile force method for estimating shrinkage induced curvatures was first suggested by Peabody (7) in 1936. In this method the concrete member is first considered to be free to shrink. For this to be so the compression and tension steels must be compressed by fictitious forces $\epsilon_{cs}A_s'E_s$ and $\epsilon_{cs}A_sE_s$ respectively, where ϵ_{cs} is the unrestrained concrete shrinkage, A_s' and A_s are the areas of compression and tension steel respectively and E_s is the Young's modulus of the steel. When these loads are released it is equivalent to eccentric tension loads $\epsilon_{cs}A_s'E_s$ and $\epsilon_{cs}A_sE_s$ applied at the steel level to the entire transformed area of the member and the bending moment produced is given by

$$M_{cs} = \epsilon_{cs}A_sE_s\,(d - x) - \epsilon_{cs}A_s'E_s\,(x - d') \quad . \; . \; . \quad (1)$$

where d and d' are the depths of the tension and compression reinforcement respectively. The curvature due to shrinkage of concrete for an uncracked section is

$$\frac{1}{r_{cs}} = \frac{(1 + \psi)\,\epsilon_{cs}E_s\left[A_s\,(d-x) - A_s'\,(x-d')\right]}{E_cI_u} \quad . \; . \; . \quad (2)$$

and for the cracked section

$$\frac{1}{r_{cs}} = \frac{(1+\psi)\,\epsilon_{cs} E_s \left[A_s\,(d-x) - A_s'\,(x-d')\right]}{E_c\, I_c} \quad . \quad . \quad . \qquad (3)$$

where ψ is the appropriate creep coefficient, E_c is the Young's modulus of the concrete, x is the depth of the centroid of the transformed section and I_u and I_c are the second moment of area of the uncracked and crack section respectively. In calculating x, I_u and I_c the reduced Young's modulus is used (8). The maximum compressive strain in the concrete, ϵ_m, and the tensile strain in the tension reinforcement, ϵ_s, are

$$\epsilon_m = \frac{x}{r} + \frac{S}{1 + \dfrac{(1+\psi)\,\rho\, E_s}{(1-\rho)\, E_c}} \quad . \quad . \quad . \quad . \quad . \quad . \quad . \qquad (4)$$

and $$\epsilon_s = \frac{d - x}{r} - \frac{S}{1 + \dfrac{(1+\psi)\,\rho\, E_s}{(1-\rho)\, E_c}} \quad . \quad . \quad . \quad . \quad . \quad . \qquad (5)$$

respectively where $\rho = (A_s + A_s')/A_c$ where A_c is the area of concrete in the case of the uncracked section and the area of uncracked concrete in the case of the cracked section. A major difficulty associated with the application of equations 1 to 5 is the choice of values for E_c and ψ. This is because the induced stresses increase continuously and as a result there is not a specific age at which the shrinkage load is applied.

In Figure 1 the predicted induced curvatures for an uncracked concrete beam are compared with experimental curvatures obtained after prolonged drying by Miller (9) and Mohamedbhai and Taylor (10) on singly reinforced beams and by Washa and Fluck (10) on singly and doubly reinforced beams. The chosen values of E_s/E_c and ψ are given on the Figure. In Figures 2 and 3 the shrinkage induced curvatures obtained using the Branson (12) and CP110 methods (13) are compared with the same experimental data. The CP110 method is a modification of Branson's method which is an empirical method based on the experimental data obtained by Miller. From Figures 1 to 3 it can be seen that the observed curvatures both for singly and doubly reinforced beams are in closest agreement with the predicted curvatures obtained using the equivalent tensile force method.

If it is assumed that the shrinkage-induced curvature is proportional to the depth of the tension reinforcement then equations 2 and 3 may be written

$$\frac{1}{r_{cs}} = \frac{\rho_{ou}\ \epsilon_{cs}}{d} \quad . \quad . \quad . \quad . \quad . \quad . \quad . \quad . \quad . \quad . \qquad (6)$$

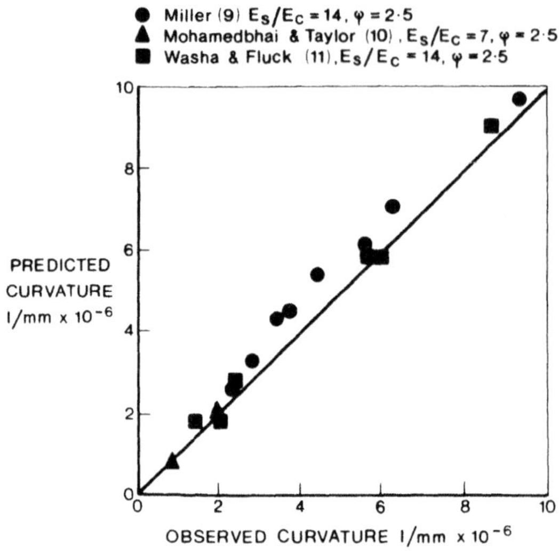

FIG 1 Comparison between predicted and observed curvatures. Equivalent tensile force method

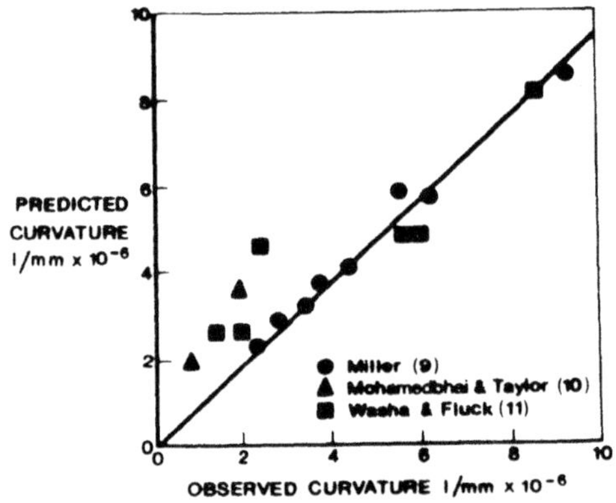

FIG 2 Comparison between predicted and observed curvatures. Branson's method

$$\text{and} \quad \frac{1}{r_{cs}} = \frac{\rho_{oc}\, \epsilon_{cs}}{d} \quad . \; . \; . \; . \; . \; . \; . \; . \; . \; . \qquad (7)$$

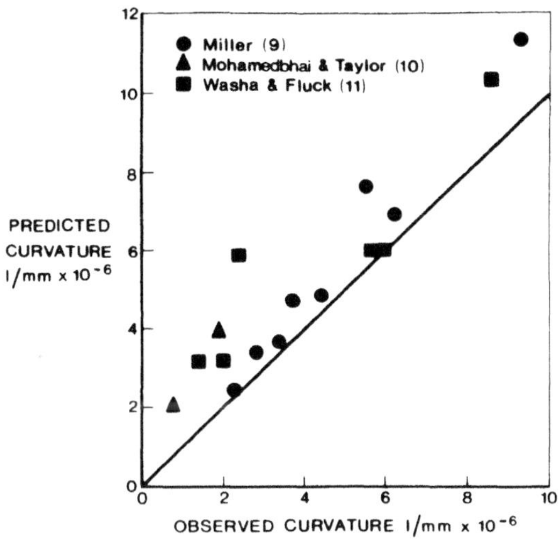

FIG 3 Comparison between predicted and observed curvatures. CP110 method

respectively. The coefficients ρ_{ou} and ρ_{oc} are given in Table 1 for $E_s/E_c = 7$, $\psi = 1.75$ and 1.0, $d/h = 0.9$, $A_s = 0.005$, 0.01, 0.02 and $A_s' = 0$, 0.01. More detailed information appears elsewhere (14). Two important points to note from the Table are firstly that the shrinkage induced curvature of a cracked member may be several times greater than that of an uncracked member and secondly that variations in the percentage of tension steel have little effect upon the shrinkage induced curvature of a cracked member.

TOTAL CURVATURE DUE TO ELASTIC MOVEMENT, CREEP AND SHRINKAGE

Equation 3 is applicable to the partially cracked section if the following simplifications are made, firstly that the long-term tensile stresses in the concrete are independent of concrete shrinkage and secondly that the depth of the centroid of the transformed section is the same as that of the cracked section.

Unfortunately it is not possible to check the shrinkage-induced curvature predictions for partially cracked beams because no experimental data are currently available. However the total curvature produced by a combination of elastic movement, creep and shrinkage can be predicted from

$$\frac{1}{r} = \frac{(1+\psi)\quad(M - M_t + M_s)}{E_c \quad I_c} \quad . \quad . \quad . \quad . \quad . \quad . \quad (8)$$

TABLE 1 Magnitude of ρ_{ou} and ρ_{oc}. E_s/E_c = 7. d/h = 0.9

Exposure condition	Creep factor ψ	Tension steel $\frac{100A_s}{bd}$	Compression steel $\frac{100A'_s}{bd}$	ρ_{ou}	ρ_{oc}
		0.5	0	0.23	1.12
		0.5	1.0	-0.17	0.68
		1.0	0	0.40	1.16
external	1	1.0	1.0	0.00	0.73
		2.0	0	0.63	1.21
		2.0	1.0	0.23	0.79
		0.5	0	0.30	1.13
		0.5	1.0	-0.20	0.61
		1.0	0	0.50	1.18
internal	1.75	1.0	1.0	0.00	0.66
		2.0	0	0.74	1.24
		2.0	1.0	0.25	0.73

and therefore the predictions can be checked indirectly. In equation 8 M is the applied bending moment and M_t the moment due to concrete tension in the cracked region. M_t may be taken as (5)

$$M_t = \frac{b\,(h - x)^2}{8} \quad f_{ct} \quad . \; . \; . \; . \; . \; . \; . \; . \; . \; . \quad (9)$$

where f_{ct} is the long term tensile of the concrete (approximately 1.5 Pa) and b is the breadth and h the depth of the beam.

In Figures 4 to 6 the curvatures obtained from equation 8 are compared with experimental curvatures obtained by Stevens (6) on a range of beams, which were loaded at 28 days, and stored internally or externally. In these Figures the curvatures are compared when the load is applied, and 1 year and 2 years after loading. The values of S and ψ used are given in Table 2. The magnitude of ψ was obtained using the CEB FIP recommendations (15) and the magnitude of the shrinkage was estimated using the method proposed by Hobbs and Parrott (16), namely

$$S = 0.125\, W_o f_1 \left[\frac{A}{V}^2 t \right] f_3 \; (\Upsilon_h) \quad . \; . \; . \; . \; . \; . \; . \quad (10)$$

where W_o is the original water content by weight (%), V/A is half the effective section thickness and Υ_h is the relative humidity. The functions f_1 and f_3 are shown plotted in Figures 7 and 8 respectively. The equation is not applicable to concrete subject to a prolonged cure.

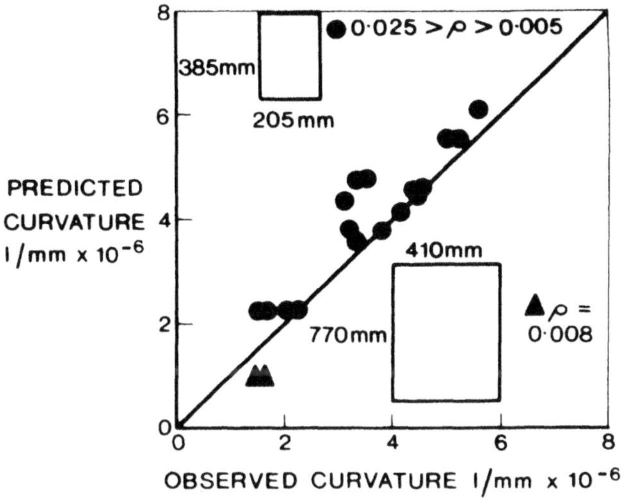

FIG 4 Comparison between predicted and observed curvatures at the time of loading

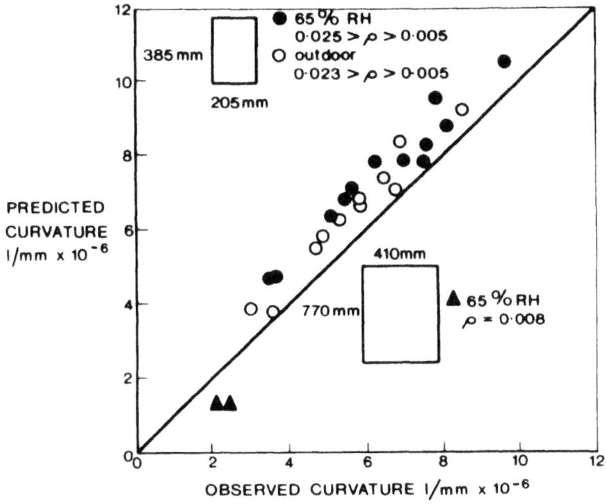

FIG 5 Comparison between predicted and observed curvatures. 1 year after loading

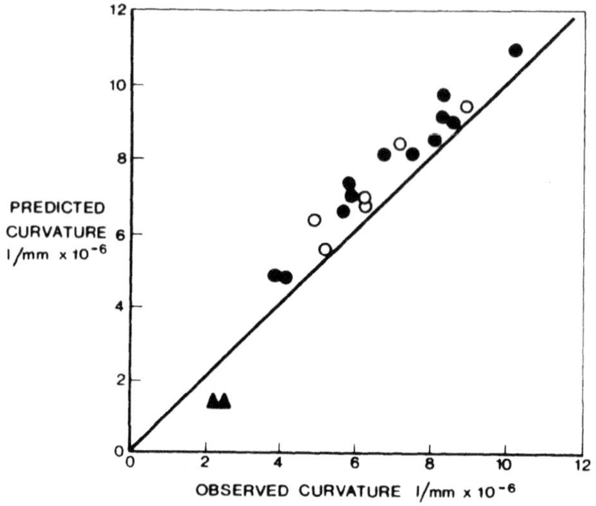

FIG 6 Comparison between predicted and observed curvatures. 2 years after loading

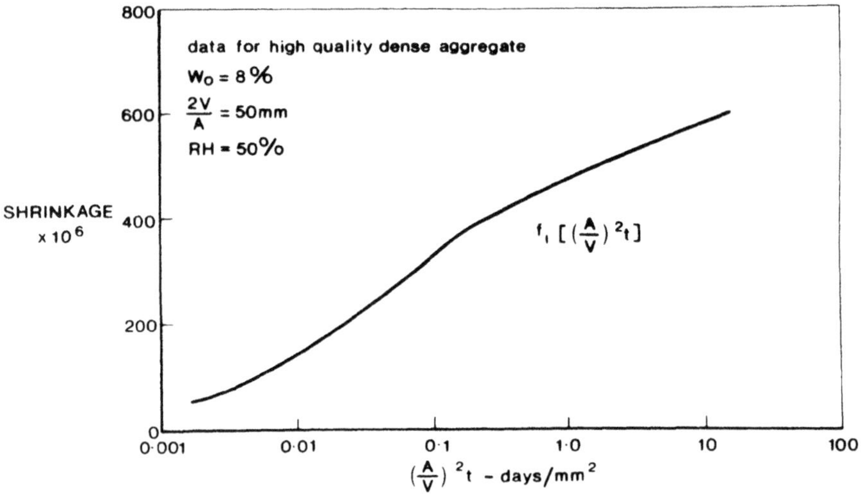

FIG 7 Shrinkage as a function of drying period and size

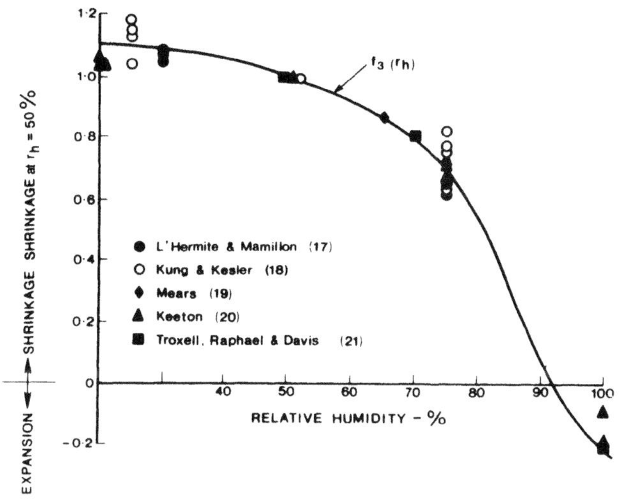

FIG 8 Influence of relative humidity on shrinkage

It is apparent from Figures 4 to 6 that equation 8 overestimates the curvature of the beams, possibly in part because the cracks are not fully developed (see Figure 4). However it is considered on the basis of Figures 4 to 6 that a reasonable upper limit for curvature can, in general, be calculated from equation 8.

The contribution of shrinkage to the total 2 year curvature of the beams is estimated to range from 11 to 24% (mean 17%) for internal drying and from 7 to 10% for external drying. The contribution of creep to the total 2 year curvature is estimated to range from 21 to 40% (mean 31%) for internal drying and from 23 to 35% for external drying.

CONCLUSIONS

1. The equivalent tensile force method is a reliable method for estimating the shrinkage-induced curvature of reinforced concrete members.

2. The equivalent tensile force method predicts that the shrinkage-induced curvature of a cracked member can be several times greater than that of an uncracked member.

3. A comparison of the predicted curvature, due to a combination

TABLE 2 Magnitude of S and ψ^*

Time under load (yrs)	Effective section thickness mm	Exposure condition RH (%)	Shrinkage S (micro-strain)	Creep factor ψ
1	134	83	142	2.1
	134	65	274	2.9
	268	65	165	1.9
2	134	83	170	2.2
	134	65	321	3.1
	268	65	214	2.2

*Cement content 300 kg/m^3, w/c 0.7, W_o 8%, E_c at the time of loading was 34.3 kPa

of elastic movement, creep and shrinkage, of partially cracked beams with the experimental curvatures determined by Stevens showed approximate agreement.

ACKNOWLEDGEMENTS

The author is grateful to Mr T Biss and Mr N Swinton for their assistance with the calculations.

REFERENCES

1. Beeby, A.W. "Modified proposals for controlling deflections by means of ratio of span to effective depth". Technical Report No. 42 456 (Cement and Concrete Association, London 1971)

2. Noble, P.W. and L.V Leech. Design Tables for Composite Steel and Concrete Beams (Constrado, London 1975)

3. Johnson, P.R and D.G.E. Smith. "Design rules for the control of deflection in composite beams". The Structural Engineer 53 (1975) 367-376.

4. Taylor, R. and F. Frodin. "A contribution to the design of composite beams to satisfy deflexion criteria". Proceedings of the Institution of Civil Engineers. Part 2 : Research and Theory 63 (1977) 105-122

5. Hobbs, D.W. "Drying shrinkage and movement of reinforced concrete". Technical Report No. 42 522 (Cement and Concrete Association 1978).

6. Stevens, R.E. "Deflexions of reinforced concrete beams". Proceedings of the Institution of Civil Engineers. Part 2 : Research and Theory 53 (1972) 207-224.

7. Peabody, D. The Design of Reinforced Concrete Structures. Second Edition (Ronald Press Company, New York 1957).

8. Kong, F.K. and R.H. Evans. Reinforced and Prestressed Concrete. (Thomas Nelson and Sons Ltd, Sunbury-on-Thames 1975).

9. Miller, A.L. "Warping of reinforced concrete due to shrinkage". Journal of the American Concrete Institute, Proceedings 54 (1958) 939-950.

10. Mohamedbhai, G.T.G. and R. Taylor. "The long-term behaviour of one-way spanning slabs under variable loading". Proceedings of the Institute of Civil Engineers. Part 2 : Research and Theory 57 (1974) 67-82.

11. Washa, G.W. and P.G. Fluck. "Effect of compressive reinforcement on the plastic flow of reinforced concrete beams". Journal of the American Concrete Institute, Proceedings 49 (1952) 89-108.

12. Branson, D.E Deformation of Concrete Structures. (McGraw-Hill International Book Company, New York 1977)

13. British Standards Institute. "The structural use of concrete" CP 110 : Part 1 (1972) 136-137.

14. Hobbs, D.W. "Shrinkage-induced curvature of reinforced concrete beams". Development Report No 4 (Cement and Concrete Association, London 1979).

15. CEB-FIP Recommendations. "International Recommendations for the Design and Construction of Concrete Structures". Second Edition (Cement and Concrete Association, London 1970).

16. Hobbs, D.W. and L.J. Parrott. "Prediction of drying and shrinkage". Concrete 13 (1979) 19-24.

17. L'Hermit, R G. and M. Mamillan. "Further results of shrinkage and creep tests". The Structure of Concrete and its Behaviour under Load. Proceedings of an International Conference, London 1965 (Cement and Concrete Association, London 1968) 423-433, 449-452.

18. Kung, S.H. and C E. Kesler. "A study of free and restrained shrinkage of mortar". Engineering Experiment Station, University of Illinois. T and AM Report No 647 (1964).

19. Mears, A.R. "The effect of aggregate size and grading on initial air-drying shrinkage". Technical Report No. 42 484 (Cement and Concrete Association, London 1970).

20. Keeton, J.R. "Study of creep in concrete, phases 1-5". Naval Engineering Laboratory, Technical Reports R33-I, II, III (Port Hueneme, California 1965).

21. Troxell, G.E., J.M. Raphael and R.W. Davis. " Long-time creep and shrinkage tests of plain and reinforced concrete". Proceedings ASTM 58 (1958) 1-20.

PRISE EN COMPTE DU FLUAGE PAR LA METHODE DES FORCES

M. Koprna

Institut de statique et structures - Béton armé et Précontraint

Ecole Polytechnique Fédérale de Lausanne

ABSTRACT

Time dependent behaviour problems of beam structures composed of viscoelastic or purely elastic elements, are treated on the basis of a generally valid system of algebraic equations, which have been derived from the force method. Specific cases studies lead to a simplified system of equations, then to the formation of independent equations. The resolution of some typical problems illustrates this method.

RESUME

Les problèmes des effets différés pour des structures linéaires formées d'éléments viscoélastiques et purement élastiques sont traités sur la base d'un système général d'équations algébriques, établi à partir de la méthode classique des forces. L'étude de cas particuliers conduit à un système simplifié d'équations, puis à la formation d'équations indépendantes. La résolution de quelques problèmes types illustre cette méthode.

ZUSAMMENFASSUNG

Die Probleme der Langzeit-Vorgänge für stabförmige, aus viskoelastischen oder rein elastischen Elementen gebildete Tragwerke werden mit Hilfe eines allgemeingültigen algebraischen Gleichungssystems, welches von der Kraftmethode hergeleitet wurde, behandelt. Die Studie von Einzelfällen führt zu einem vereinfachten System von Gleichungen, dann zur Bildung von voneinander unabhängigen Gleichungen. Die Lösung einiger typischer Probleme illustriert diese Methode.

1. INTRODUCTION

Le comportement rhéologique du béton influence les déformations et par conséquent provoque une redistribution des efforts; il représente un intérêt particulier, surtout dans les états-limites d'utilisation, qui déterminent souvent le dimensionnement des structures.

La méthode des forces, bien connue des ingénieurs, peut aisément être élargie afin de tenir compte des effets différés dans une structure linéaire quelconque, composée d'éléments viscoélastiques et purement élastiques.

Récapitulons tout d'abord la méthode des forces sans tenir compte des effets différés.
Les conditions de compatibilité écrites pour le système de base isostatique donnent le système d'équations suivant:

$$[\alpha_{ik}]\{X_{kEL}\} + \{a_{iq}\} = \{\Delta_i\} + [R_{ri}]^T\{\Delta_r\} \tag{1}$$

avec

$$\alpha_{ik} = a_{ik} + \delta_{ik}.c_i \tag{2}$$

où

X_{kEL} - $k^{\text{ième}}$ hyperstatique, correspondant à la solution purement élastique, due à la totalité des sollicitations (q, Δ),

a_{ik}, a_{iq} - déformation en *i* due à $X_{kEL} = 1$, resp. aux charges q:

$$a_{ik} = a_{ik\varphi} + \sum_r R_{ri}.R_{rk}c_r$$
$$a_{iq} = a_{iq\varphi} + \sum_r R_{ri}.R_{rq}c_r \tag{3a}$$

avec

$$a_{ik\varphi} = \int \frac{M_i.M_k}{EI} dx$$
$$a_{iq\varphi} = \int \frac{M_i.M_q}{EI} dx \tag{3b}$$

R_{rk}, R_{rq} - effort en *r* dû à $X_{kEL} = 1$, resp. aux charges q,

c_i, c_r - constante de ressort en *i*, resp. en *r*,

Δ_i, Δ_r - déformation imposée en *i*, resp. en *r*,

δ_{ik} - delta de Kronecker:

$$\delta_{ik} = 1 \quad \text{pour } i = k$$
$$= 0 \quad \text{pour } i \neq k,$$

i, k - liaisons correspondant aux hyperstatiques $(i,k = 1 \div n)$,

r - liaisons à considérer autres que celles correspondant aux hyperstatiques $(r = 1 \div m)$.

2. BASES THEORIQUES

Pour considérer les effets différés, admettons l'hypothèse de linéarité et l'hypothèse de superposition formulées par l'équation intégrale de Volterra:

$$\varepsilon(t) = \varepsilon_{tot}(t) - \varepsilon_s(t) = \Phi(t,t_o).\sigma_o + \int_{t_o}^{t} \Phi(t,\tau).d\sigma(\tau) \tag{4}$$

ceci pour les contraintes

$$\sigma(t) = 0 \quad \text{pour } t < t_o$$
$$= \sigma_o \quad \text{pour } t = t_o$$
$$= \sigma(\tau) \quad \text{pour } t_o < \tau \leqslant t$$

avec

$\varepsilon(t)$ - déformation relative sans l'effet du retrait au temps t,

$\varepsilon_{tot}(t)$ - déformation relative totale,

$\varepsilon_s(t)$ - déformation relative due au retrait,

$\Phi(t,t_o)$ - fonction fluage:

$$\Phi(t,t_o) = \frac{1}{E_o}(1 + \varphi) \tag{5}$$

$\varphi(t,t_o) \equiv \varphi$ - coefficient de fluage,

t_o - âge du béton au moment de la mise en charge.

L'introduction du coefficient de vieillissement $\chi(t,t_o) \equiv \chi$ nous permet d'écrire l'éq.(4) sous la forme (voir {1}):

$$\varepsilon(t) = \frac{\sigma_o}{E_o}(1 + \varphi) + \frac{\Delta\sigma(t)}{E_o}(1 + \chi\cdot\varphi) \tag{6}$$

avec

$$\Delta\sigma(t) = \sigma(t) - \sigma_o$$

$$\chi(t,t_o) = \frac{1}{\varphi(t,t_o)}\left(\frac{E_o}{\Delta\sigma(t)}\int_{t_o}^{t}\Phi(t,\tau).d\sigma(\tau) - 1\right)$$

$$= \frac{\int_{t_o}^{t}\varphi(t,\tau).d\sigma(\tau)}{\varphi(t,t_o).\Delta\sigma(t)} \tag{7}$$

Note: Le coefficient de vieillissement χ *dépend, dans le cas général, de l'histoire de la contrainte. La solution du problème, conduisant à la résolution numérique de l'équation intégrale de Volterra, est relativement difficile.*

Toutefois, dans le cas où la déformation relative varie selon la loi

$$\varepsilon(t) = a + b.\varphi \quad \text{pour } t \geqslant t_o$$
$$= 0 \quad \text{pour } t < t_o$$

a, b *étant des constantes arbitraires (affinité entre* ε *et* φ*), on peut prouver que le coefficient de vieillissement* χ *devient indépendant de l'histoire de la contrainte et qu'il est défini par (voir {2}):*

$$\chi(t,t_o) = \frac{E_o}{E_o - r(t,t_o)} - \frac{1}{\varphi(t,t_o)} \tag{8}$$

$r(t,t_o)$ *représentant la fonction relaxation.*
Cette affinité facilite le calcul du coefficient de vieillissement et permet sa tabularisation.

Il existe en outre des formules simplifiées permettant une estimation du coefficient χ.

L'éq.(6) nous permet de modifier le système d'éqs.(1) afin de tenir compte des effets différés:
En expriment $\varepsilon(t)$, σ_o et $\Delta\sigma(t)$ en fonction des hyperstatiques, nous obtenons, après quelques transformations élémentaires, le sys-

tème d'équations suivant:

$$[\alpha_{ik}+a^{**}_{ik}]\{X_{kt}\} + \{a^{*}_{iq}\} = [\alpha_{ik}]\{X_{kEL}\} - [a^{*}_{ik}-a^{**}_{ik}]\{X_{ko}\} \tag{9a}$$

avec

$X_{kt} = X_{ko} + \Delta X_k$ - $k^{ième}$ hyperstatique au temps t,

X_{ko} - valeur de X_{kt} au temps $t = t_o$,

ΔX_k - augmentation de X_{ko} dans l'intervalle de temps $t - t_o$,

$$a^{*}_{ik} = \int \varphi \frac{M_i M_k}{EI} dx \tag{10a}$$

$$a^{**}_{ik} = \int \chi \cdot \varphi \frac{M_i M_k}{EI} dx \tag{10b}$$

$$a^{*}_{iq} = \int \varphi \frac{M_i M_q}{EI} dx \tag{10c}$$

Le système d'éqs.(9a) est le système général d'équations algébriques permettant de traiter les problèmes des effets différés pour des structures linéaires formées d'éléments viscoélastiques et purement élastiques dans le cadre des hypothèses susmentionnées.

Traitons ici deux cas particuliers:

A. Si les propriétés viscoélastiques des éléments viscoélastiques sont identiques:

$a^{*}_{ik} = \varphi . a_{ik\varphi}$

$a^{**}_{ik} = \chi\varphi . a_{ik\varphi}$

$a^{*}_{iq} = \varphi . a_{iq\varphi}$

et si, de plus, $c_r = 0$ pour $r = 1 \div m$:

$a_{ik} = a_{ik\varphi}$

$a_{iq} = a_{iq\varphi}$.

En introduisant

$\{a_{iq}\} = -[\alpha_{ik}]\{X_{kELq}\}$

où

X_{kELq} - $k^{ième}$ hyperstatique due aux charges q sans prise en considération des effets différés, le système d'éqs.(9a) se simplifie et s'écrit:

$$[\alpha_{ik}+\chi\varphi a_{ik}]\{X_{kt}\} = [\alpha_{ik}]\{X_{kEL}+\varphi X_{kELq}\} - (\varphi-\chi\varphi)[a_{ik}]\{X_{ko}\} \quad (9b)$$

B. Si, en outre, les coefficients de rigidité

$$[\xi_{ik}] = [\alpha_{ik}]^{-1}[a_{ik}]$$

forment une matrice diagonale, c'est-à-dire

$$[\xi_{ik}] = [\delta_{ik}\xi_k]$$

(par exemple, pour une structure formée uniquement d'éléments viscoélastiques, $\xi_k = 1$ pour $k = 1 \div n$), les équations du système (9b) deviennent indépendantes:

$$(1+\chi\varphi\xi_k)X_{kt} = X_{kEL} + \varphi X_{kELq} - (\varphi-\chi\varphi)\xi_k X_{ko} \quad (9c)$$

3. APPLICATIONS

Appliquons les systèmes d'éqs.(9) à la résolution de quelques problèmes types:

A. Modification du système statique

La modification du système statique, entraînée par une ou plusieurs liaisons effectuées ultérieurement, provoque la redistribution des efforts par suite du fluage.

En introduisant dans les systèmes d'éqs.(9) les conditions

$$X_{kEL} = X_{ko}$$

nous obtenons pour les hyperstatiques X_{kt} au temps t:

éqs.(9a): $[\alpha_{ik}+a^{**}_{ik}]\{X_{kt}\} + \{a^{*}_{iq}\} = [\alpha_{ik}-a^{*}_{ik}+a^{**}_{ik}]\{X_{ko}\}$

éqs.(9b): $[\alpha_{ik}+\chi\varphi a_{ik}]\{X_{kt}\} = \varphi[\alpha_{ik}]\{X_{kELq}\} + [\alpha_{ik}-(\varphi-\chi\varphi)a_{ik}]\{X_{ko}\}$

éqs.(9c): $X_{kt} = X_{ko} + \dfrac{\varphi}{1+\chi\varphi\xi_k}(X_{kELq}-\xi_k X_{ko})$

B. Effet des charges agissant sur le système définitif

Même dans le système définitif, le fluage peut provoquer une redistribution des efforts.
En introduisant dans les systèmes d'éqs.(9) les conditions

$$\boxed{X_{kEL} = X_{ko} = X_{kELq}}$$

nous obtenons:

éqs.(9a): $[\alpha_{ik}+a^{**}_{ik}]\{X_{kt}\} + \{a^{*}_{iq}\} = [\alpha_{ik}-a^{*}_{ik}+a^{**}_{ik}]\{X_{kEL}\}$

éqs.(9b): $[\alpha_{ik}+\chi\varphi a_{ik}]\{X_{kt}\} = [(1+\varphi)\alpha_{ik} - (\varphi-\chi\varphi)a_{ik}]\{X_{kEL}\}$

éqs.(9c): $X_{kt} = (1 + \dfrac{\varphi(1-\xi_k)}{1+\chi\varphi\xi_k})X_{kEL}$

C. Déformations imposées instantanées

Ce problème, représentant souvent la relaxation, est lié aux conditions

$$\boxed{\begin{array}{l} X_{kEL} = X_{ko} \\ a^{*}_{iq} = X_{kELq} = 0 \end{array}}$$

ce qui donne:

éqs.(9a): $[\alpha_{ik}+a^{**}_{ik}]\{X_{kt}\} = [\alpha_{ik}-a^{*}_{ik}+a^{**}_{ik}]\{X_{kEL}\}$

éqs.(9b): $[\alpha_{ik}+\chi\varphi a_{ik}]\{X_{kt}\} = [\alpha_{ik} - (\varphi-\chi\varphi)a_{ik}]\{X_{kEL}\}$

éqs.(9c): $X_{kt} = (1 - \dfrac{\varphi\xi_k}{1+\chi\varphi\xi_k})X_{kEL}$

D. Déformations imposées lentes

Pour un développement quelconque des déformations imposées, le coefficient de vieillissement χ est dépendant de l'histoire des contraintes et doit être déterminé séparément pour chaque cas.

Un développement des déformations affine à celui du coefficient de fluage φ permet l'utilisation de diagrammes préétablis (voir {3}, {4}).

Les conditions

$$X_{ko} = 0$$

$$a^*_{iq} = X_{kELq} = 0$$

introduites dans les systèmes d'éqs.(9) donnent:

éqs.(9a): $[\alpha_{ik}+a^{**}_{ik}]\{X_{kt}\} = [\alpha_{ik}]\{X_{kEL}\}$

éqs.(9b): $[\alpha_{ik}+\chi\varphi a_{ik}]\{X_{kt}\} = [\alpha_{ik}]\{X_{kEL}\}$

éqs.(9c): $X_{kt} = \frac{1}{1+\chi\varphi\xi_k} X_{kEL}$

Références:

1. *H.Trost,*
   Auswirkungen des Superpositionsprinzips auf Kriech- und Relaxationsprobleme bei Beton und Spannbeton.
   Beton- und Stahlbetonbau. 1967.

2. *Z.P.Bazant,*
   Prediction of concrete creep effects using age-adjusted effective modulus method.
   Journal of the American Concrete Institute. 1972.

3. *R.Favre, M.Koprna, A.Radojicic,*
   Effets différés, fissuration et déformation.
   Ecole Polytechnique fédérale de Lausanne. 1980.
   Editions Georgi, Saint-Saphorin.

4. *CEB, Manuel de calcul,*
   Effets structuraux du fluage et des déformations différées. 1980.
   Bulletin d'information No 136.

COMPORTEMENT D'UNE STRUCTURE CONTINUE REALISEE A L'AIDE D'ELEMENTS PREFABRIQUES

René J. Suter, Dr.
Institut de statique et structures
Ecole Polytechnique Fédérale de Lausanne, Suisse

ABSTRACT

A continuous structure realised from prefabricated elements by means of connections over the supports shows up two particularities:
- the transformation of static system resulting in a redistribution of forces
- the presence of joints which diminish the continuity and the transmission of forces over the supports.

RESUME

Une structure continue, réalisée par la liaison sur appuis entre des éléments préfabriqués, présente deux particularités:
- la transformation du système statique ayant pour conséquence une redistribution des efforts
- la présence de joints qui diminuent la continuité et la transmission des efforts sur appuis.

ZUSAMMENFASSUNG

Zwei Faktoren beeinflussen das Tragverhalten eines Durchlaufträgers, der aus vorgefertigten, über den Auflagern verbundenen Einfeldträgern hergestellt wurde:
- die Aenderung des statischen Systems bewirkt eine Schnittkraftumlagerung im Laufe der Zeit
- die Verbindungen beeinträchtigen die Durchlaufwirkung über den Auflagern.

Une série de n éléments préfabriqués en béton armé est mise en place sous forme de poutres simples, librement appuyées, et assemblées ultérieurement (figure 1). Le système initialement isostatique se transforme, après la réalisation des liaisons, en un système hyperstatique, - poutre continue à n travées -.

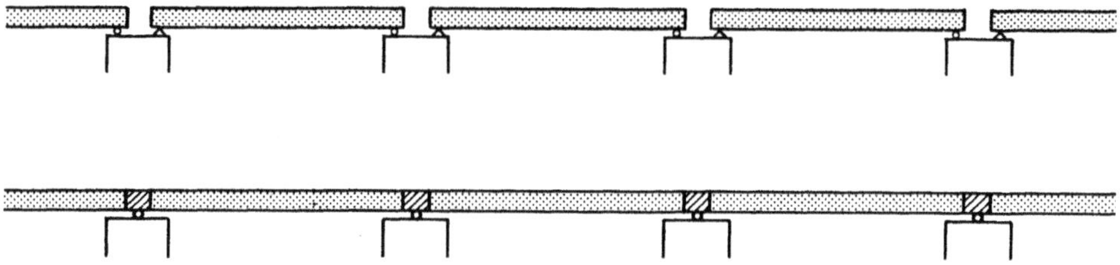

FIG 1

Principe de construction

Une telle construction se distingue d'une structure monolithique par:

- un procédé de construction particulier
- la présence de joints sur appuis.

Ces deux particularités ont une influence capitale sur la répartition des efforts le long de la structure.

Considérons en premier les effets du procédé de construction. Par suite des déformations dues au fluage des éléments sous l'effet de leur poids propre, un moment se crée sur les appuis intermédiaires. Ce phénomène s'explique par des déformations empêchées par la liaison, provoquant ainsi, dans un système hyperstatique, des sollicitations supplémentaires.

La difficulté qu'introduit le changement du système statique peut facilement être résolu à l'aide d'une transformation du mode de charge, illustré pour une poutre à deux travées (figure 2). La répartition des moments au temps $t = t_o$ peut être imaginée comme une superposition de deux cas de charge sur un système continu:

- la surcharge répartie g
- un abaissement de l'appui médian, compensant le moment négatif sur appui.

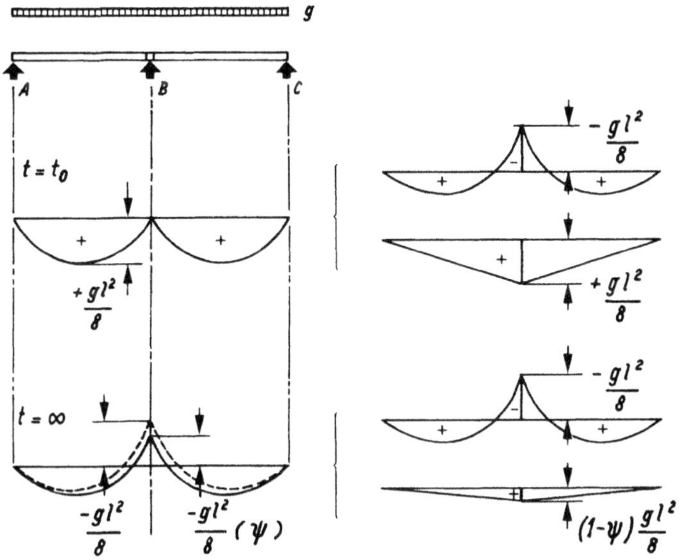

FIG 2

Transformation du système statique (deux travées)

Dans le temps, la répartition des moments dus à la charge ne varie pas, tandis que les moments engendrés par une déformation imposée sont réduits en fonction des propriétés viscoélastiques du béton.

Au temps $t > t_o$, le moment sur appui vaut:

$$X_B(t) = X_B^{mo}.\psi(t,t_o)$$

$\psi(t,t_o)$: coefficient de relaxation au temps t, partant du temps t_o

Différents auteurs se sont déjà préoccupés de ce problème et de l'évolution dans le temps des efforts sur appuis. Ainsi, Dischinger a proposé, en 1938 déjà:

$$X_B(t) = X_B^{mo} . (1 - e^{-\varphi})$$

Selon Rüsch, le moment vaut:

$$X_B(t) = X_B^{mo} . (1 - c_r)$$

$c_r(\varphi)$: coefficient décrivant l'influence du fluage pour une déformation imposée instantanée

En introduisant un coefficient de vieillissement ρ selon Trost, on obtient:

$$X_B (t) = X_B^{mo} . \frac{\varphi}{1 + \rho\varphi}$$

Le coefficient de vieillissement a été repris par Bazant, le CEB et Favre sous le symbole de χ. Sa présentation plus différenciée permet de mieux tenir compte de différentes caractéristiques de béton et de différents âges de mise en charge.

Il ressort néanmoins de ces théories que le moment qui se crée sur appui, par suite du fluage des éléments, est relativement élevé et voisin du moment total de continuité d'une poutre monolithique.

Les considérations théoriques ci-avant sur la redistribution des efforts se basent sur des systèmes statiques idéalisés, passant d'appuis simples à des liaisons parfaites. C'est en effet cette rigidité absolue des liaisons qui explique un accroissement aussi élevé du moment négatif sur appui. Toutefois, des liaisons rigides ne sont exécutées que pour des constructions d'une certaine importance, par exemple des ponts, et souvent au moyen d'une précontrainte.

Pour des bâtiments ou des constructions industrielles, où les portées sont courtes ou moyennes et les épaisseurs des éléments relativement faibles, la réalisation d'une précontrainte sur appuis n'est généralement pas envisagée, pour des raisons économiques et constructives. Il en résulte une certaine souplesse de la liaison introduisant une discontinuité dans le comportement d'ensemble de la structure continue.

En effet, les imperfections constructives des liaisons influencent les phénomènes de redistribution mentionnés et les moments qui se créent sur appuis sont diminués par rapport aux valeurs théoriques.

Bien que l'influence des liaisons constitue la partie essentielle et originale de ce projet de recherche [1], nous n'allongerons pas dans ce contexte les considérations théoriques y relatives. On se concentre plutôt sur le thème principal de cette session, c'est-à-dire sur les répercussions du fluage sur les structures en béton.

Dans cette optique est présentée brièvement l'étude expérimentale réalisée au Centre d'Etude du Béton Armé et Précontraint [2]. Afin de tenir compte des effets contradictoires causés par les différents phénomènes, cette étude expérimentale a été subdivisée en deux parties (figure 3):

- une première partie, qui a pour but la détermination des caractéristiques mécaniques de la liaison sous charges de courte durée;
- une seconde partie, dans laquelle est examiné le comportement d'ensemble d'une telle structure sous charge de longue durée, compte tenu de la redistribution des efforts par suite du fluage.

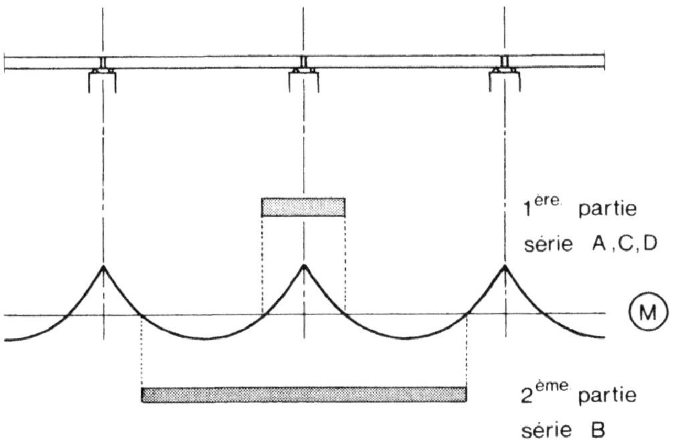

FIG 3

Principe des études expérimentales

Nous nous limitons à quelques explications concernant la seconde partie, où l'on a procédé à deux essais sur des dalles unidirectionnelles et continues à deux travées.

FIG 4

Vue générale des essais: série B

Pour chaque essai, trois dalles identiques de 3.75 m environ de longueur, de 0.80 m de largeur et de 0.14 m d'épaisseur ont été bétonnées et placées, à l'âge de 14 jours, sur leurs appuis en tant que poutres simples. A l'âge de 28 jours, deux de ces dalles ont été reliées, pendant que la troisième restait en tant que dalle de référence (figure 4). Durant une année, on a ensuite observé les variations de la réaction médiane, les flèches et les déformations relatives dans les sections déterminantes.

Les deux montages d'essai, B1 et B2, se distinguent uniquement par la réalisation de la liaison:

- pour B1, on a essayé de simuler le cas théorique d'une liaison rigide par soudure des armatures contre un profilé métallique (figure 5);
- pour B2, on a réalisé une liaison à l'aide de boucles d'armatures dans un joint bétonné.

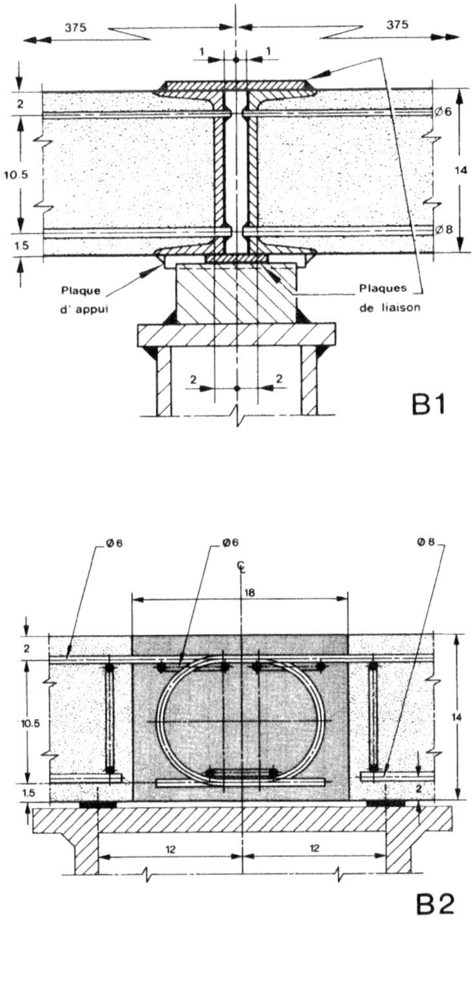

FIG 5

Détails des liaisons sur appuis: montages B1 et B2

A la figure 6 est représentée l'évolution des moments sur l'appui médian, ou plus précisément le moment observé par rapport au moment d'appui d'une poutre continue monolithique sur deux travées ($gl^2/8$). En tant que comparaison est également indiquée la courbure théorique de relaxation $\psi(t,t_o)$, obtenue à partir des valeurs de $\varphi(t,t_o)$ mesurées sur échantillons et sur la dalle de référence.

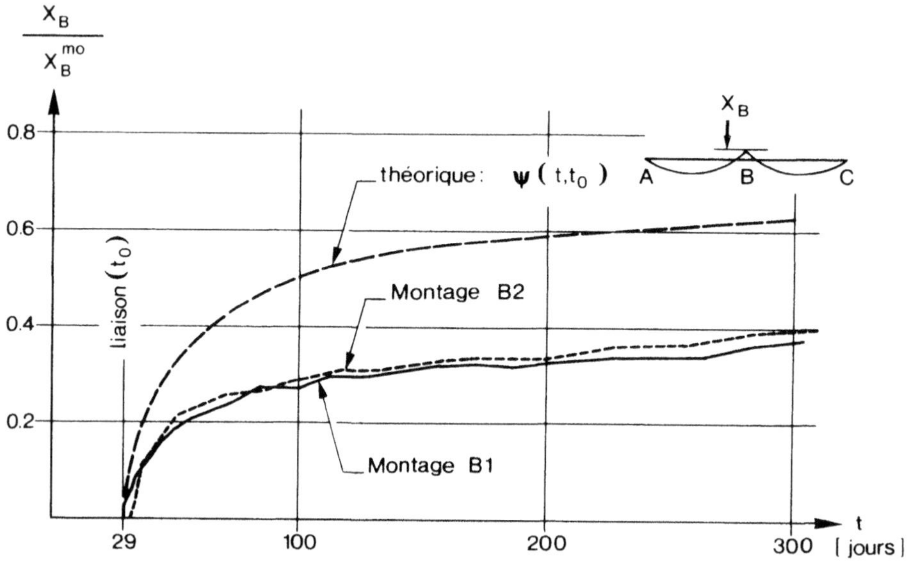

FIG 6

Evolution des moments sur l'appui médian

L'analyse des courbes donne lieu à deux constatations:

- la redistribution des efforts pour les deux montages B1 et B2 est pratiquement identique;
- les valeurs observées correspondent aux 60% environ des valeurs théoriques.

Cela signifie que la réalisation d'une liaison rigide pour le montage B1 n'a pas réussi, malgré le soin que l'on a apporté à l'exécution de celle-ci. La discontinuité qui se fait généralement au droit des joints de bétonnage, s'est en effet produite juste derrière les profilés métalliques où l'adhérence avec le béton peut pratiquement être admise comme nulle.

Une fois le moment effectif $X_B(t=300j)$ connu, on détermine facilement la répartition des moments le long des dalles assemblées (figure 7). A titre de comparaison, on a également reporté la répartition théorique pour une liaison absolument rigide.

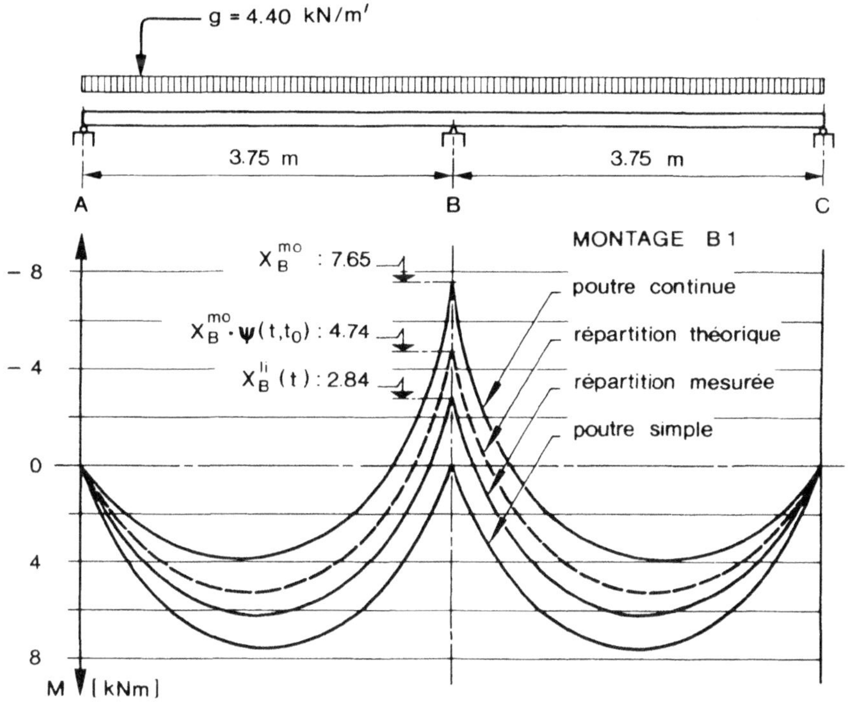

FIG 7

Répartition des moments au temps t = 300 j

Il est très intéressant aussi d'étudier la déformée des dalles à l'âge de 300 jours à partir de la liaison, représentée à la figure 8. On y remarque bien l'angle de discontinuité γ sur l'appui médian de la dalle assemblée qui explique la différence entre les valeurs des moments observés et théoriques.

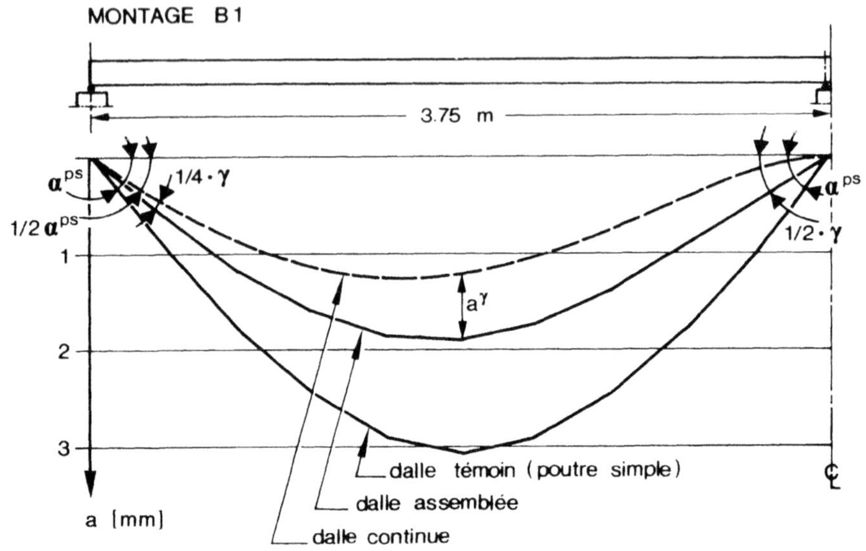

FIG 8

Allure de la déformée au temps t = 300 j

On donne encore brièvement, à l'aide de l'exemple du montage B2, l'introduction du fluage et de la discontinuité sur appui dans le calcul statique. Dans ce but, on établit les équations algébriques de compatibilité des angles sur l'appui médian, pour des charges permanentes appliquées avant la réalisation des joints (dans ce cas le poids propre des éléments).

Si la liaison est parfaitement rigide dès sa réalisation, on a:

$$\alpha(t) = \alpha_o . \varphi(t,t_o) + \alpha_1 . \Delta X_B(t) . [1+\chi(t,t_o) . \varphi(t,t_o)] = 0$$

$$\Delta X_B(t) = \frac{-\alpha_o . \varphi(t,t_o)}{\alpha_1 . [1+\chi(t,t_o) . \varphi(t,t_o)]} = X_B^{mo} . \frac{\varphi(t,t_o)}{1+\chi(t,t_o) . \varphi(t,t_o)}$$

α_o: angle dû aux charges

α_1 : angle dû à un moment unitaire $X_B = 1$

Si la liaison n'est pas parfaite, on introduit deux termes supplémentaires:

- un facteur c^{II}, proportionnel au moment sur appui pour tenir compte d'une discontinuité mécanique;
- un angle $\gamma(\varepsilon_s)$ qui tient compte d'un éventuel retrait du béton à l'intérieur de la liaison.

On obtient ainsi:

$$\alpha(t) = \alpha_o . \varphi(t,t_o) + \alpha_1 . \Delta X_B(t) . [1+\chi(t,t_o). \varphi(t,t_o)] +$$

$$+ c^{II}. \Delta X_B(t) - \gamma(\varepsilon_s) = 0$$

$$\Delta X_B(t) = - \frac{\alpha_o . \varphi(t,t_o) - \gamma(\varepsilon_s)}{1+\chi(t,t_o). \varphi(t,t_o)+ c^{II}}$$

Il faut encore préciser que le ressort c^{II} finalement retenu dans ce travail de recherche constitue un ressort purement élastique, mais valable pour des charges et surcharges de courte et de longue durée. Toute viscoélasticité ou plasticité de la liaison a pu ainsi être éliminée.

Pour l'exemple de l'essai B2, la concordance entre valeurs observées et calculées est bonne.

REFERENCES

1. R. Suter, "Influence des liaisons sur le comportement d'une structure continue réalisée à l'aide d'éléments préfabriqués", Thèse de doctorat, EPFL 1980.
2. R. Favre, R. Suter, J.-P. Jaccoud, "Comportement d'une structure continue réalisée à l'aide d'éléments préfabriqués", EPFL-CEBAP, Rapports d'essai C141-04 (1977), C141-05 (1978), C141-06 (1980).
3. CEB, "Code-modèle CEB-FIP pour les structures en béton", Bulletin d'information No 124-125/1978.
4. CEB, "Manuel de calcul: Effets structuraux du fluage et des déformations différés", Bulletin d'information No 136/1980.
5. R. Favre, M. Koprna, A. Radojicic, "Effets différés, fissuration et déformations des structures en béton", EPFL, Editions GEORGI, Saint-Saphorin, 1980.

DEFORMATION DES DALLES EN BETON
METHODE BILINEAIRE POUR LE CALCUL DES FLECHES

Jean-Paul Jaccoud
ingénieur à
l'Insitut de statique et structures -
Béton Armé et Précontraint (IBAP)
de l'Ecole Polytechnique Fédérale de Lausanne, Suisse

ABSTRACT

A bilinear model is proposed for the behaviour at the serviceability state of concrete slabs, taking into account cracking and long-term effects. It permits the calculation of the intermediate value of the probable deflection situated between the extreme values in the uncracked and cracked states. The comparison with the results of long-term experiments is satisfactory.

RESUME

Un modèle bilinéaire est proposé pour le comportement à l'état de service des dalles en béton, compte tenu de la fissuration et des effets différés. Il permet de calculer la valeur intermédiaire de la flèche probable située entre les valeurs extrêmes en stades homogène et fissuré. La comparaison avec les résultats d'essais de longue durée est satisfaisante.

ZUSAMMENFASSUNG

Um das Verhalten von Massivplatten im Gebrauchszustand zu erfassen, wird ein bilineares Modell abgeleitet, wobei Rissebildung sowie die Einflüsse von Kriechen und Schwinden berücksichtigt werden. Dieses Modell erlaubt die Berechnung der wahrscheinlichen Durchbiegung als Zwischenwert innerhalb der zwei Extremwerte des homogenen und des völlig gerissenen Zustandes. Die Uebereinstimmung mit gemessenen Werten aus Langzeitversuchen ist gut.

1. INTRODUCTION

Le problème des déformations des structures en béton armé et précontraint préoccupe les ingénieurs depuis de nombreuses années [1, 2, 3]. C'est une évidence de dire que les déformations des structures en béton dépendent d'une multitude de paramètres, souvent difficiles à maîtriser lors du calcul, comme les propriétés mécaniques et rhéologiques du béton, les conditions d'ambiance, la cure et l'histoire de la construciton. Les déformations sont fortement accrues lors de la fissuration du béton et elles augmentent dans le temps à cause du fluage et du retrait (FIG. 1).

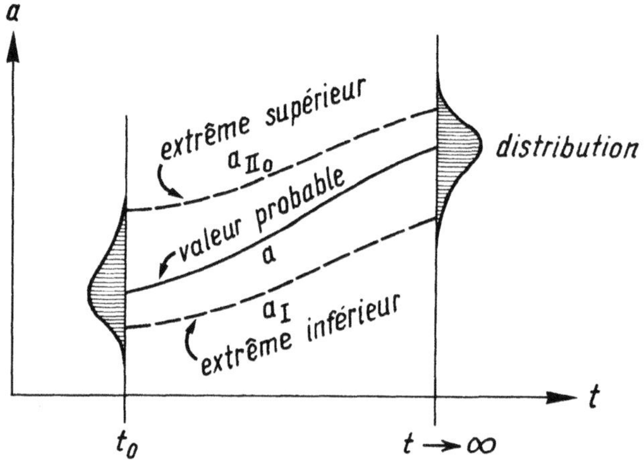

FIG. 1
Représentation schématique de l'évolution des flèches dans le temps

Une part importante des dégâts constatés dans les constructions en service est due à des déformations excessives. De plus, ce problème a paradoxalement pris davantage d'importance ces dernières années à cause des progrès réalisés tant dans les matériaux et techniques de construction que de ceux réalisés dans les concepts et méthodes de calcul, qui permettent un dimensionnement plus sûr et plus économique, ceci malheureusement souvent au détriment de la rigidité des éléments porteurs. Des efforts considérables ont été entrepris sur le plan international afin d'améliorer les connaissances sur le comportement des ouvrages à l'état d'utilisation d'une part, et dans le but de mettre à disposition des ingénieurs des normes ainsi que des méthodes pratiques de calcul, d'autre part [4, 6, 7, 8]. Notre institut et plus particulièrement son directeur, le professeur Renaud Favre, sont étroitement associés à ces efforts. Le but du présent est de donner un bref aperçu des travaux théoriques et expérimentaux entrepris dans ce sens à notre institut.

2. METHODE DE CALCUL

La méthode bilinéaire découle, moyennant certaines simplifications, de la méthode générale de calcul des flèches par intégration des courbures décrite dans [7]. La méthode bilinéaire permet d'estimer la valeur intermédiaire de la flèche probable a entre les valeurs extrêmes a_I et a_{IIo} (FIG. 1 et 2), calculées respectivement en stade I (béton homogène; c = 0) et en stade II-nu (fissuré, sans participation du béton tendu: c = 1.0). Elle découle également de la constatation empirique que la relation flèche-moment peut être approchée à l'état de service par une fonction bilinéaire (FIG. 2).

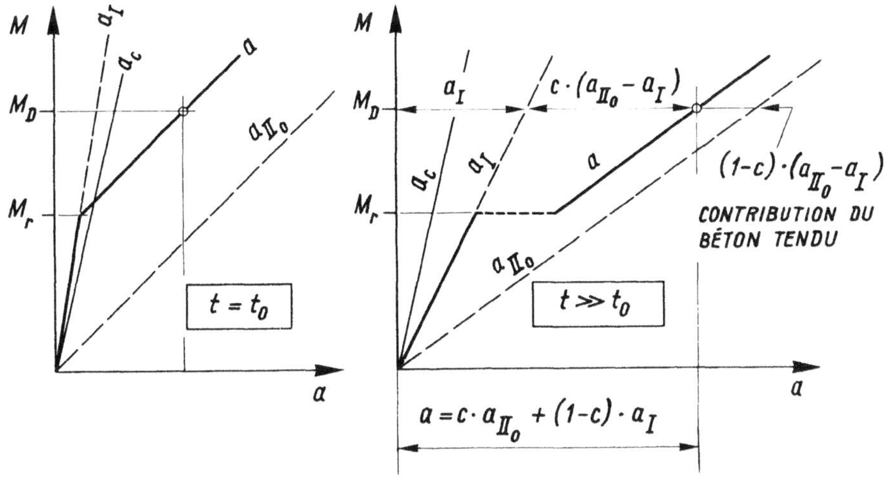

FIG. 2 Approximation bilinéaire de la relation flèche-moment

Le principe de la méthode réside dans l'évaluation des flèches extrêmes a_I et a_{IIo} en corrigeant la flèche de base a_c (calculée élastiquement en admettant le béton homogène) afin de tenir compte de la fissuration, du fluage et du retrait du béton. Ces corrections sont effectuées au moyen de coefficients k_A, respectivement k_φ et k_s calculés uniquement dans une section déterminante. La méthode est exposée de manière plus détaillée dans la littérature et l'on y trouvera notamment des aides et abaques donnant les valeurs numériques des coefficients de correction [4, 6, 7].

$$a = a_I + c.(a_{IIo} - a_I)$$

où

$$a_I = \int \frac{M}{EI_I} .\bar{M}.dx$$

$$\cong \underbrace{k_A^I.a_c}_{\text{partie instantanée}} + \underbrace{k_A^I.k_\varphi^I.\varphi.a_{cg}}_{\text{partie due au fluage}} + \underbrace{\alpha_s.k_s^I.|\varepsilon_s|.\frac{\ell^2}{d}}_{\text{partie due au retrait}}$$

$$a_{IIo} = \int \frac{M}{EI_{IIo}}.\bar{M}.dx$$

$$\cong k_A^{II}.a_c + k_A^{II}.k_\varphi^{II}.\varphi.a_{cg} + \alpha_s.k_s^{II}.|\varepsilon_s|.\frac{\ell^2}{d}$$

avec

a - flèche probable;

a_c - flèche de base, résultant d'un calcul élastique avec $EI = E_c I_c$;

a_{cg} - flèche de base, due à la fraction de la charge agissant de manière permanente;

$c = 0$ pour $M_D \leqslant M_r$ (stade I),

$c = 1 - \beta_1.\beta_2.\frac{M_r}{M_D}$ pour $M_D > M_r$;

M_r - moment de fissuration dans la section déterminante:

$M_r = W_c.\ (f_{ct} - \frac{N}{A_c})$;

f_{ct} - résistance à la traction du béton;

M_D - moment total dans la section déterminante;

β_1 - coefficient caractérisant les qualités d'adhérence des barres d'armature:

β_1 = 1.0 pour les barres à adhérence augmentée,
= 0.5 pour les barres lisses, selon [5, 7];

β_2 - coefficient représentant la durée d'application et la répétition des charges:

β_2 = 1.0 au premier chargement,
= 0.5 pour les charges de longue durée ou répétées, selon [5, 7];

La section déterminante est la section où est calculée la flèche en travée, ou la section d'encastrement pour une console.

3. COMPARAISON AVEC LES ESSAIS

Dès la fin 1977, trois séries expérimentales de longue durée ont été réalisées:

La série A, comprenant 5 dalles unidirectionnelles identiques (FIG. 3), pour lesquelles le paramètre variable fut les conditions d'ambiance lors du bétonnage et les conditions de cure (TABLEAU 1). Celles-ci ont joué un rôle considérable sur les flèches et fissures observées, le niveau de charge ayant été intentionnellement fixé au voisinage du moment de fissuration (FIG. 3).

La série B, comprenant 3 dalles carrées simplement appuyées sur leur pourtour et avec les angles ancrés (FIG. 4); le paramètre variable fut, d'une part, le niveau de charge et, d'autre part, la distribution de l'armature. Pour le niveau de charge le plus élevé - correspondant à 60% de la charge ultime - deux dispositions furent réalisées: armature isotropre et armature orthotrope. Les essais ont montré un comportement identique en ce qui concerne les flèches, dicté par la moyenne de l'armature dans les deux directions perpendiculaires.

La série C, comprenant 7 dalles unidirectionnelles identiques (FIG. 5), pour lesquelles le paramètre variable est le niveau de charge variant de 20 à 60% de la charge ultime. Les mesures relatives à cette dernière série ne sont pas encore terminées. Cependant, elles permettent déjà de vérifier l'excellente concordance entre les valeurs mesurées et celles calculées par la méthode bilinéaire, ceci aussi bien pour les charges de courte durée (FIG. 6) que pour les charges de longue durée (FIG. 7) pour lesquelles la précision est encore meilleure.

Les résultats indiqués sur les FIGURES 8 à 11 permettent de porter un jugement plus général sur la précision de la méthode bilinéaire pour le calcul des flèches. Pour chacune des dalles d'essai, on a reporté le rapport entre la flèche mesurée et celle calculée à partir des caractéristiques mécaniques et rhéologiques du béton, mesurées sur éprouvettes ayant subi la même cure que les dalles; et ceci en fonction de l'âge réel $t-t_o$ compté à partir de l'instant t_o de la mise en charge.

On constate que l'écart entre les valeurs mesurées et celles calculées est le plus élevé pour les dalles chargées au voisinage du moment de fissuration (toutes les dalles de la série A, dalles B1, C12 et C22). Pour ces dalles, les flèches dépendent fortement des conditions d'ambiance lors du bétonnage et de la cure (TABLEAU 1). Pour autant cependant que l'on connaisse ces conditions et les caractéristiques des matériaux qui en découlent, le modèle de calcul proposé permet d'en tenir compte de manière satisfaisante.

Cet écart diminue quelque peu pour les âges élevés (FIG. 10 et 11). Cela est dû à l'apparition parfois tardive et progressive de la fissuration (après plusieurs jours à quelques mois), phénomène dont il n'est qu'imparfaitement tenu compte dans le modèle de calcul au moyen du coefficient β_2.

En conclusion, on peut dire que la précision de la méthode bilinéaire de calcul des flèches est de l'ordre de 15 à 20% pour des âges élevés, exceptés les éléments sollicités au voisinage du moment de fissuration où elle peut n'être plus que de 30 à 40%.

4. EXEMPLE NUMERIQUE

Illustrons la méthode bilinéaire au moyen d'une application numérique, à savoir le calcul de la flèche de la dalle C14 à l'âge $t-t_o = 6$ mois. Toutes les indications nécessaires sont données aux FIGURES 5 et 10 ainsi qu'au TABLEAU 1.

La flèche de base résultant d'un calcul élastique avec

$$E_c I_c = 28{,}27.10^9 .\ 0{,}000256 = 7{,}237.10^6 \text{ Nm}^2 \text{ vaut:}$$

$$a_c = a_c(g) + a_c(P) = 2{,}28 \text{ mm}$$

Les valeurs extrêmes de la flèche peuvent être calculées à partir des coefficients de correction k tirés de [6] et avec $\alpha_s = 1/8$ pour une poutre simple:

$$a_I = k_A^I.(1+k_\varphi^I.\varphi).\ a_c + \alpha_s.\ k_s^I.\ |\varepsilon_s|.\frac{\ell^2}{d}$$

$$= 0.955.(1+o,893.2,17).2,28 + \frac{1}{8}.0,218.362.10^{-6}.\frac{3100^2}{131}$$

$$= 7,13\ \text{mm}$$

$$a_{IIo} = k_A^{II}.(1+k_\varphi^{II}.\varphi).a_c + \alpha_s.k_s^{II}.|\varepsilon_s|.\frac{\ell^2}{d}$$

$$= 5,15.(1+0,159.2,17).2,28 + \frac{1}{8}.\ 1,086.362.10^{-6}.\frac{3100^2}{131}$$

$$= 19,40\ \text{mm}$$

Le moment de fissuration et le moment total dans la section déterminante valent respectivement:

$$M_r = W_c.\ f_{ct} = 0,0032\ .\ 3,04.\ 10^6 = 9'728\ \text{Nm}$$

$$M = M(g) + M(P) = 16'035\ \text{Nm}$$

D'où, avec $\beta_1 = 1.0$ et $\beta_2 = 0.5$, $c = 1-\beta_1.\beta_2.\ \frac{M_r}{M} = 0,697$ et la valeur probable de la flèche:

$$a = 7,13 + 0,697.\ (19,40 - 7,13) = \underline{\underline{15,7\ \text{mm}}}$$

REFERENCES

1. H. Mayer, "Die Berechnung der Durchbiegung von Stahlbetonteilen", DAfSt, Heft 194, 1967.

2. CEB, "Calcul des flèches", Bulletin d'information No 90, 1973.

3. U. Kraemer, G. Thielen und E. Grasser, "Berechnung der Durchbiegung von biegebeanspruchten Stahlbetonbauteilen unter Gebrauchslast", Beton- und Stahlbetonbau 4/1975.

4. E. Grasser und G. Thielen, "Hilfsmittel zur Berechnung der Schnittgrössen und Formänderungen von Stahlbetontragwerken". DAfSt, Heft 240, 2. Auflage 1978.

5. CEB, "Code-Modèle CEB-FIP pour les structures en béton", Bulletin d'information No 124/125F, 1978.

6. R. Favre, M. Koprna et A. Radojicic, "Effets différés, fissuration et déformations des structures en béton", Ecole Polytechnique Fédérale de Lausanne, Editions GEORGI, Saint-Saphorin (Suisse), 1980.
7. CEB, commission V serviciabilité, "Manuel déformations", en préparation.
8. SIA, commission 162, "Groupe de travail 2: déformations", nouvelle norme SIA 162, en préparation.

NOTE:

Toutes les caractéristiques mécaniques et rhéologiques du béton indiquées aux FIG. 8 à 11 ainsi qu'au TAB. 1 ont été mesurées sur éprouvettes prismatiques de 120 x 120 x 360 mm ayant subi la même cure que les différentes dalles. La signification des symboles utilisés est la suivante:

+ conditions de bétonnage et de cure favorables,
0 conditions de bétonnage et de cure normales,
- conditions de bétonnage et de cure défavorables,
G protection durant la cure au moyen d'une natte Guritherm;

$t_{o,corr} = \alpha . \sum_i \frac{T_i + 10}{30} . \Delta t_i$ âge corrigé, avec pour ciment portland utilisé (CPN): $\alpha = 1,5$;

β_{wm} résistance moyenne à la compression sur cube (½ prisme);

f_{ctm} résistance moyenne à la traction (= ½ résistance à la traction par flexion mesurée sur prisme);

E_{cm} module de déformation initiale, mesuré 5 minutes après la mise en charge.

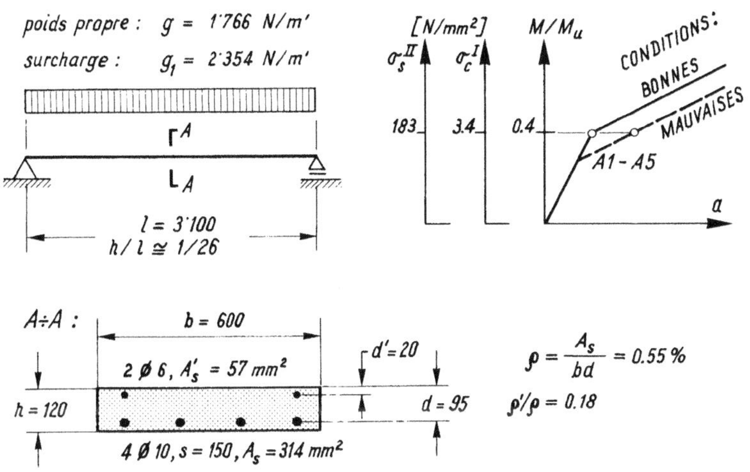

FIG. 3 Dalles d'essais série A.

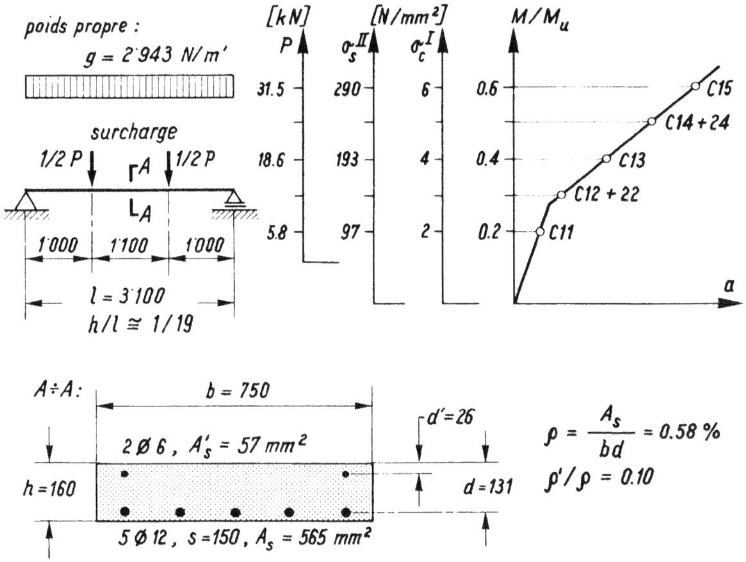

FIG. 5 Dalles d'essais série C.

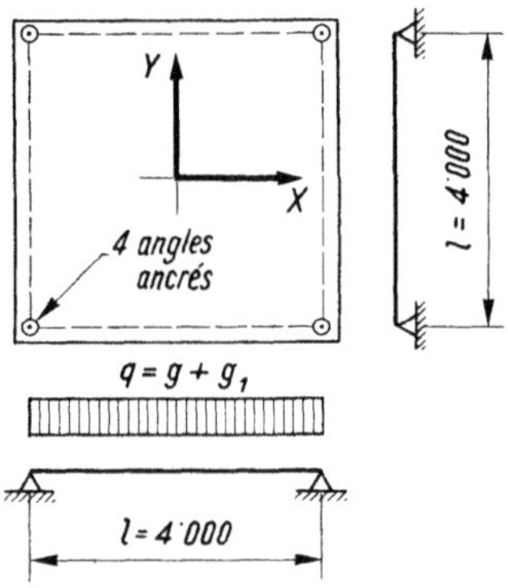

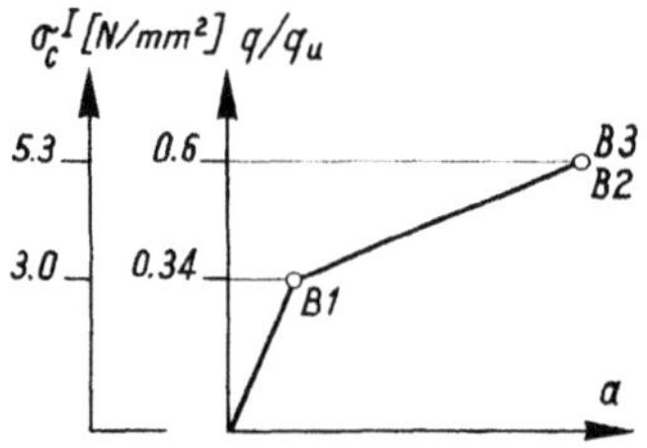

épaisseur : $h = 120$ mm

$h/l \cong 1/33$

poids propre : $g = 2'943$ N/m²

dans le champ : $\rho'/\rho = 0$

dans les angles (ancrés) : $\rho' = \rho$

	DALLE N°	B1	B2	B3
	g_1 [N/m²]	7'907	15'814	15'814
Direction X	Ø [mm]	10	10	12
	s [mm]	150	150	150
	d [mm]	100	100	99
	ρ [%]	0.52	0.52	0.76
	σ_s^{II} [N/mm²]	155	267	190
Direction Y	Ø [mm]	10	10	8
	s [mm]	150	150	150
	d [mm]	90	90	89
	ρ [%]	0.58	0.58	0.38
	σ_s^{II} [N/mm²]	172	297	456 !
$\bar{\rho} = 1/2(\rho_x + \rho_y)$ [%]		0.55	0.55	0.57

FIG. 4 Dalles d'essais série B.

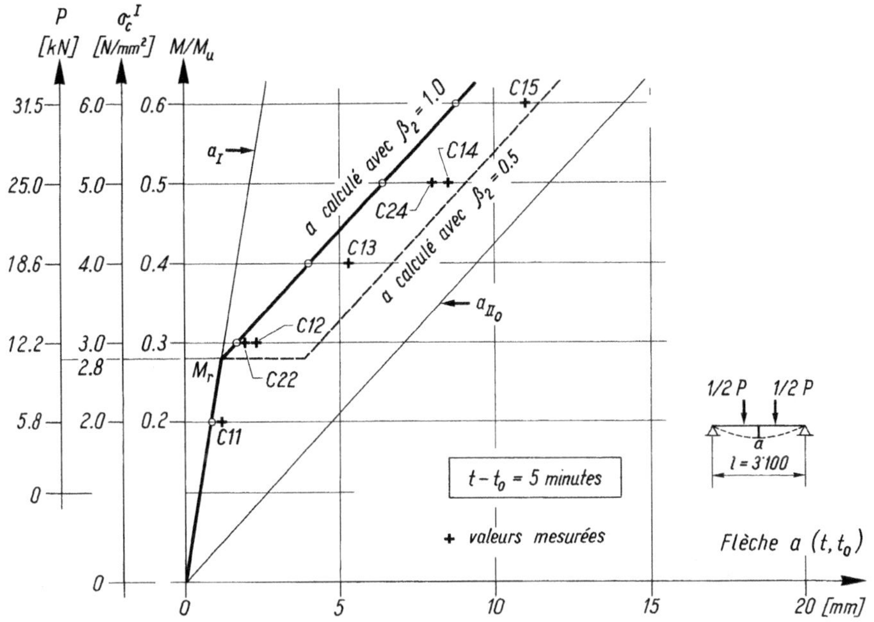

FIG. 6 Flèches après 5 minutes - dalles série C.

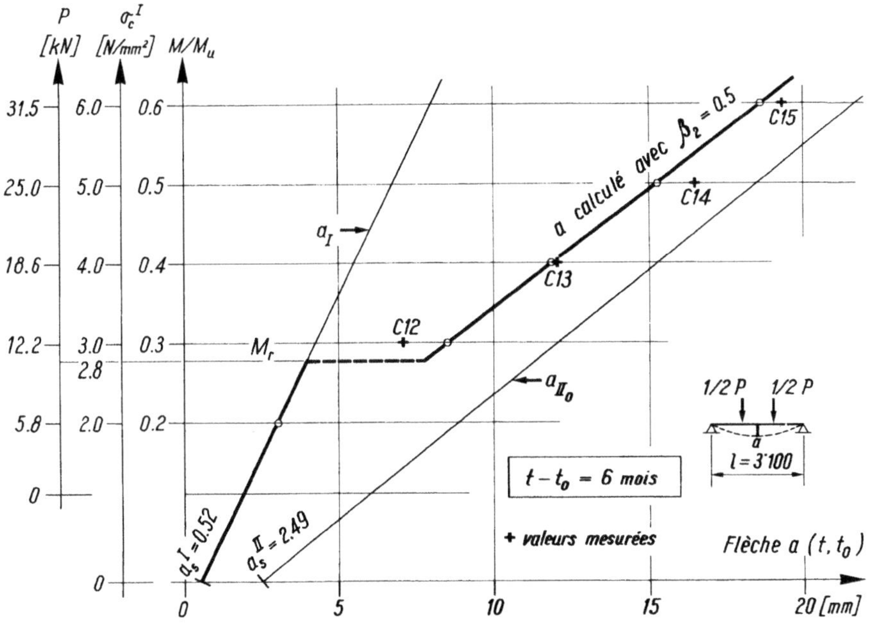

FIG. 7 Flèches après 6 mois - dalles série C.

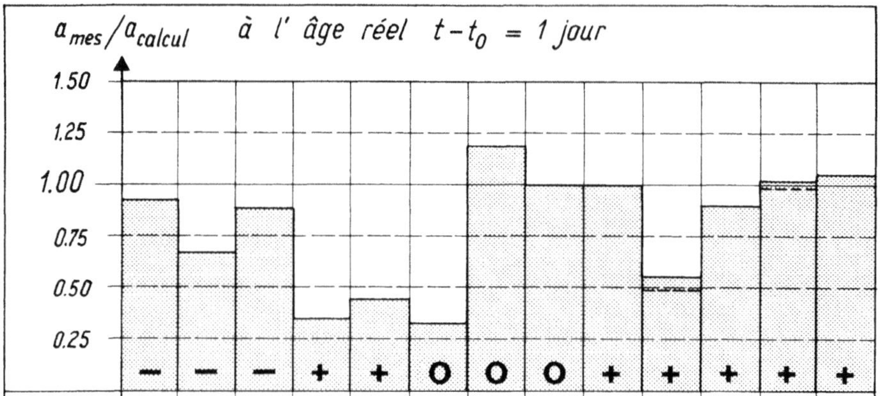

DALLE N°	A1	A2	A3	A4	A5	B1	B2	B3	C11	C12 C22	C13	C14 C24	C15
φ	0.35	0.46	0.38	0.35	0.39	0.43	0.38	0.40	–	0.32	–	0.42	0.32
$-\varepsilon_s\ [10^{-6}]$	14	10	10	0	2	11	8	11	–	12	–	1	0

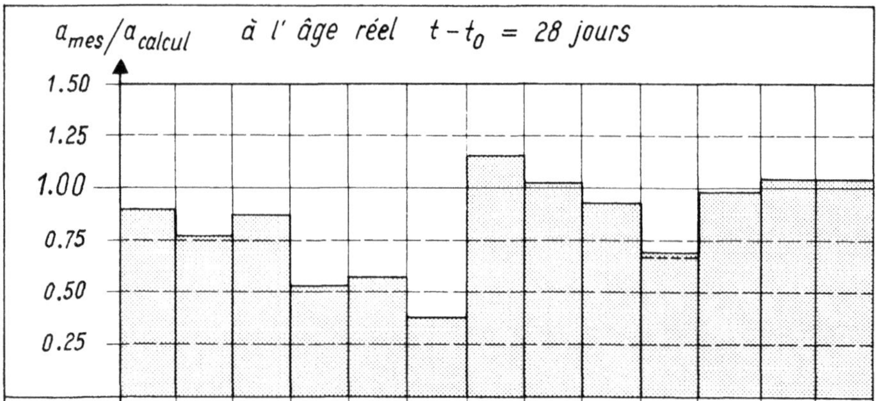

DALLE N°	A1	A2	A3	A4	A5	B1	B2	B3	C11	C12 C22	C13	C14 C24	C15
φ	1.34	1.55	1.31	1.38	1.42	1.45	1.30	1.34	–	1.19	–	1.23	1.10
$-\varepsilon_s\ [10^{-6}]$	212	203	203	150	158	180	151	149	–	78	–	119	88

FIG. 8 & 9

Rapport des flèches mesurées aux flèches calculées par la méthode bilinéaire. Coefficient de fluage et retrait spécifique déduits des mesures.

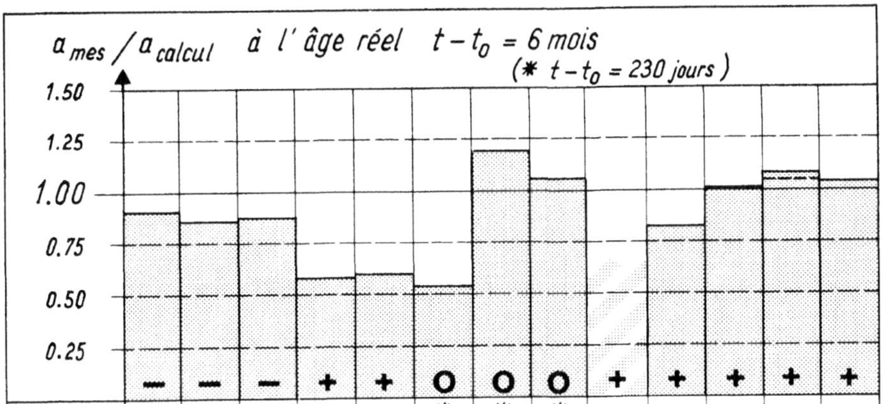

DALLE N°	A1	A2	A3	A4	A5	* B1	* B2	* B3	C11	C12 C22	C13	C14 C24	C15
φ	2.38	2.47	2.31	2.11	1.96	2.50	2.21	2.31	–		–	2.17	1.92
$-\varepsilon_s\,[10^{-6}]$	448	462	429	386	379	407	354	432	–		–	362	242

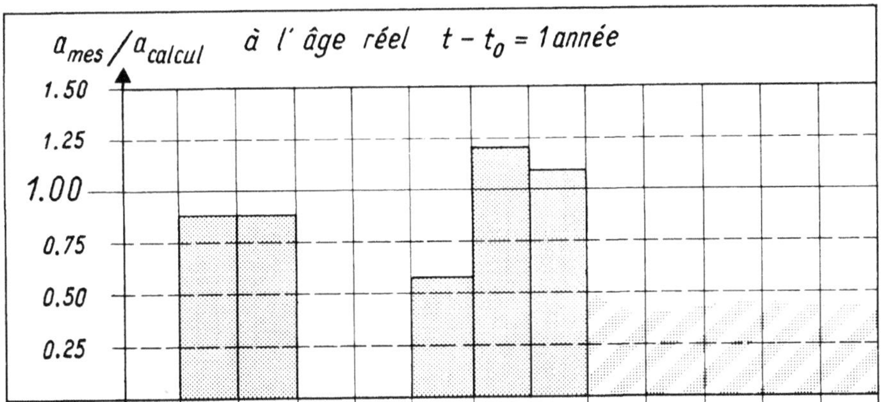

DALLE N°	A1	A2	A3	A4	A5	B1	B2	B3	C11	C12 C22	C13	C14 C24	C15
φ	–	2.60	2.44	–	–	2.65	2.56	2.43	–		–		
$-\varepsilon_s\,[10^{-6}]$	–	533	505	–	–	457	388	456	–		–		

FIG. 10 & 11

Rapport des flèches mesurées aux flèches calculées par la méthode bilinéaire. Coefficient de fluage et retrait spécifique déduits des mesures.

	Caractéristiques mesurées à l' âge réel t_0 de mise en charge							Climat halle d' essais		
DALLE N°	Cond. cure	t_0 [j]	$t_{0,corr.}$ [j]	β_{wm} [N/mm² = MPa]	f_{ctm}	$f_{ct,min.}$	E_{cm} [GPa]	Climatisation	T [°C]	HR [%]
A 1	–	15	20.4	24.1	2.07	2.01	22.37	oui	20 (±1)	65 (±5)
A 2	– G	15	15.7	28.4	2.15	2.14	23.59			
A 3	–	15	17.8	23.2	1.97	1.90	24.09			
A4	+ G	15	22.5	42.5	3.1	2.7	33.82	non	23 (+10/-5)	46 (±20)
A5	+	15	21.2	38.2	2.7	2.33	32.67			
B1	0 G	14	21.6	35.8	3.16	3.0	28.39	non	23 (+10/-5)	46 (±20)
B2	0 G	14	22.5	39.8	3.14	2.75	29.72			
B3	0 G	14	22.5	35.1	2.84	2.65	25.93			
C11	+ G	28	~42	34.4	2.56	2.27	28.97	oui	20 (±1)	65 (±5)
C12 C22				32.6 36.8	3.06 2.89	3.04 2.75	28.27 30.77			
C13				36.9	2.82	2.55	29.23			
C14 C24				C14 idem C12 37.3	 2.56	 2.25	 30.87			
C15				33.9	2.9	2.73	28.73			

TABLEAU 1

Caractéristiques mécaniques du béton mesurées sur éprouvettes. Données relatives aux conditions de bétonnage et aux conditions climatiques durant les essais.

ANALYSIS OF LINEAR VISCO-ELASTIC STRUCTURES SUBJECTED TO DELAYED RESTRAINTS

by Mario Alberto CHIORINO[1] and Franco MOLA[2]

[1] Professor of Structural Analysis, Technical University of Turin, Italy

[2] Professor of Design of Structures, Department of Structural Engineering, Technical University of Milan, Italy

Abstract

The static analysis of linear visco-elastic homogeneous structures subjected to additional delayed restraints introduced after the application of loads can be easily performed if the additional restraints are introduced all together. If the delayed restraints are introduced at different times, the state of stress in the structure has on the contrary to be determined following step-by-step the variation in time of the structural scheme. The paper presents the general solution of the problem under the hypothesis of elastic modulus constant in time and assuming that the initial structural scheme is statically determined. The solution is applied for the evaluation of the state of stress in a simple structure with two delayed restraints.

Résumé

L'analyse statique des structures homogènes viscoélastiques avec liaisons additionnelles appliquées après leur chargement n'est pas difficile si les liaisons sont appliquées toutes dans le meme instant. Si par contre les liaisons sont appliquées en temps différents, l'analyse devient complexe car il faut considérer les successives variations du schéma structural. La solution generale du probleme est presentée dans ce memoir sous l'hypothèse de module elastique constant dans le temps et en supposant que la structure soit initialement isostatique. La solution est appliquée à l'analyse statique d'une simple structure avec deux liaisons imposées en temps différents.

1. Introduction

Reinforced concrete structures are frequently subjected in structural design practice to modifications of their statical system particularly during their construc-

tion phase and/or during their service life through the imposition of one or more delayed additional restraints. Some typical examples of this are: prefabricated beams or slabs initially simply supported and then made continuous, cantilever built bridges in which a solidarization of cantilever edge sections is introduced at a later stage, prefabricated beams with a R.C. flange cast on them and so on. When one imposes a restraint at a point of a viscoelastic structure (time t^*) after the application of loads (time t_o), the creep deformations in that point are prevented, so a redundant action arises in the imposed restraint. It is interesting to investigate this fact and give a general solution of the problem in order to obtain the exact state of stress and strain of the structure at any time. When all the additional restraints are imposed at $t^* = t_o^+$ immediately after the loads, the solution is very simple to calculate because the redundant actions vary in time in an affine way and are proportional to those evaluated elastically on the structure in its final statical scheme. The coefficient of affinity is represented /1/ by the well known expression $[1-r(t,t_o)/E(t_o)]$, $r(t,t_o)$ being the relaxation function corresponding to the given creep function $f(t,t_o)$. When the delayed restraints are applied at different times $t_j^* > t_o$, $(j=1,2,....,n)$, the redundant actions are no longer affine and the problem becomes more involved. A particular solution is presented in /1/, where all the delayed restraints are imposed at the same time $t_1^* > t_o$, $(n=1)$. Although in this case the redundants are still affine, the affinity coefficient is different from the preceding one and assumes the form of an integral operator: $\xi(t,t_o,t_1^*) = \int_{t_1^*}^{t} (\partial f(\tau,t_o)/\partial\tau) \cdot r(t,\tau)\, d\tau$, which, apart some minor computational concerns (for normally refined creep functions a numerical integration is mandatory) may be readily evaluated if – as in the preceding case – the relaxation function $r(t,\tau)$ is available or has been previously been determined (from the corresponding given creep function $f(t,\tau)$).
In the present paper we solve the general problem when the delayed restraints are imposed at different times $t_j^* > t_o$ and determine the recurrent expressions which allow to obtain the redundant actions at any time. The solution, which is obtained for sake of semplicity in the case of constant elastic modulus and with reference to a homogeneous structure initially statically determined, can be easily generalized taking into account also structural members of different creep behaviour. A numerical example related to three simply supported beams made continuous at different times shows the practical application of the solution, and its analytical expression is determined in the case of hereditary or rate-of-creep behaviour of the material. These two solutions may be considered as the bounds at the interior of which lie the solutions corresponding to more refined creep models.

2. Solution of the Problem

Let's consider the homogeneous structure of Fig. 1. The structure is assumed to be initially statically determined (system (o)) and subjected to static actions constant in time applied at time t_o. To the structure are imposed n delayed restraints at times $t_o \leqslant t_1^* \leqslant t_2^* \ldots \leqslant t_j^* \leqslant \ldots \leqslant t_n^*$ and we indicate with $X_j(t)$, $(j=1,2,...,n)$ the redundant actions that arise in the delayed restraints for $t_{j+1}^* > t \geqslant t_j^*$. For the evaluation of $X_j(t)$, we have to follow step-by-step the structural behaviour, writing for $t \geqslant t_j^*$ the compatibility equation related to j-th delayed restraint in the structure having j restraints (system (j)).
Beginning from system (1) and thinking the elastic modulus constant in time, the

compatibility equation is:

$$X_1(t)\cdot\delta_{11}^{(1)} + \int_{t_1^*}^{t} X_1(\tau)\cdot\delta_{11}^{(1)}\cdot\phi(t,\tau)\,d\tau = -\delta_{10}^{(0)}\,[f(t,t_0)-f(t_1^*,t_0)]\cdot E \qquad 1)$$

where $f(t,\tau)$ and $\phi(t,\tau) = -E(\partial f(t,\tau)/\partial\tau)$ are respectively the creep function and the creep kernel, $\delta_{11}^{(1)}$ and $\delta_{10}^{(0)}$ are the usual elastic influence coefficients related to the unit redundant action and to external loads, evaluated respectively in system (1) and system (0), Fig. 2.
Indicating by $r(t,\tau)$ the relaxation function, the solution of eq. 1) is /1,2/:

$$X_1(t) = -\frac{\delta_{10}^{(0)}}{\delta_{11}^{(1)}}\int_{t_1^*}^{t}\frac{\partial f(\tau,t_0)}{\partial\tau}\cdot r(t,\tau)\,d\tau, \qquad 2)$$

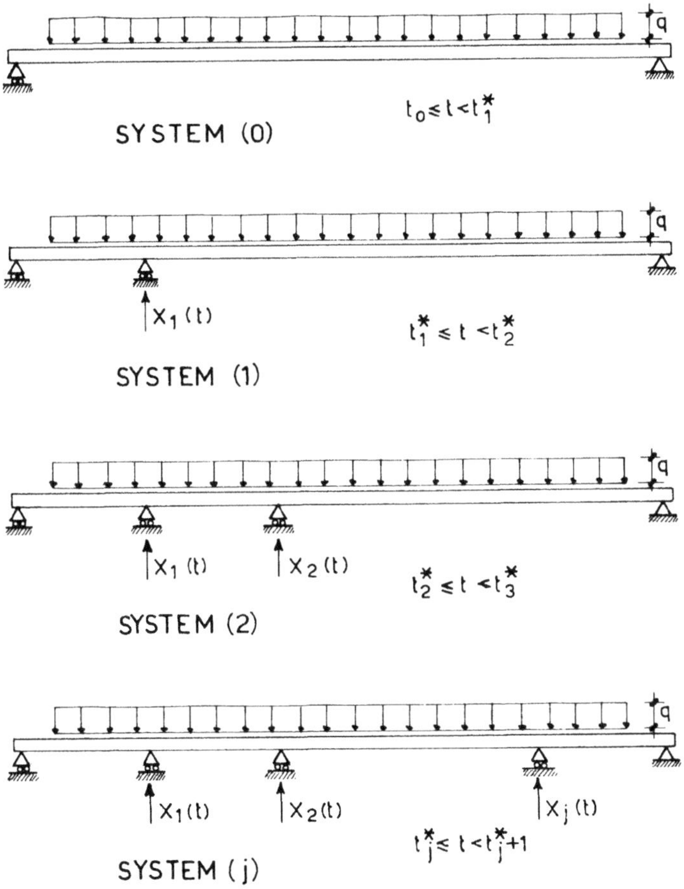

Fig. 1 - Schematisation of the structure and its delayed restraints

and can be put in the simpler form:

$$X_1(t)=\overline{X}_{1e}\cdot x_1(t),(t_1^*\leqslant t<t_2^*) \qquad 3)$$

setting:

$$\overline{X}_{1e} = -\frac{\delta_{10}^{(0)}}{\delta_{11}^{(1)}}$$

$$x_1(t) = \int_{t_1^*}^{t} \frac{\partial f(\tau,t_O)}{\partial \tau}\cdot r(t,\tau)\,d\tau.$$

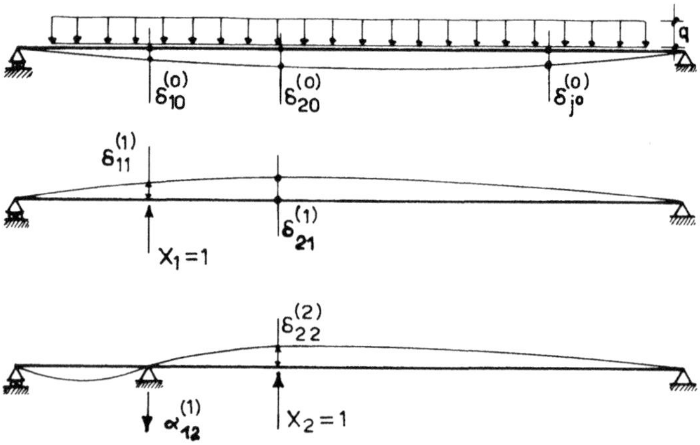

Fig. 2 - Schemes for the evaluation of elastic influence coefficients

We now consider system (2) and calculate for $t_3^* > t \geqslant t_2^*$ the displacement $\eta_2(t)-\eta_2(t_2^*) = \overline{\eta}_2(t)$ in the direction of $X_2(t)$ due to external actions and $X_1(t)$. We obtain:

$$\overline{\eta}_2(t)=\overline{X}_{1e}\,(x_1(t)-x_1(t_2^*))\cdot\delta_{21}^{(1)}+\int_{t_1^*}^{t}\overline{X}_{1e}\,x_1(\tau)\,\delta_{21}^{(1)}\,\phi(t,\tau)\,d\tau-$$

$$-\int_{t_1^*}^{t_2^*}\overline{X}_{1e}\,x_1(\tau)\,\delta_{21}^{(1)}\,\phi(t_2^*,\tau)\,d\tau+\delta_{20}^{(0)}\,(f(t,t_O)-f(t_2^*,t_O))\cdot E,$$

or shortly, putting:

$$\int_{t_1^*}^{t} x_1(\tau)\,\phi(t,\tau)\,d\tau = x_{1\phi}(t),$$

we have:

$$\overline{\eta}_2(t)=\overline{X}_{1e}\,[(x_1(t)+x_{1\phi}(t))-(x_1(t_2^*)+x_{1\phi}(t_2^*))]\,\delta_{21}^{(1)}+\delta_{20}^{(0)}\,(f(t,t_O)-f(t_2^*,t_O))\cdot E \qquad 4)$$

so that the compatibility equation in the second delayed restraint becomes:

$$X_2(t)\delta_{22}^{(2)} + \int_{t_2^*}^{t} X_2(\tau)\,\delta_{22}^{(2)}\,\phi(t,\tau)\,d\tau = -\,\bar{\eta}_2(t), \qquad 5)$$

where $\delta_{22}^{(2)}$ is the elastic influence coefficient related to unit redundant action and e-valuated in system (2), (Fig. 2).
The solution of eq. 5), remembering eq. 4) is:

$$X_2(t) = -\frac{\delta_{20}^{(0)}}{\delta_{22}^{(2)}} \int_{t_2^*}^{t} \frac{\partial f(\tau,t_0)}{\partial \tau}\, r(t,\tau)\,d\tau - \bar{X}_{1e}\,\frac{\delta_{21}^{(1)}}{\delta_{22}^{(2)}} \int_{t_2^*}^{t} \frac{\partial}{\partial \tau}[x_1(\tau) + x_{1\phi}(\tau)]\,\frac{r(t,\tau)}{E}\,d\tau \qquad 6)$$

and can be put in the simpler form:

$$X_2(t) = \bar{\bar{X}}_{2e} \cdot x_2(t)\,, \quad (t_2^* \leqslant t < t_3^*) \qquad 7)$$

setting:

$$\bar{X}_{2e} = -\frac{\delta_{20}^{(0)}}{\delta_{22}^{(2)}}\,, \quad \bar{\bar{X}}_{2e} = \bar{X}_{2e} - \bar{X}_{1e}\,\frac{\delta_{21}^{(1)}}{\delta_{22}^{(2)}}$$

$$x_2(t) = \frac{1}{\bar{\bar{X}}_{2e}}\Big[\bar{X}_{2e} \int_{t_2^*}^{t} \frac{\partial f(\tau,t_0)}{\partial \tau} \cdot r(t,\tau)\,d\tau - \bar{X}_{1e}\,\frac{\delta_{21}^{(1)}}{\delta_{22}^{(2)}} \int_{t_2^*}^{t} \frac{\partial}{\partial \tau}(x_1(\tau) + x_{1\phi}(\tau) \cdot \qquad 8)$$

$$\cdot \frac{r(t,\tau)\,d\tau}{E}\Big].$$

It is important to observe that the displacements produced by external actions and $X_1(t)$ together are consistent with the first delayed restraint, so that $X_2(t)$ can be obtained independently of $X_1(t)$ provided that its influence coefficient is evaluated in system (2). In this way $X_2(t)$ causes a variation $\Delta X_1(t)$ of $X_1(t)$ which can be calculated in an elastic way according to the first principle of linear viscoelasticity, for the restraint 1 is preceding the arise of $X_2(t)$.
The variation $\Delta X_1(t)$ can be briefly written:

$$\Delta X_1(t) = \bar{\bar{X}}_{2e} \cdot x_2(t) \cdot \alpha_{12}^{(1)}\,,$$

where $\alpha_{12}^{(1)}$ is the elastic action in the first restraint when a unit action $X_2 = 1$ is ap-plied in system (1), (Fig. 2).
At time $t_3^* > t \geqslant t_2^*$ we therefore have:

$$X_2(t) = \bar{\bar{X}}_{2e} \cdot x_2(t)$$

$$X_1(t) = \bar{X}_{1e} \cdot x_1(t) + \bar{\bar{X}}_{2e} \cdot x_2(t) \cdot \alpha_{12}^{(1)}. \qquad 9)$$

Proceeding in an analogous way for the remaining times t_j^* the expression 9) can be generalized as follows:

$$X_j(t) = \bar{\bar{X}}_{je}\, x_j(t)\,,\quad (t_j^* \leq t < t_{j+1});\quad (j = 1,2,\dots,n)$$

$$X_k(t) = \bar{\bar{X}}_{ke} \cdot x_k(t) + \sum_{m=k+1}^{j} \bar{\bar{X}}_{me} \cdot x_m(t) \cdot \alpha_{km}^{(k)}\,;\quad (1 \leq k \leq j-1) \tag{10}$$

with:

$$x_j(t) = \frac{1}{\bar{\bar{X}}_{je}}\Big[\bar{X}_{je} \int_{t_j^*}^{t} \frac{\partial f(\tau,t_o)}{\partial \tau}\, r(t,\tau)\, d\tau - \sum_{k=1}^{j-1} \bar{\bar{X}}_{ke} \frac{\delta_{jk}^{(k)}}{\delta_{jj}^{(j)}} \int_{t_j^*}^{t} \frac{\partial}{\partial \tau}\big(x_k(\tau)+x_{k\phi}(\tau)\big) \frac{r(t,\tau)}{E}\, d\tau$$

$$\bar{X}_{je} = -\frac{\delta_{jo}^{(0)}}{\delta_{jj}^{(j)}} \tag{11}$$

$$\bar{\bar{X}}_{je} = \bar{X}_{je} - \sum_{k=1}^{j-1} \bar{\bar{X}}_{ke} \frac{\delta_{jk}^{(k)}}{\delta_{jj}^{(j)}}\,,\quad (\bar{\bar{X}}_{1e} = \bar{X}_{1e})$$

$$x_{j\phi}(t) = \int_{t_j^*}^{t} x_j(\tau)\, \phi(t,\tau)\, d\tau\,,\quad (t_{j+1}^* > t \geq t_j^*);\quad (j = 1,2,\dots,n).$$

The recurrent expression 10) and 11) represent the general solution of the problem and their calculation is reduced to the iterative integration of combinations of functions f, ϕ, r.
An important particular form of expression 10) and 11) takes place when all delayed restraints are imposed at time t_o^*, immediately after the application of external loads. In this case, we have $t_o = t_1^* = t_2^* = \dots = t_j^* = \dots = t_n^*$ so that, remembering the fundamental relations of linear viscoelasticity /2/ we obtain:

$$x_1(t) = \int_{t_o}^{t} \frac{\partial f(\tau,t_o)}{\partial \tau}\, r(t,\tau)\, d\tau = 1 - \frac{r(t,t_o)}{E} \tag{12}$$

$$x_{1\phi}(t) = -\int_{t_o}^{t} \Big(1 - \frac{r(\tau,t_o)}{E}\Big) E \frac{\partial f(t,\tau)}{\partial \tau}\, d\tau. \tag{13}$$

Integrating eq. 13) by parts we have:

$$x_{1\phi}(t) = -2 + \frac{r(t,t_o)}{E} + E\, f(t,t_o)\,,$$

and therefore:

$$\int_{t_o}^{t} \frac{\partial}{\partial \tau}\big(x_1(\tau) + x_{1\phi}(\tau)\big) \frac{r(t,\tau)}{E}\, d\tau = \int_{t_o}^{t} \frac{\partial f(\tau,t_o)}{\partial \tau}\, r(t,\tau)\, d\tau = 1 - \frac{r(t,t_o)}{E}\,,$$

so that, remembering eqs. 8) we obtain ($t_2^* = t_o$):

$$x_2(t) = 1 - \frac{r(t,t_o)}{E}.$$

Proceeding in the same way for the remaining $t_j^* = t_o$ we obtain finally the solution:

$$\begin{cases} X_j(t) = \bar{\bar{X}}_{je}\,(1 - \frac{r(t,t_o)}{E}), \quad (t \geq t_o); \; (j = 1,2,...,n) \\ X_k(t) = [\bar{\bar{X}}_{ke} + \sum_{m=k+1}^{j-1} \bar{\bar{X}}_{me}\,\alpha_{km}^{(k)}]\,(1 - \frac{r(t,t_o)}{E}); \; (1 \leq k \leq j-1) \end{cases} \qquad 14)$$

coinciding with the well known solution wich states that all redundant actions are related in an affine way to the elastic solution calculated as the restraints were preceding the application of external loads. This solution represents the so-called "principle of acquisition of the modified system stress distribution", /11/.

3. Calculation of solution

The expression 10) and 11) generally cannot be calculated in a direct way for the functions f, ϕ, and r have not an analytic form at all or their analytic form does not make possible a closed form calculation of the integrals appearing in the expression of the $X_j(t)$. It is necessary to proceed in a numerical way by means of computer programs so that the problem becomes rather complex and the expression 10) and 11) cannot be used in everyday practice.

Nevertheless, it is possible to calculate the solution in a simple way, in two particular cases of creep behaviour of concrete. In fact, when the creep behaviour is of hereditary type, i.e. as that described by the Kelvin Voigt /3/ model, we have /4/:

$$\begin{cases} f(t,\tau) = f(t-\tau) = \frac{1}{E}[1 + \varphi_\infty(1 - e^{-(\frac{t-\tau}{\tau^*})})] \\ \phi(t,\tau) = \frac{\varphi_\infty}{\tau^*} \cdot e^{-(\frac{t-\tau}{\tau^*})} \\ r(t,\tau) = \frac{E}{(1+\varphi_\infty)}[1 + \varphi_\infty e^{-(\frac{t-\tau}{\tau^{*\prime}})}], \; \tau^{*\prime} = \frac{\tau^*}{1+\varphi_\infty} \end{cases} \qquad 15)$$

and the expression 10) and 11) can be easily calculated as they contain combinations of exponential functions of simple integration.

Analogously, when the creep behaviour is rate-of-creep type, i.e. as that described by the Dischinger model /15/ we have:

$$\begin{cases} f(t,\tau) = \frac{1}{E}[1 + (\beta(t) - \beta(\tau)) \cdot \varphi_\infty] \\ \phi(t,\tau) = \phi(\tau) = \varphi_\infty \frac{d\beta(\tau)}{d\tau} \\ r(t,\tau) = E\, e^{-\varphi_\infty(\beta(t) - \beta(\tau))} \end{cases} \qquad 16)$$

and the expression 10) and 11). can be still easily calculated for they contain exact differential forms.
Moreover, the two models represent two limiting cases of creep behaviour as the first assumes that the delayed deformation is fully reversible, while the second considers this deformation completely irreversible. The results that we obtain from the two models besides to be of simple evaluation give the upper and lower bounds at the interior of which lies the solution when more refined creep models are used. As there are not great differences between the results obtained considering an hereditary or an a rate-of-creep behaviour, for everyday practice it is generally sufficient to know the results of these two models as it will be shown in the following example.

4. Numerical example

Let's consider the structure of Fig. 3, formed by three simply supported beams of constant rigidity (EJ) and subjected at time t_o to a constant load q. At times t_1^* and $t_2^* \geqslant t_1^*$ we impose two additional restraints to the structure making it continuous on supports 1 and 2. Denoting with $X_1(t)$ and $X_2(t)$ the bending moments which arise in the two delayed restraints we want to calculate the expression of $X_1(t)$ and $X_2(t)$ in the hypotesis of creep behaviour according to hereditary model or rate-of-creep model.
Preliminarly we calculate the elastic influence coefficients.
From system (o), Fig. 4, we have:

$$\delta_{10}^{(0)} = 2\,\frac{ql^3}{24EJ}$$

$$\delta_{20}^{(0)} = 2\,\frac{ql^3}{24EJ}$$

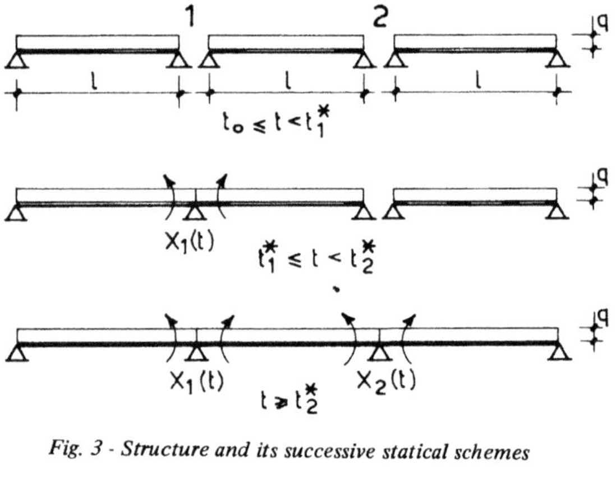

Fig. 3 - Structure and its successive statical schemes

From system (1), Fig. 4, we have:

$$\delta_{11}^{(1)} = \frac{2}{3}\,\frac{1}{EJ}$$

$$\delta_{21}^{(1)} = \frac{1}{6EJ}$$

$$\overline{X}_{1e} = -\frac{2}{24}\cdot\frac{3}{2}\,ql^2 = -\frac{1}{8}\,ql^2\ ;\ \alpha_{12}^{(1)} = -\frac{1}{6EJ}\cdot\frac{3EJ}{21} = -\frac{1}{4}\,.$$

Finally, from system (2), Fig. 4, we obtain:

$$\delta_{22}^{(2)} = \frac{21}{3EJ} - \frac{1}{4}\,\frac{1}{6EJ} = \frac{15}{24}\,\frac{1}{EJ}$$

$$\overline{X}_{2e} = -\frac{2}{24}\,\frac{ql^3}{EJ}\,\frac{24}{15}\,\frac{EJ}{l} =$$

$$= -\frac{2}{15}\,ql^2$$

$$\overline{\overline{X}}_{2e} = -\frac{2}{15}\,ql^2 - (1/8\,ql^2)\,\cdot$$

$$\cdot\,\frac{1}{6EJ}\,\frac{24EJ}{15l} = -\,ql^2/10.$$

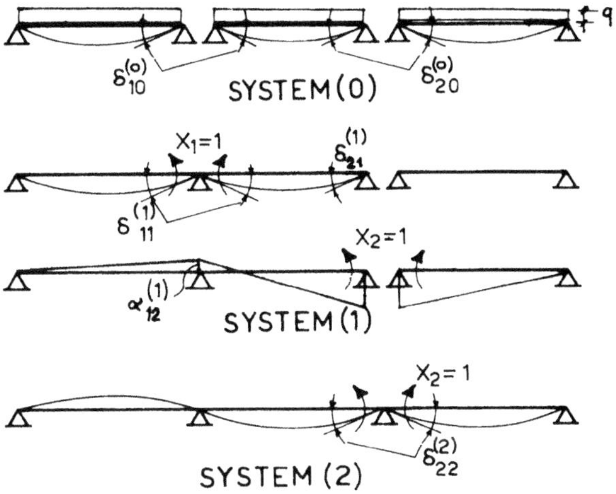

Fig. 4 - Calculation of elastic influence coefficients

4.1. Hereditary Model (H.M.)

Using relations 15) we obtain:

$$x_1(t) = \frac{\varphi_\infty}{\tau^*(1+\varphi_\infty)} \int_{t_1^*}^{t} e^{-\left(\frac{\tau - t_O}{\tau^*}\right)} \cdot \left(1+\varphi_\infty e^{-\left(\frac{t-\tau}{\tau^{*'}}\right)}\right) d\tau =$$

$$= \frac{\varphi_\infty}{1+\varphi_\infty}\, e^{-\left(\frac{t_1^* - t_O}{\tau^*}\right)} \cdot \left(1 - e^{-\left(\frac{t-t_1^*}{\tau^{*'}}\right)}\right)$$

$$x_{1\phi}(t) = \frac{\varphi_\infty}{1+\varphi_\infty}\, e^{-\left(\frac{t_1^* - t_O}{\tau^*}\right)} \int_{t_1^*}^{t} \left(1 - e^{-\left(\frac{\tau - t_1^*}{\tau^{*'}}\right)}\right) \frac{\varphi_\infty}{\tau^*}\, e^{-\left(\frac{t-\tau}{\tau^*}\right)} d\tau =$$

$$= \frac{\varphi_\infty}{1+\varphi_\infty}\, e^{-\left(\frac{t_1^* - t_O}{\tau^*}\right)} \left[e^{-\left(\frac{t-t_1^*}{\tau^{*'}}\right)} - (1+\varphi_\infty)\, e^{-\left(\frac{t-t_1^*}{\tau^*}\right)} \right]$$

$$\frac{\partial}{\partial \tau}\left(x_1(\tau) + x_{1\phi}(\tau)\right) = \frac{\varphi_\infty}{\tau^*}\; e^{-\left(\frac{\tau - t_O}{\tau^*}\right)}$$

$$x_2(t) = \frac{\varphi_\infty}{\tau^*(1+\varphi_\infty)} \int_{t_2^*}^{t} e^{-\left(\frac{\tau - t_O}{\tau^*}\right)} \left(1+\varphi_\infty\, e^{-\left(\frac{t-\tau}{\tau^{*'}}\right)}\right) d\tau = \frac{\varphi_\infty}{1+\varphi_\infty}\, e^{-\left(\frac{t_2^* - t_O}{\tau^*}\right)} \cdot$$

$$\cdot \left(1 - e^{-\left(\frac{t-t_2^*}{\tau^{*'}}\right)}\right),$$

so that expressions 9) are:

$$X_2(t) = -\frac{ql^2}{10} \cdot \frac{\varphi_\infty}{1+\varphi_\infty}\; e^{-\left(\frac{t_2^* - t_O}{\tau^*}\right)} \left(1 - e^{-\left(\frac{t-t_2^*}{\tau^{*'}}\right)}\right)$$

$$X_1(t) = -\frac{ql^2}{10} \cdot \frac{\varphi_\infty}{1+\varphi_\infty} \left(1 - e^{-\left(\frac{t-t_1^*}{\tau^{*'}}\right)}\right) \left[1.25\, e^{-\left(\frac{t_1^* - t_O}{\tau^*}\right)} - 0.25\, e^{-\left(\frac{t_2^* - t_O}{\tau^*}\right)}\right].$$

In Fig. 5 are reported the ratios $X_1(t)/(ql^2/10)$ and $X_2(t)/(ql^2/10)$ for $t_1^* = t_0$ and different values of $t_2^* - t_1^*$, under the assumptions:

$$\varphi_\infty = 3; \quad \tau^* = 1000 \text{ days.}$$

4.2. Rate-of-Creep Model (R.C.M.)

Using relations 16) we obtain:

$$x_1(t) = \int_{t_1^*}^{t} \frac{d\beta(\tau)}{d\tau}\, e^{-(\beta(t)-\beta(\tau))}\, d\tau = 1 - e^{-(\beta(t)-\beta(t_1^*))}$$

$$x_{1\phi}(t) = \int_{t_1^*}^{t} \left(1 - e^{-(\beta(\tau)-\beta(t_1^*))}\right) \frac{d\beta(\tau)}{d\tau}\, d\tau = \beta(t) - \beta(t_1^*) - 1 + e^{-(\beta(t)-\beta(t_1^*))}$$

$$\frac{\partial}{\partial\tau}\left(x_1(\tau) + x_{1\phi}(\tau)\right) = \frac{d\beta(\tau)}{d\tau}$$

$$x_2(t) = \int_{t_2^*}^{t} \frac{d\beta(\tau)}{d\tau}\, e^{-(\beta(t)-\beta(\tau))}\, d\tau = 1 - e^{-(\beta(t)-\beta(t_2^*))},$$

so that expressions 9) are:

$$X_2(t) = -\frac{ql^2}{10}\left(1 - e^{-(\beta(t)-\beta(t_2^*))}\right)$$

$$X_1(t) = -\frac{ql^2}{10}\left[1.25\left(1 - e^{-(\beta(t)-\beta(t_1^*))}\right) - 0{,}25\left(1 - e^{-(\beta(t)-\beta(t_2^*))}\right)\right].$$

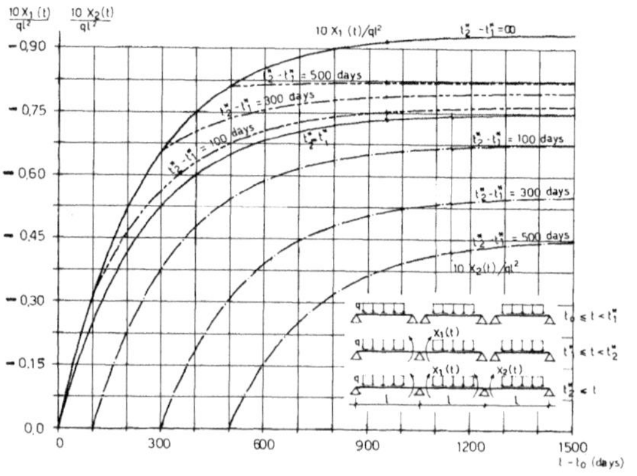

Fig. 5 - Continuity bending moments according to Hereditary Model (H.M.)

In Fig. 6 are reported the ratios $X_1(t)/(ql^2/10)$, $X_2(t)/(ql^2/10)$ obtained assuming for $\beta(t)$ the exponential form:

$$\beta(t) = (1 - e^{-\gamma(t - t_o)}) \cdot \varphi_\infty \; ; \; \varphi_\infty = 3 \; ; \; \gamma = 1/\tau^*.$$

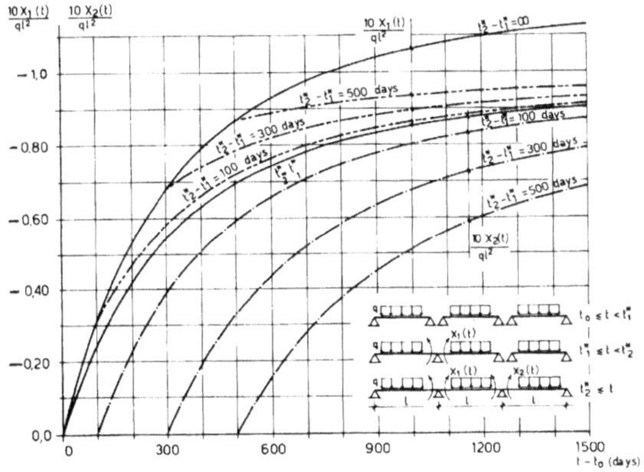

Fig. 6 - Continuity bending moments according to Rate-of-Creep Model (R.C.M.)

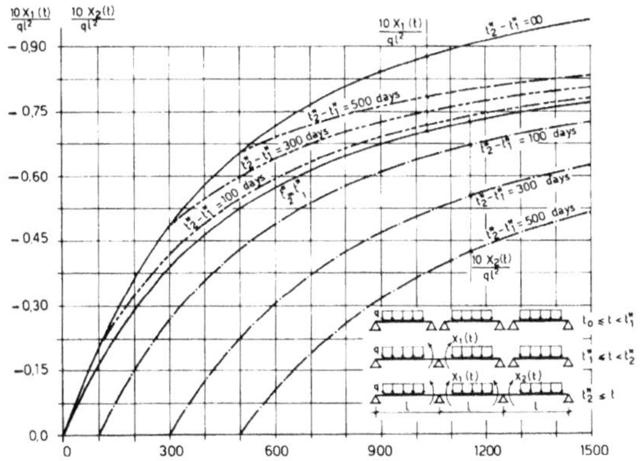

Fig. 7 - Continuity bending moments according to Improved Dischinger Model (I.D.M.)

As we can easily see, the results deriving from rate-of-creep behaviour are always greater than those pertaining to the hereditary behaviour but the difference is sufficiently limited. As a comparison, in Fig. 7 are shown the results obtained considering an intermediate creep behaviour according to improved Dischinger model (I.D.M.), (Rüsch - Jungwirth Model /6/). The diagrams and their asymptotic values reported in Table 1 show that the results of Rüsch - Jungwirth model lie between those pertaining to hereditary and rate-of-creep models.

TABLE 1

	H.M.		R.C.M.		I.D.M.	
$t_2^* - t_1^*$ (days)	$-\frac{10X_1}{ql^2}$	$-\frac{10X_2}{ql^2}$	$-\frac{10X_1}{ql^2}$	$-\frac{10X_2}{ql^2}$	$-\frac{10X_1}{ql^2}$	$-\frac{10X_2}{ql^2}$
0	0.750	0.750	0.950	0.950	0.844	0.844
100	0.810	0.679	0.954	0.934	0.852	0.814
300	0.833	0.556	0.965	0.892	0.868	0.747
500	0.852	0.455	0.978	0.838	0.886	0.676
∞	0.937	0.000	1.187	0.000	1.055	0.000

References

1. M.A. Chiorino and P. Napoli, F. Mola, M. Koprna, *Manuel de Calcul: Effects Structuraux du Fluage et des Déformations Différées*, Bulletin d'Information CEB n. 136, (1980)

2. F. Mola, *General and Approximate Method for the Analysis of Inhomogeneous Linear Viscoelastic Structures*, Studi e Ricerche, Corso di Perfezionamento per le Costruzioni in Cemento Armato F.lli Pesenti, Politecnico di Milano, n. 1, pp. 101-143, (1979)

3. W. Nowacki, *Théorie du Fluage*, Eyrolles, Paris (1965)

4. A. Migliacci, *Applicazioni dei principi di viscosità*, Tamburini, Milano, (1971)

5. F. Dischinger, *Elastische und Plastische Verformungen der Eisenbetontragwerke und insbesondere der Bogenbrücken*, Der Bauingenieur H. 20, pp. 53-63, 286-294, 426-437, 563-572 (1939)

6. H. Rüsch and D. Jungwirth, H. Hilsdorf, *Kritische Sichtung der Verfahren zur Berücksichtigung der Einflüsse von Kriechen*, Beton und Stahlbetonbau H. 68, pp. 49-60, 76-86, 152-158 (1973).

GENERAL AND APPROXIMATE METHOD FOR THE ANALYSIS OF LINEAR VISCO-ELASTIC STRUCTURES SENSITIVE TO SECOND ORDER EFFECTS

by Franco MOLA[1] and Giuseppe CREAZZA[2]

[1] Professor of Design of Structures, Department of Structural Engineering, Technical University of Milan, Italy

[2] Professor of Strenght of Materials, University of Venice, Italy

Abstract

Creep causes a variation of the state of stress in redundant slender R.C. structures because the deformations of the single parts are not related by affinity and depend on the applied axial load. The general method of analysis of slender R.C. frame structures in which lateral displacements are prevented leads to a set of systems of integro-differential Volterra equations and is of rather complex application. For this reason an approximate procedure is presented, based on the Age Adjusted Effective Modulus Method (AAEMM) which allows to solve the problem through a particular elastic analysis especially recommended for practical uses. Two numerical examples show the application of the Method.

Résumé

Le fluage donne lieu à une variation de l'état de contrainte dans les structures élancées en béton armé car les déformations des éléments structuraux ne sont pas liées par affinité et dépendent de l'intensité de l'action normale appliquée. La méthode générale d'analyse des portiques élancés en béton armé avec les noeuds qui ne sont pas déplaçables conduit à la résolution de systhèmes d'équations intégro-différentielles de Volterra et pourtant est de très difficile application. Pour cette raison est ici présenté un procédé approché de calcul basé sur la Méthode AAEMM (Age Adjusted Effective Modulus Method) qui permet de reconduire la solution du problème à une analyse élastique particulière. Deux exemples numériques montrent l'application du procédé qui par sa simplicité se prête bien à un usage pratique.

1. Introduction

The effects of concrete creep on the behaviour of R.C. structures are quite different depending on whether or not the second order phenomena are considered. In the absence of such phenomena, the static behaviour of a hyperstatic structure subject to static actions (forces) does not vary because of creep and coincides with the behaviour that may be calculated in the elastic phase according to the first principle of linear visco-elasticity.
This fact, which is rigorously effective for homogeneous structures, cannot be valid for R.C. structures since their cross sections are made up of different materials. However, the small percentages of reinforcement present in the sections ensure that the structural behaviour, as a whole, is little different from the one foreseen by the first principle of linear visco-elasticity. Consequently, in practical applications, with reference only to the calculation of internal actions, it is possible to classify the R.C. structures as homogeneous.
As for the calculation of deformations we should not always disregard the presence of reinforcement which, even if generally limited to small values, can notably reduce the creep.
When the structural members present high values of slenderness, in the evaluation of the static behaviour it is necessary to take into account the second order effects and in this case the first principle of linear visco-elasticity cannot be applied any more.
In fact, the state of stress depends on the state of deformation, so that the increase of deformation due to creep causes a variation in the state of stress.
Since it is necessary to assess carefully the state of deformation, the reinforcement present in cross sections must be taken into account, which makes the structural analysis very complex. This is shown in /1, 2, 3/, where the study of single bar simply supported at the ends and subject to a transversal load and to a normal force, constant in time, leads to an integro-differential equation in terms of partial derivatives.
When the structure is hyperstatic, the compatibility equations are difficult to solve, since we have no a priori knowledge of the relationship existing between the applied actions and the deformation of the geometrical axis of the structural members. As regard frame structures having no-sway connections and formed by constant section bars, the solution of the compatibility equation system can be obtained by an eigen-function expansion of the quantities involved in the problem, so that the determination of hyperstatic unknowns can be obtained by solving an infinite set of systems of Volterra integral equations.
This way of carrying on, reported in detail in /4, 5/, is general and gives exact solutions but its complexity makes it feasible only by means of computer programs and therefore cannot be adopted in everyday practice. The above method can be applied also in handling way, as reported in /5/, when the creep behaviour of the material depends only on ageing. This process based on the principle of superposition in time still involves a remarkable amount of calculation and, even if conceptually easy, cannot be used as usual procedure in the structural design. Therefore, since in the service state a variation of static behaviour takes place, it seems necessary to have a quite feasible method of calculation which enables to estimate, with sufficient approximation, the state of stress in the structure. This aim can be achieved only if a constitutive visco-elastic law algebraically-expressed is adopted for concrete, since it is then possible to obtain at a generic time the

relationship existing between the state of deformation and the applied stresses. In such a way the compatibility equations are reduced to linear algebraic equations. Among the algebraically-expressed constitutive laws, the most reliable one is that proposed by Trost and Bazant /6, 7/, which, as is demostrated in /8/, includes, as special cases, the most usual algebraically-expressed laws adopted in literature. The law of Trost and Bazant led to the so-called AAEMM method (Age Adjusted Effective Modulus Method) which has already been applied with satisfactory results to the solution of a single isostatic bar with an initial curvature /9/.
Further on, such method will be applied to the calculation of hyperstatic structures sensitive to second order phenomena.

2. Constitutive Equation of a bar according to the AAEMM Method

Let's consider the bar of Fig. 1, having the cross section with symmetrical reinforcement subject at time t_o to a transverse load $q(z)$ and to a normal force P. Assuming for concrete and steel the constitutive laws:

$$\begin{aligned} \epsilon_c(t) &= \frac{\sigma_c(t_o)}{E_c}(1+\varphi(t,t_o)) + \frac{\sigma_c(t)-\sigma_c(t_o)}{E_c}(1+\chi(t,t_o)\varphi(t,t_o)) \\ \epsilon_s(t) &= \frac{\sigma_s(t)}{E_s}, \end{aligned} \qquad 1)$$

according to the law of plane sections, the longitudinal strains caused by bending are:

$$\begin{aligned} \epsilon_c(z,t,y) &= -\eta''(z,t)\cdot y \\ \epsilon_s(z,t,y) &= -\eta''(z,t)\cdot y. \end{aligned} \qquad 2)$$

If we introduce equations 2) into eqs. 1), solve them with respect to stresses and neglect for brevity to indicate the dependence of χ and φ from t and t_o, we obtain:

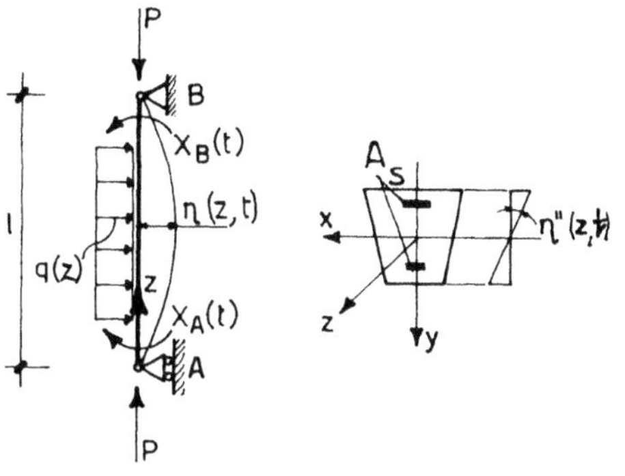

Fig. 1 - R.C. slender bar and its constant cross section.

$$\begin{aligned} \sigma_c(z,t,y) &= -E'_c\cdot y\cdot[\eta''(z,t) - \eta''_o(z)\varphi(1-\chi)] \\ \sigma_s(z,t,y) &= -E_s\cdot y\cdot\eta''(z,t) \end{aligned} \qquad 3)$$

with: $E'_c = E_c/(1+\chi\varphi)$.

Once the eqs. 3) are introduced into the equilibrium equation:

$$\int_{A_c}\sigma_c(z,t,y)\cdot y\cdot dA_c + \int_{A_s}\sigma_s(z,t,y)\cdot y\cdot dA_s = M(z,t) \qquad 4)$$

they give:

$$\eta''(z,t) = -\frac{M(z,t)}{E_c' J^{*'}} - \frac{M_o(z)}{E_c' J^{*'}} \cdot \frac{J_c}{J^*} \cdot \frac{\varphi(1-\chi)}{1+\chi\varphi} \qquad 5)$$

where:

$$J_c = \int_{A_c} y^2\, dA_c;\; J^* = J_c + \frac{E_s}{E_c} \int_{A_s} y^2\, dA_s;\; J^{*'} = J_c + \frac{E_s}{E_c'} \int_{A_s} y^2\, dA_s.$$

Since for the analyzed bar it results:

$$\left|\begin{array}{l} M_o(z) = M_q(z) + X_A(t_o)\cdot f_A(z) + X_B(t_o)\cdot f_B(z) + P\cdot \eta_o(z) \\ M(z,t) = M_q(z) + X_A(t)\cdot f_A(z) + X_B(t)\cdot f_B(z) + P\cdot \eta(z,t) \end{array}\right. \qquad 6)$$

being:

$$f_A(z) = 1 - \frac{z}{l};\; f_B(z) = \frac{z}{l};\; M_q(z) = \frac{z}{l}\cdot [\int_o^l q(\xi)\cdot (l-\xi)d\xi] - \int_o^z q(\xi)(z-\xi)d\xi,$$

the inserting of eqs. 6) in eq. 5) gives:

$$\eta''(z,t) + \frac{P\cdot\eta(z,t)}{E_c' J^{*'}} = -\frac{M_q(z)}{E_c' J^{*'}} \cdot (1+K) - \frac{[X_A(t_o)\cdot f_A(z) + X_B(t_o)\cdot f_B(z)]}{E_c' J^{*'}} \cdot K -$$

$$- \frac{[X_A(t)\cdot f_A(z) + X_B(t) f_B(z)]}{E_c' J^{*'}} - \frac{P\cdot \eta_o(z)}{E_c' J^{*'}} \cdot K, \qquad 7)$$

setting:

$$K = \frac{J_c}{J^*} \cdot \frac{\varphi(1-\chi)}{1+\chi\varphi}.$$

Eq. 7) is the constitutive equation of the bar and its solution enables to express the displacements as a function of the state of stress existing at the generic time t. The derivative of these displacements with respect to z gives the rotations (in particular those at the ends), to be used in compatibility equations.

3. Slope-Deflection Relationships

To integrate eq. 7) with respect to z, it is necessary, above all, to know the function $\eta_o(z)$, which represents the elastic solution of the problem at the time t_o. This function in particular case of $q(z)=q=\text{const.}$ is:

$$\eta_o(z) = \frac{ql^4}{E_c J^*} \cdot a(z,\alpha) + X_A(t_o)\cdot \frac{l^2}{E_c J^*} \cdot b(z,\alpha) + X_B(t_o)\cdot \frac{l^2}{E_c J^*} \cdot c(z,\alpha) \qquad 8)$$

being respectively /10/:

$$a(z,\alpha)=\frac{1}{\pi^4\alpha^2}\left[\frac{\cos(\pi\sqrt{\alpha}/2-\pi z\sqrt{\alpha}/l)}{\cos(\pi\sqrt{\alpha}/2)}-1\right]-\frac{1}{2\pi^2\alpha}\cdot\frac{z}{l}\cdot\left(1-\frac{z}{l}\right)$$

$$b(z,\alpha)=\frac{1}{\pi^2\alpha}\left[\frac{\sin(\pi\sqrt{\alpha}(1-z/l))}{\sin(\pi\sqrt{\alpha})}-\left(1-\frac{z}{l}\right)\right] \qquad 9)$$

$$c(z,\alpha)=\frac{1}{\pi^2\alpha}\left[\frac{\sin(\pi\sqrt{\alpha}\,z/l)}{\sin(\pi\sqrt{\alpha})}-\frac{z}{l}\right],\ \alpha=P\cdot l^2/(E_cJ^*\pi^2).$$

Introducing eq. 8) into eq. 7) and integrating twice as regard z under the condition:

$$\eta(o)=\eta(l)=o,$$

we obtain:

$$\eta(z,t)=\frac{ql^4}{E_c'J^{*\prime}}[(1+K)\cdot a(z,\alpha')+K\cdot d(z,\alpha,\alpha')]+X_A(t)\frac{l^2}{E_c'J^{*\prime}}\cdot b(z,\alpha')+$$

$$+X_B(t)\frac{l^2}{E_c'J^{*\prime}}\cdot c(z,\alpha')+X_A(t_o)\frac{l^2}{E_c'J^{*\prime}}\cdot K\cdot[b(z,\alpha')+e(z,\alpha,\alpha')]+$$

$$+X_B(t_o)\frac{l^2}{E_c'J^{*\prime}}\cdot K\cdot[c(z,\alpha')+g(z,\alpha,\alpha')], \qquad 10)$$

being respectively:

$$d(z,\alpha,\alpha')=\frac{1}{\pi^4}\left[\frac{(\alpha+\alpha')}{\alpha\alpha'^2}+\frac{\pi^2}{2\alpha'}\cdot\frac{z}{l}\cdot\left(1-\frac{z}{l}\right)+\frac{\alpha}{\alpha'^2(\alpha'-\alpha)}[\cos(\pi\sqrt{\alpha'}z/l)+\right.$$

$$\left.+\frac{1-\cos(\pi\sqrt{\alpha'})}{\sin(\pi\sqrt{\alpha'})}\cdot\sin(\pi\sqrt{\alpha'}z/l)]-\frac{1}{\alpha(\alpha'-\alpha)}\cdot\frac{\cos(\pi\sqrt{\alpha}/2-\pi\sqrt{\alpha}z/l)}{\cos(\pi\sqrt{\alpha}/2)}\right]$$

$$e(z,\alpha,\alpha')=\frac{1}{\pi^2}\left[\left(1-\frac{z}{l}\right)\frac{1}{\alpha'}+\frac{1}{\alpha-\alpha'}\frac{\sin(\pi\sqrt{\alpha}(1-z/l))}{\sin(\pi\sqrt{\alpha})}-\frac{\alpha}{\alpha'(\alpha-\alpha')}\cdot\frac{\sin(\pi\sqrt{\alpha'}(1-z/l))}{\sin(\pi\sqrt{\alpha'})}\right]$$

$$g(z,\alpha,\alpha')=\frac{1}{\pi^2}\left[\frac{z}{l}\cdot\frac{1}{\alpha'}+\frac{1}{\alpha-\alpha'}\frac{\sin(\pi\sqrt{\alpha}z/l)}{\sin(\pi\sqrt{\alpha})}-\frac{\alpha}{\alpha'(\alpha-\alpha')}\cdot\frac{\sin(\pi\sqrt{\alpha'}z/l)}{\sin(\pi\sqrt{\alpha'})}\right], \qquad 11)$$

$$\alpha'=\alpha(1+\chi\varphi).$$

Therefore the slope-deflection relationships at time t are:

$$\vartheta_A(t)=\eta'(o,t)=\frac{ql^3}{E'_c J^{*'}}[(1+K)\rho_a(\alpha')+K\rho_d(\alpha,\alpha')]+X_A(t)\frac{1}{E'_c J^{*'}}\cdot\rho_b(\alpha')+$$
$$+X_B(t)\frac{1}{E'_c J^{*'}}\cdot\rho_c(\alpha')+X_A(t_o)\frac{1}{E'_c J^{*'}}\cdot K\cdot[\rho_b(\alpha')+\rho_e(\alpha,\alpha')]+$$
$$+X_B(t_o)\frac{1}{E'_c J^{*'}}\cdot K\cdot[\rho_c(\alpha')+\rho_g(\alpha,\alpha')] \qquad 12)$$
$$\vartheta_B(t)=-\eta'(l,t)=\frac{ql^3}{E'_c J^{*'}}[(1+K)\rho_a(\alpha')+K\rho_d(\alpha,\alpha')]+X_A(t)\frac{1}{E'_c J^{*'}}\cdot\rho_c(\alpha')+$$
$$+X_B(t)\frac{1}{E'_c J^{*'}}\rho_b(\alpha')+X_A(t_o)\frac{1}{E'_c J^{*'}}\cdot K\cdot[\rho_c(\alpha')+\rho_g(\alpha,\alpha')]+$$
$$+X_B(t_o)\frac{1}{E'_c J^{*'}}\cdot K\cdot[\rho_b(\alpha')+\rho_e(\alpha,\alpha')],$$

being respectively:

$$\rho_a(\alpha')=l\cdot a'(o,\alpha')=-l\cdot a'(l,\alpha')=\frac{2\sqrt{\alpha'}\,tg(\pi\sqrt{\alpha'}/2)-\pi\alpha'}{2\pi^3\alpha'^2}$$
$$\rho_b(\alpha')=l\cdot b'(o,\alpha')=-l\cdot c'(l,\alpha')=\frac{1-\pi\sqrt{\alpha'}\,ctg(\pi\sqrt{\alpha'})}{\pi^2\alpha'}$$
$$\rho_c(\alpha')=l\cdot c'(o,\alpha')=-l\cdot b'(l,\alpha')=\frac{\pi\sqrt{\alpha'}-\sin(\pi\sqrt{\alpha'})}{\pi^2\alpha'\sin(\pi\sqrt{\alpha'})}$$
$$\rho_d(\alpha,\alpha')=l\cdot d'(o,\alpha,\alpha')=-l\cdot d'(l,\alpha,\alpha')=\frac{1}{\pi^3}[\frac{\pi}{2\alpha'}+\frac{\alpha\sqrt{\alpha'}}{\alpha'^2(\alpha'-\alpha)}\frac{1-\cos(\pi\sqrt{\alpha'})}{\sin(\pi\sqrt{\alpha'})}-$$
$$-\frac{\sqrt{\alpha}\,tg(\pi\sqrt{\alpha}/2)}{\alpha(\alpha'-\alpha)}] \qquad 13)$$
$$\rho_e(\alpha,\alpha')=l\cdot e'(o,\alpha,\alpha')=-l\cdot g'(l,\alpha,\alpha')=\frac{1}{\pi^2}[-\frac{1}{\alpha'}-\frac{\pi\sqrt{\alpha}\,ctg(\pi\sqrt{\alpha})}{(\alpha-\alpha')}+$$
$$+\frac{\alpha\pi\sqrt{\alpha'}\,ctg(\pi\sqrt{\alpha'})}{\alpha'(\alpha-\alpha')}]$$
$$\rho_g(\alpha,\alpha')=l\cdot g'(o,\alpha,\alpha')=-l\cdot e'(l,\alpha,\alpha')=\frac{1}{\pi^2}[\frac{1}{\alpha'}+\frac{\pi\sqrt{\alpha}}{(\alpha-\alpha')\sin(\pi\sqrt{\alpha})}-\frac{\alpha\pi\sqrt{\alpha'}}{\alpha'(\alpha-\alpha')\sin(\pi\sqrt{\alpha'})}]$$

The functions ρ, which are reported in Figs. 2÷5, enable to express the relationship between the rotations at the ends of the bar and the actions applied to it. It is interesting to point out that, if we put $\chi = 1$, we obtain the slope-deflection relationships valid for the Effective Modulus Method (EMM). In this case we have K= 0 and eqs. 12) are simply:

$$\begin{aligned} \vartheta_A(t) &= \frac{ql^3}{E'_c J^{*\prime}} \rho_a(\alpha') + X_A(t) \frac{l}{E'_c J^{*\prime}} \rho_b(\alpha') + X_B(t) \frac{l}{E'_c J^{*\prime}} \rho_c(\alpha') \\ \vartheta_B(t) &= \frac{ql^3}{E'_c J^{*\prime}} \rho_a(\alpha') + X_A(t) \frac{l}{E'_c J^{*\prime}} \rho_c(\alpha') + X_B(t) \frac{l}{E'_c J^{*\prime}} \rho_b(\alpha') \end{aligned} \qquad 14)$$

and coincide with the well known elastic relationships of a bar made by a concrete of modulus $E'_c = E_c/(1+\varphi)$. Finally, if we have $\alpha = \alpha' = 0$ (absence of second order effects) eqs. 12) are:

$$\begin{aligned} \vartheta_A(t) &= \frac{ql^3}{24E'_c J^{*\prime}}(1+K) + (2X_A(t_o) + X_B(t_o))\frac{Kl}{6E'_c J^{*\prime}} + (2X_A(t) + X_B(t))\frac{l}{6E'_c J^{*\prime}} \\ \vartheta_B(t) &= \frac{ql^3}{24E'_c J^{*\prime}}(1+K) + (X_A(t_o) + 2X_B(t_o))\frac{Kl}{6E'_c J^{*\prime}} + (X_A(t) + 2X_B(t))\frac{l}{6E'_c J^{*\prime}} \end{aligned} \qquad 15)$$

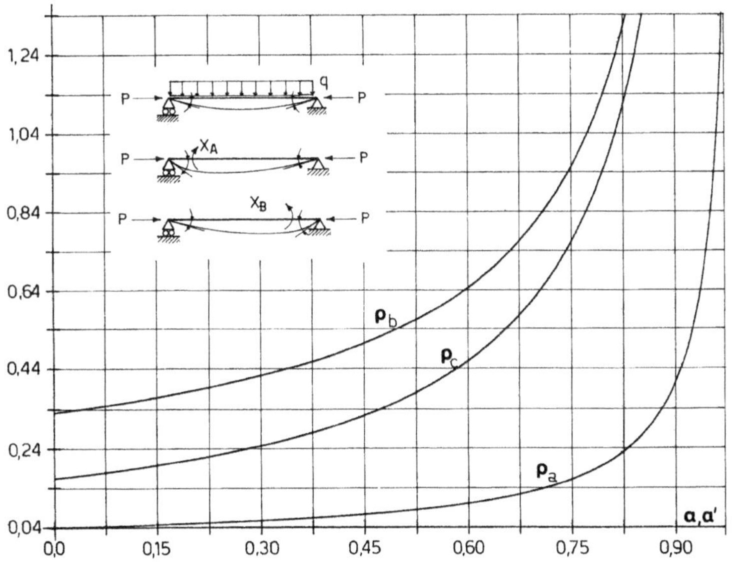

Fig. 2 - Functions ρ_a, ρ_b, ρ_c

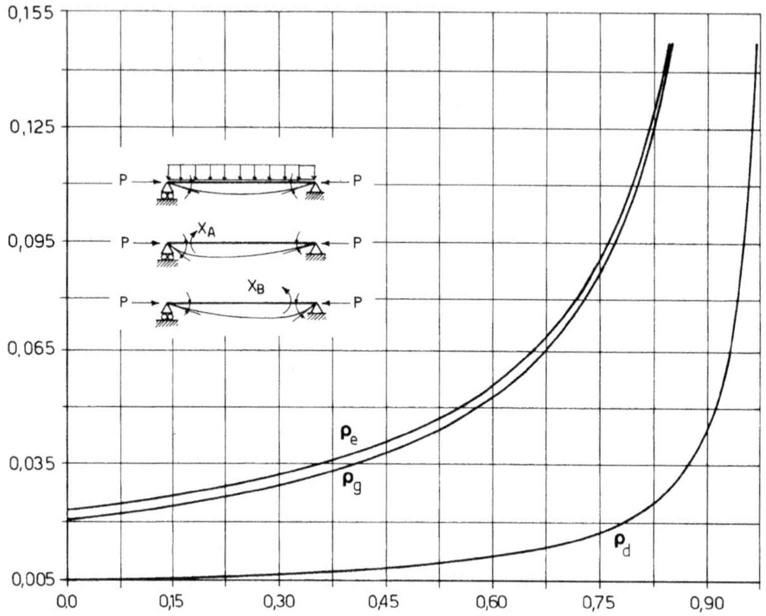

Fig. 3 - Functions ρ_d, ρ_e, ρ_g (α=0.10)

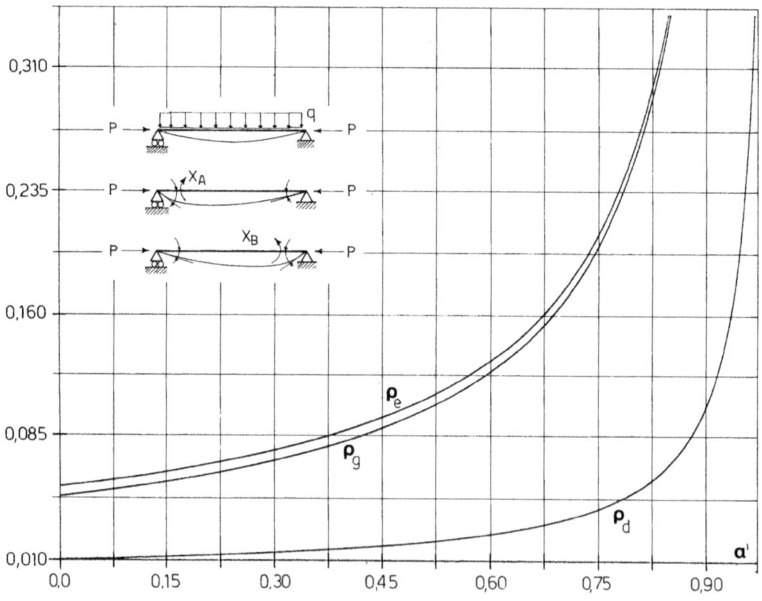

Fig. 4 - Functions ρ_d, ρ_e, ρ_g (α=0.20)

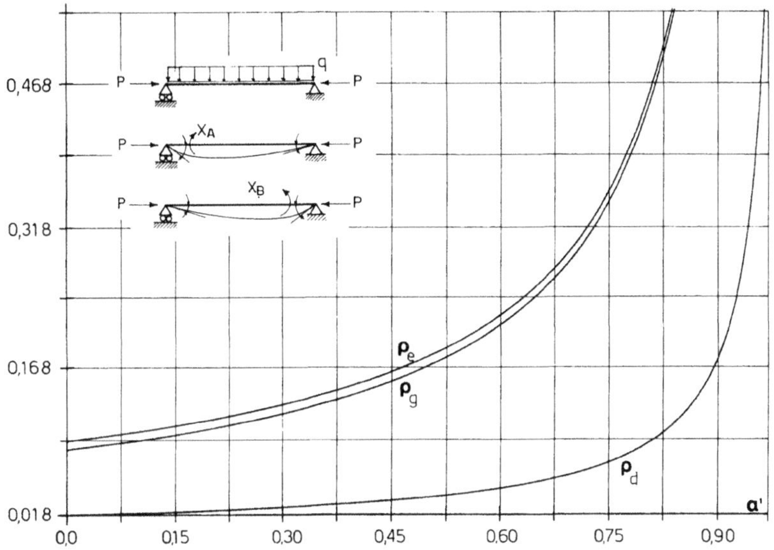

Fig. 5 - Functions ρ_d, ρ_e, ρ_g (α=0.30)

4. Numerical Examples

Eqs. 12), 14), 15) have been used for the solution of two typical problems which often appear in structural analysis and wich are affected by creep in a completely different way. The first of them, Fig. 6, refers to the calculation of the fixed-end moment $X_A(t)$ of a beam built-in at one end and supported at the other.
Applying the first of eqs. 12), the compatibility equation at the time t is:

$$\vartheta_A(t) = 0,$$

hence:

$$X_A(t) = -\frac{ql^2}{\rho_b(\alpha')}[(1+K)\rho_a(\alpha') + K\rho_d(\alpha,\alpha') -$$

$$-\frac{K\rho_a(\alpha)}{\rho_b(\alpha)}(\rho_b(\alpha') + \rho_e(\alpha,\alpha'))],$$

being, in fact:

$$X_A(t_o) = -\frac{\rho_a(\alpha)}{\rho_b(\alpha)} \cdot ql^2 .$$

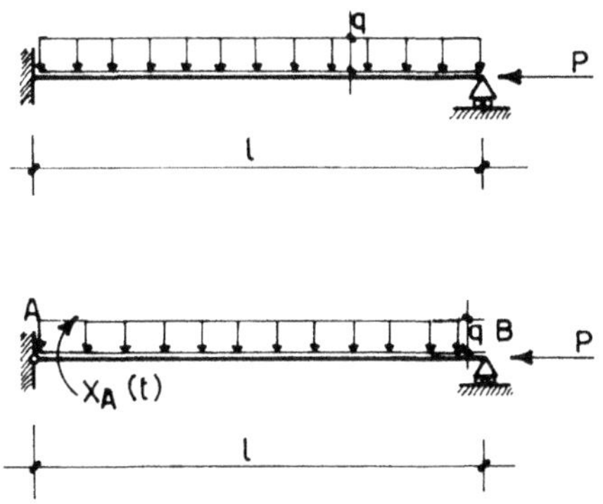

Fig. 6 - Hyperstatic bar and its statically determined scheme

In the example a rate-of-creep behaviour has been assumed by setting:

$$\varphi(t,t_o) = \varphi_\infty(1 - e^{-\gamma(t-t_o)});\ \chi(t,t_o) = \frac{1}{1-e^{-\varphi(t,t_o)}} - \frac{1}{\varphi(t,t_o)};\ \varphi_\infty = 3;$$

γ = 0,005 $days^{-1}$, and the structure has been considered as homogeneous by assuming: Jc/J^* = 1. In Fig. 7, the ratios $X_A(t)/X_A(t_o)$ have been reported for different values of α.

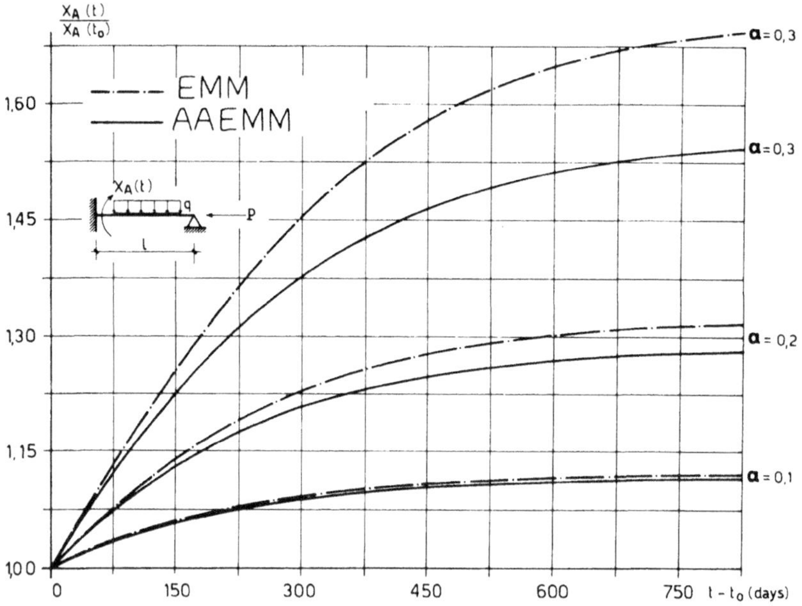

Fig. 7 - Fixed-end bending moment versus time

These ratios which turn out to increase as α rises, show a good agreement with the calculated values following the exact method, reported in /11/. In the same figure, to make a comparison, also the results which are obtained by applying, eq. 14) are reported, according to (EMM) method.

The second example refers to the calculation of the moment of continuity X(t) existing at a beam-column connection (Fig. 8). By applying the first eqs. 12) and the second of eqs. 15) we have:

$$\vartheta_B^{(1)}(t) + \vartheta_A^{(2)}(t) = 0,$$

hence:

$$X(t)\left[\frac{1}{3E'_c J_1^{*'}} + \frac{h\rho_b(\alpha')}{E'_c J_2^{*'}}\right] + X(t_o)\left[\frac{K_1 l}{3E'_c J_1^{*'}} + \frac{K_2 h}{E'_c J_2^{*'}}(\rho_b(\alpha') + \rho_e(\alpha,\alpha'))\right] + \frac{ql^3}{24E'_c J_1^{*'}}(1+K_1) = 0,$$

being $X(t_o)$ given by the equation:

$$X(t_o)\left[\frac{1}{3E_c J_1^*} + \frac{h\rho_b(\alpha)}{E_c J_2^*}\right] + \frac{ql^3}{24E_c J_1^*} = 0.$$

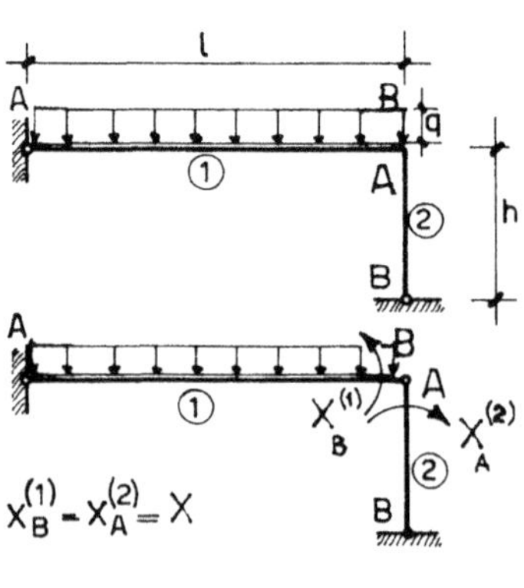

Fig. 8 - Beam-column connection and its statically determined scheme

In Fig. 9, the values of $X(t)/X(t_0)$ are reported, for $\alpha = 0$ and $\alpha = 0.25$, for different ratios Jc_2/J_2^* setting $Jc_1/J_1^* = 0.75, Jc_1 = Jc_2$, $l = h$, and keeping the same values of φ and χ as the previous example.
It is also reported the case of homogeneous structure $(Jc_1/J_1^* = Jc_2/J_2^* = 1)$. These diagrams show the progressive loss of rigidity of the column as a result of the deformation induced by second order effects, and how this effect is reduced by the increase of the percentage of reinforcement present in the cross sections of the column itself.
When second order effects are not present ($\alpha = 0$), the structural behaviour is quite different and we can have an increase of $X(t)$ when the percentage of reinforcement in the cross sections of the beam is smaller than that present in the cross sections of the column.

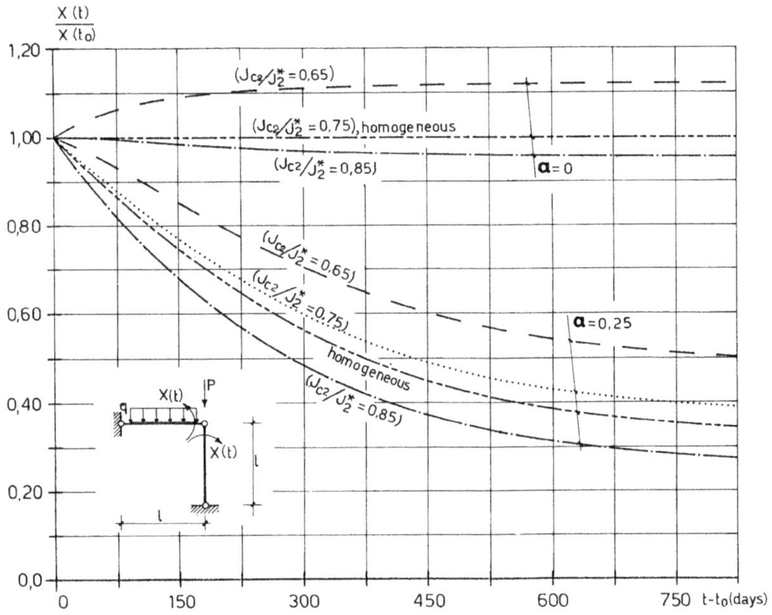

Fig. 9 - Continuity bending moment versus time

References

1. F. Levi and G. Pizzetti, *Fluage, Plasticité, Précontrainte,* Dunod, Paris, (1951)

2. G. Creazza and F. Mola, *Il Calcolo delle Deformazioni di Elementi Snelli di Calcestruzzo in Fase Elasto-viscosa Lineare,* Tech, Rep. n. 1, Department of Structural Engineering, Technical University of Milan, (1978)

3. G. Creazza and F. Mola, *L'Influenza delle Armature sulle Deformazioni di Elementi Snelli in Cemento Armato,* Atti dell'Istituto di Scienza delle Costruzioni, Politecnico di Torino n. 398 (1978)

4. F. Mola and I. Iori, *Effetti della Viscosità e del Ritiro nelle Travi Continue in Cemento Armato in Presenza di Forze Assiali,* Costruzioni in Cemento Armato, Studi e Rendiconti n. 14, pp. 179-185, Masson Italia, Milan, (1977)

5. G. Creazza and F. Mola, *Effetti del Secondo Ordine nelle Strutture Iperstatiche a Comportamento Elasto-Viscoso Lineare,* Atti dell'Istituto di Scienza delle Costruzioni, Politecnico di Torino n. 399 (1978)

6. H. Trost, *Auswirkungen des Superpositionsprinzips auf Kriech und Relaxation Probleme bei Beton und Spannbeton,* Beton und Stahlbetonbau H. 10, pp. 230-238, 261-269 (1967)

7. Z.P. Bazant, *Prediction of Concrete Creep Effects Using Age-Adjusted Effective Modulus Method,* ACI Journal n. 4, pp. 212-217 (1972)

8. P. Napoli, *Sintesi Critica dei Metodi Approssimati per il Calcolo degli Effetti Strutturali della Viscoelasticità del Calcestruzzo,* L'Industria Italiana del Cemento n. 12, pp. 831-837 (1976)

9. Z.P. Bazant and L. Najjar, *Comparison of Approximate Linear Methods For Concrete Creep,* Journal of ASCE, Struct. Div. n. 9, pp. 1851-1874 (1973)

10. S.P. Timoshenko and J.M. Gere, *Theory of Elastic Stability,* Mc Graw-Hill, New York (1961)

11. F. Mola and I. Iori, *Sul Comportamento Elasto-Viscoso Lineare di Travi Continue di Calcestruzzo in Presenza di Forze Assiali,* Tech. Rep. n. 42, Department of Structural Engineering, Technical University of Milan (1975)

Panel Discussions

Panel Discussion 1

MICROSTRUCTURAL ASPECTS, MECHANISMS AND CONSTITUTIVE MODELLING

Chairman: F.H. Wittmann

Recorder: L.J. Parrott

Panelists: J.F.Young, J.W.Dougill, S.E.Pihlajavaara, Z.P.Bazant, E.Cinclair, C.D.Pomeroy and H.K.Hilsdorf

F.H. WITTMANN The panel discussion was introduced by Professor Wittmann. He indicated that one purpose of the meeting was to bridge the gap between materials scientists who test concrete and engineers who design and build concrete structures. The panelists were then asked to give short statements relating to the completed sessions on the materials science and property modelling aspects of creep and shrinkage.

J.F. YOUNG Professor Young suggested that mathematical modelling of time dependent deformation normally involves some empiricism. It was the task of materials scientists to provide a link between microstructure and mathematical modelling so that empiricism is minimised. Professor Young cited recent work in which reductions in irrecoverable creep and shrinkage with heat treatment were correlated with heat-induced polysilicate formation in cement hydrates. He stated that such correlations could be used in mathematical modelling and thus help to bridge the gap referred to by Professor Wittmann.

J.W. DOUGILL The improvements in our physical picture of the components of cement paste encouraged Professor Dougill and he thought that they would assist the application of mechanics at the microstructural level. He felt that more attention should be paid to the effects of aggregates and aggregate/matrix interfaces. Professor Dougill thought that there was a growing awareness of the limitations of linear stress/strain models and foresaw more general models being developed in mechanics.

S.E. PIHLAJAVAARA It was stressed by Dr Pihlajavaara that simple approaches were required for describing properties. As an example he cited the schematic figures and the nomograms that he had developed for predicting the drying of concrete.

Z.P. BAZANT Professor Bazant thought that test specimen behaviour should be modelled in terms of constitutive equations and mechanics. In this way it was possible to allow for the non-uniformity of stress and of other parameters within the test specimen. Constitutive equations should, in turn, be linked to microstructural features such as porosity and polymerisation.

E. CINCLAIR Professor Cinclair posed the question "Is variability important?". If the answer was yes he stated that the techniques of probability are easily learnt and used. He emphasised that engineering problems are partly deterministic and partly probablistic in nature.

C.D. POMEROY It was pointed out by Dr Pomeroy that the predictions of concrete properties will depend upon the input data that is used and that each input parameter will be the subject of variability. An additional point was that input data is very limited at the design stage. Such limitations were recognised and taken into account when methods of predicting concrete deformation were recently developed in the United Kingdom.

H.K. HILSDORF The various levels of sophistication for prediction of creep and shrinkage were discussed by Professor Hilsdorf and it was asked how they could be linked to the mechanisms of these properties. It was mentioned that there seemed to be some justification for assuming that the mechanisms of recoverable and irrecoverable creep differed and that their separation in predictive methods was probably justifiable. Professor Hilsdorf suggested that there should be a range of predictive methods available that reflect the accuracy needed in design and that stochastic methods might be helpful in putting bounds upon the predictions.

S.E. PIHLAJAVAARA Dr Pihlajavaara showed a series of pictures that illustrated how the processes of education and communication can, with the best intentions, produce unfortunate consequences.

B.L. MEYERS It was pointed out by Professor Meyers that an additional element between microstructural data and prediction methods was required: it was necessary to study the effects, such as cracking, that cause discrepancies between the behaviour of in-situ and laboratory concrete.

J.F. YOUNG Professor Young replied that microstructural considerations may not yet provide a full solution for prediction requirements but they will certainly help to develop more fundamentally based predictive methods. He illustrated his point by suggesting that the connections between polysilicate formation and creep or shrinkage could be developed to provide a predictive method.

Z.P. BAZANT It was emphasised by Professor Bazant that specimen response should be modelled using constitutive equations that were adjustable so that a good fit to experimental data could be obtained.

C.D POMEROY The suggestion was made by Dr Pomeroy that we should not confine our attention to the prediction of properties but should also consider how our knowledge can be used to control concrete behaviour.

R. FAVRE Professor Favre indicated that there were serious difficulties in eluminating stress and temperature gradients in a concrete sample and he suggested that materials scientists preferred to model the behaviour of materials by theoretical models rather than by laboratory measurement.

F.H. WITTMANN Professor Wittmann agreed that the difficulties were indeed difficult to avoid but micromechanics could be used to disentangle the various components of the response of the system. He cited the paper of Mr Acker as an example of how the overall response of a concrete specimen could be formulated in terms of reasonable assumptions regarding the basic concrete properties. An inverse type of analysis could then be used to obtain the true materials parameters.

S.E. PIHLAJAVAARA The possibility of modifying the response of concrete by using additives or special cements and by controlling drying conditions was mentioned by Dr Pihlajavaara.

J.M. ILLSTON Dr Illston thought that future improvements in predictive methods will be empirical in nature but he emphasised that formulations should be devised that would allow new, scientific information to be incorporated as it becomes available. The development of creep models in terms of rate process theory was quoted as an example of such an approach.

P. ACKER
Mr Acker spoke of the difficulties that were encountered in measuring and computing the creep behaviour of microconcrete. The test samples had stresses and moisture contents that varied spacially. Computational mechanics normally involved the unjustified assumptions of linearity and superposition so

Mr Acker's work had moved towards the development of incremental models. He cited a practical example of a prestressed structure in which non-linear behaviour was suspected and suggested that the behaviour was possibly due to cracking.

S.E. PIHLAJAVAARA It was pointed out by Dr Pihlajavaara that the distribution of moisture in concrete can be checked experimentally using published methods.

D. RAVINA Dr Ravina believed that a greater emphasis should have been placed upon experimental studies of the mechanisms of creep and shrinkage. He thought that results from such basic studies could be more easily incorporated into analyses.

C.D. POMEROY Research at the Cement and Concrete Association in which it had been possible to measure in-situ behaviour of concrete and to conduct parallel, basic science studies were mentioned by Dr Pomeroy. Such studies were helpful, for example, in understanding the causes of seasonal strain variations in reinforced concrete structures. It was also pointed out by Dr Pomeroy that seasonal variations of relative humidity can produce shrinkage variations at $\pm$ 40% about the mean value. Thus predictive methods that imply an exponential type of shrinkage development should be used with caution. He thought that future refinements of predictive methods will take more account of in-situ data.

T. JAVOR Dr Javor was pessimistic about current trends and felt that research should be directed towards providing simple answers to practical questions.

F.H. WITTMANN In reply Professor Wittmann stated that simple answers were not possible if it was necessary to model complicated environmental effects.

Professor Wittmann ended the panel discussion by summarising the topics that had been covered. The topics included the application of microstructural data to creep and shrinkage interpretation, the relationship of micromechanics to specimen behaviour and the relationship of specimen behaviour to structural response. He felt that these topics should be considered very carefully when planning research so that comprehensive theories of creep and shrinkage can be developed.

Panel Discussion 11

CONSTITUTIVE MODELLING AND STRUCTURAL CREEP ANALYSIS

Chairman: Z.P. Bazant

Recorder: D.W. Hobbs

Panelists: C.A. Anderson, T.Javor, M.A. Chiorino, R. Favre, B.L. Meyers and K. Willam

Z.P. BAZANT The panel discussion was opened by Professor Bazant. He pointed out that structural analysis and applications should be based on the concepts of materials science. In his view constitutive models were basically of two types - those for routine analysis of structures, namely codes, and those which draw on maximum knowledge and are intended for special structures such as nuclear ones. Professor Bazant then posed a number of questions which he hoped the panelists and audience would consider in their discussions. How can cracking, drying and its interaction with temperature, cyclic loading, and statistical aspects be included in a constitutive model ? What is the constitutive law and how can it be applied ? How good are the simplified models ? What are the variabilities, uncertainties and errors of our predictions ?

M.A CHIORINO Professor Chiorino suggested that for normal structures methods exist, such as the single step numerical method of the Trost-Bazant type, which lead to satisfactory accuracy. He pointed out that values of the relaxation function may now be obtained by numerical exact procedures or simple algebraic approximations. In the case of structures which were complex due to the history of loading, geometry or restraint Professor Chiorino concluded from the Symposium that the rate of creep formulations to describe material behaviour were possibly more suitable.

R. FAVRE In the view of Professor Favre it was essential that the gap between Material Scientists and Engineers should be narrowed. He cited two important subject areas which had been ignored by the Symposium. These were the deformation and deflection of slabs which are subject to two-way directional bending moments and the movement of ground foundations.

C.A. ANDERSON The general engineering designer was according to Mr Anderson unlikely to use a 3 dimensional finite element analysis, in part, because of inadequate knowledge concerning the behaviour of concrete and, in part, because of the complexity which the codes would necessarily embody. He stated that in most instances an engineering solution was adequate but he hoped that approximate solutions will increasingly be based on a finite element analysis.

K. WILLAM Dr Willam stressed that most models are simple equations which can be used in any finite element code. He emphasised that since a single material was being considered a single unified descriptive model was required. Dr Willam thought that the conventional method of simply adding up the contributions due to elastic, visco-elastic and delayed elastic movement was unwise because this may ignore important interactions. He concluded that further development would involve considering creep as a process which evolves with time and taking into account triaxial effects.

B.L. MEYERS Professor Meyers pointed out that the various aspects of creep which had been discussed namely material problems, mechanical problems and engineering problems, were in fact the same. He expressed some concern regarding the communications gap which existed between the workers in these areas and felt that it was not generally known that methods exist which can adequately predict the movement of 95 to 98% of our structures.

T. JAVOR Dr Javor stressed that only simple equations are required for use in practice and that these equations should be based to a large extent on structural observations. He suggested that more attention should be given to the influence of quality of concrete, aggregate and additives upon creep.

H. TROST Professor Trost indicated that for structural predictions it was possible to determine the effect of variations in creep and shrinkage upon the upper and lower bounds. However, he felt that to make structural predictions it was essential to understand the mechanical problem. He thought that a level 1 or 2 approach was suitable for most normal structures but suggested that improvements at the level 2 or 3 were desirable for more complex structures.

L.F. NIELSON, B.L. MEYERS, H.K. HILSDORF Because there are shortcomings in our understanding of both material properties and concrete structures Mr Nielson questioned whether a level 3 approach was justified. Professor Meyer agreed but suggested that a level 3 approach would improve the understanding of the problem. However, Mr Hilsdorf stressed that for certain types of structures such as nuclear ones only the best analytical models would do.

C.D. POMEROY At this stage in the discussion Dr Pomeroy stated that perhaps our knowledge should be used to modify the properties of concrete to ensure that the movements and deflections were acceptable to the designer.

H. ASCHL In response to a question raised by Professor Bazant concerning creep for a 3 dimensional model Mr Aschl mentioned the high cost of the experimental work. He went on to suggest that the analysts job may be simplified if a number of experimentalists carried out tests on the same concrete.

L.J. PARROTT Dr Parrott agreed with Dr Javor that measurements must be made both in the laboratory and on in-situ concrete. He pointed out firstly that in-situ data is currently being used to assist the planning of laboratory tests and secondly that environmental variations can produce larger strain variations within a single year than loading causes in the first 10 years after construction.

W. HAAS In reply to a question raised by Professor Chiorino concerning the necessity of using rate type formulations for special structures Dr Haas described the attempts he and his colleagues had made since 1969 in the field of computerized creep analysis. He reported that they had first tried simple algebraic equations, based on creep and relaxation curves but when they applied these equations to actual structural creep problems with complicated loading histories and construction phases, they found that these methods were not suitable. According to Dr Haas this was due to the fact that the creep and relaxation curves were special solutions (constant stress and constant strain) of the actual rate-type stress-strain law for concrete creep. He concluded that methods based on creep curves or relaxation curves must therefore be tailored to the particular structural creep problem and as a result a general purpose creep analysis program must be based on a rate type creep law.

R.F. WARNER, L.J. PARROTT The suggestion was made by Professor Warner that one of the conference objectives had not been met since too little attention had been given to the common ground which may exist between the microstructural and structural analysts. In Professor Warner's view attempts should be made to establish compatibility between the spectrum of models which had been proposed. In a partial reply to this comment Dr Parrott stated that rate processes are a good way of describing the basic mechanics of the material. As an example he quoted polymerisation as a microstructural parameter which could be correlated with creep rate.

C.D. POMEROY Dr Pomeroy supported Dr Javor's request that more measurements be made on in-situ concrete. He mentioned that material property gradients could exist in in-situ concrete due to heat of hydration effects and he expressed the opinion that material scientists and structural engineers would come together through a better understanding of the properties and performance of in-situ concrete.

Z.P. BAZANT Professor Bazant then closed the discussion.

NAME INDEX*

* Underlined pages indicate contributions to this volume.

SUBJECT INDEX*

* Underlined page numbers indicate that corresponding keywords are used in the title of a contribution.

GPSR Compliance

The European Union's (EU) General Product Safety Regulation (GPSR) is a set of rules that requires consumer products to be safe and our obligations to ensure this.

If you have any concerns about our products, you can contact us on ProductSafety@springernature.com

In case Publisher is established outside the EU, the EU authorized representative is:

Springer Nature Customer Service Center GmbH
Europaplatz 3
69115 Heidelberg, Germany

Batch number: 10816188

Printed by Printforce, the Netherlands